Lecture Notes in Mathematics

Volume 2274

This series reports on new developments in all areas of mathematics and their applications - quickly, informally and at a high level. Mathematical texts analysing new developments in modelling and numerical simulation are welcome. The type of material considered for publication includes:

1. Research monographs
2. Lectures on a new field or presentations of a new angle in a classical field
3. Summer schools and intensive courses on topics of current research.

Texts which are out of print but still in demand may also be considered if they fall within these categories. The timeliness of a manuscript is sometimes more important than its form, which may be preliminary or tentative.

More information about this series at http://www.springer.com/series/304

Jorge Almeida • Alfredo Costa •
Revekka Kyriakoglou • Dominique Perrin

Profinite Semigroups
and Symbolic Dynamics

 Springer

Jorge Almeida
CMUP, Departamento de Matemática,
Faculdade de Ciências
Universidade do Porto
Porto, Portugal

Alfredo Costa
CMUC, Departamento de Matemática,
Faculdade de Ciências e Tecnologia
Universidade de Coimbra
Coimbra, Portugal

Revekka Kyriakoglou
LIGM
Universite Gustave Eiffel
Marne La Vallee Cedex 2, France

Dominique Perrin
LIGM
Universite Gustave Eiffel
Marne La Vallee Cedex 2, France

ISSN 0075-8434   ISSN 1617-9692   (electronic)
Lecture Notes in Mathematics
ISBN 978-3-030-55214-5   ISBN 978-3-030-55215-2   (eBook)
https://doi.org/10.1007/978-3-030-55215-2

Mathematics Subject Classification: Primary: 20E18; Secondary: 37B10

This Springer imprint is published by the registered company Springer Nature Switzerland AG.
The registered company address is: Gewerbestrasse 11, 6330 Cham, Switzerland

Contents

1   **Introduction**   1

2   **Prelude: Profinite Integers**   7
     2.1   Introduction   7
     2.2   Profinite Integers   8
     2.3   Profinite Natural Integers   13
     2.4   Zero Set of a Recognizable Series   15
     2.5   Odometers   18
     2.6   Exercises   18
         2.6.1   Section 2.2   18
         2.6.2   Section 2.3   20
         2.6.3   Section 2.4   20
     2.7   Solutions   20
         2.7.1   Section 2.2   20
         2.7.2   Section 2.3   22
         2.7.3   Section 2.4   22
     2.8   Notes   23

3   **Profinite Groups and Semigroups**   25
     3.1   Introduction   25
     3.2   Topological and Metric Spaces   25
         3.2.1   Topological Spaces   26
         3.2.2   Metric Spaces   29
         3.2.3   Compact Spaces   30
     3.3   Topological Semigroups   32
         3.3.1   Semigroups and Monoids   32
         3.3.2   Interplay Between Algebra and Topology   34
         3.3.3   Generating Sets   36
     3.4   Topological Groups   38
     3.5   Quotients of Compact Semigroups   39
     3.6   Ideals and Green's Relations   41
         3.6.1   Green's Relations   41

		3.6.2	Stable Semigroups	43
		3.6.3	The Schützenberger Group	44
	3.7		The ω-Power	46
	3.8		Projective Limits	49
	3.9		Profinite Semigroups: Definition and Characterizations	53
	3.10		Profinite Groups and Profinite Monoids	56
	3.11		The Profinite Distance	57
	3.12		Profinite Monoids of Continuous Endomorphisms	58
	3.13		Exercises	60
		3.13.1	Section 3.2	60
		3.13.2	Section 3.3	63
		3.13.3	Section 3.5	63
		3.13.4	Section 3.6	64
		3.13.5	Section 3.7	67
		3.13.6	Section 3.8	67
		3.13.7	Section 3.9	67
		3.13.8	Section 3.12	68
	3.14		Solutions	68
		3.14.1	Section 3.2	68
		3.14.2	Section 3.3	73
		3.14.3	Section 3.5	74
		3.14.4	Section 3.6	75
		3.14.5	Section 3.7	80
		3.14.6	Section 3.8	80
		3.14.7	Section 3.9	81
		3.14.8	Section 3.12	83
	3.15		Notes	84
4	**Free Profinite Monoids, Semigroups and Groups**			**87**
	4.1		Introduction	87
	4.2		Free Monoids and Semigroups	88
	4.3		Free Groups	93
	4.4		Free Profinite Monoids and Semigroups	96
	4.5		Pseudowords as Operations	104
	4.6		Free Profinite Groups	106
	4.7		Presentations of Profinite Semigroups	111
	4.8		Profinite Codes	115
	4.9		Relatively Free Profinite Monoids and Semigroups	118
	4.10		Exercises	121
		4.10.1	Section 4.2	121
		4.10.2	Section 4.3	122
		4.10.3	Section 4.4	123
		4.10.4	Section 4.5	123
		4.10.5	Section 4.6	124
		4.10.6	Section 4.7	124

	4.10.7	Section 4.8	125
	4.10.8	Section 4.9	125
4.11	Solutions		125
	4.11.1	Section 4.2	125
	4.11.2	Section 4.3	127
	4.11.3	Section 4.4	128
	4.11.4	Section 4.5	130
	4.11.5	Section 4.6	131
	4.11.6	Section 4.7	132
	4.11.7	Section 4.8	134
	4.11.8	Section 4.9	134
4.12	Notes		135

5   Shift Spaces ... 139

5.1	Introduction		139
5.2	Factorial Sets		140
5.3	Shift Spaces		140
5.4	Block Maps and Conjugacy		143
5.5	Substitutive Shift Spaces		145
	5.5.1	Primitive Substitutions	145
	5.5.2	Matrix of a Substitution	147
	5.5.3	Recognizable Substitutions	150
5.6	The Topological Closure of a Uniformly Recurrent Set		153
	5.6.1	Uniformly Recurrent Pseudowords	153
	5.6.2	The $\mathcal{J}$-Class of a Uniformly Recurrent Set	157
5.7	Generalization to Recurrent Sets		158
5.8	Exercises		160
	5.8.1	Section 5.2	160
	5.8.2	Section 5.3	160
	5.8.3	Section 5.5	160
	5.8.4	Section 5.6	161
	5.8.5	Section 5.7	162
5.9	Solutions		162
	5.9.1	Section 5.2	162
	5.9.2	Section 5.3	163
	5.9.3	Section 5.5	164
	5.9.4	Section 5.6	165
	5.9.5	Section 5.7	167
5.10	Notes		168

6   Sturmian Sets and Tree Sets ... 169

6.1	Introduction		169
6.2	Return Words		170
	6.2.1	Left and Right Return Words	170
6.3	Neutral Sets		172
6.4	Episturmian Words		175

6.5   Tree Sets 179
6.6   Sequences of Return Sets... 184
6.7   Limit Return Sets 187
6.8   Exercises 189
       6.8.1   Section 6.2... 189
       6.8.2   Section 6.3... 189
       6.8.3   Section 6.4... 190
       6.8.4   Section 6.5... 190
6.9   Solutions 190
       6.9.1   Section 6.2... 190
       6.9.2   Section 6.3... 191
       6.9.3   Section 6.4... 193
       6.9.4   Section 6.5... 193
6.10   Notes 194

7   **The Schützenberger Group of a Minimal Set**.......................... 195
7.1   Introduction 195
7.2   Invariance of the Schützenberger Group........................... 196
7.3   A Sufficient Condition for Freeness 199
7.4   Groups of Substitutive Shifts....................................... 201
       7.4.1   Proper Substitutions 207
7.5   Exercises 208
       7.5.1   Section 7.2... 208
       7.5.2   Section 7.3... 209
       7.5.3   Section 7.4... 209
7.6   Solutions 210
       7.6.1   Section 7.2... 210
       7.6.2   Section 7.3... 211
       7.6.3   Section 7.4... 212
7.7   Notes 213

8   **Groups of Bifix Codes** 215
8.1   Introduction 215
8.2   Prefix Codes in Factorial Sets...................................... 216
       8.2.1   Prefix Codes.. 217
       8.2.2   Maximal Prefix Codes... 217
       8.2.3   Minimal Automata of Prefix Codes 219
8.3   Bifix Codes in Recurrent Sets...................................... 221
       8.3.1   Group Codes 221
       8.3.2   Parses 222
       8.3.3   Complete Bifix Codes 224
8.4   Bifix Codes in Tree Sets.. 229
       8.4.1   Cardinality Theorem 229
       8.4.2   The Finite Index Basis Theorem 230

8.5	The Syntactic Monoid of a Recognizable Bifix Code	232
	8.5.1 The F-Minimum $\mathcal{J}$-Class	232
	8.5.2 The F-Group as a Permutation Group	234
	8.5.3 The Minimal Automaton of a Recognizable Bifix Code	238
8.6	The Charged Code Theorem	240
	8.6.1 Charged Codes	241
	8.6.2 The Charged Code Theorem: Statement and Examples	242
8.7	Bifix Codes in the Free Profinite Monoid	246
8.8	Proof of the Charged Code Theorem	248
	8.8.1 A Byproduct of the Proof of the Charged Code Theorem	253
8.9	Exercises	255
	8.9.1 Section 8.2	255
	8.9.2 Section 8.3	255
	8.9.3 Section 8.5	257
	8.9.4 Section 8.6	257
	8.9.5 Section 8.7	258
	8.9.6 Section 8.8	258
8.10	Solutions	258
	8.10.1 Section 8.2	258
	8.10.2 Section 8.3	259
	8.10.3 Section 8.5	261
	8.10.4 Section 8.6	262
	8.10.5 Section 8.7	262
	8.10.6 Section 8.8	262
8.11	Notes	263
Bibliography		265
Subject Index		269
Index of symbols		275

Chapter 1
Introduction

The theory of profinite groups originates in the theory of infinite Galois groups and p-adic analysis (see Ribes and Zalesskii (2010); Fried and Jarden (2008)).

The corresponding theory for semigroups was initially motivated by the theory of varieties of languages and semigroups, put in correspondence by Eilenberg's variety theorem (see Almeida (2005a) for a short introduction). Indeed, this theory relates families of languages and varieties of finite semigroups, sometimes called pseudovarieties. The latter are not defined by identities but by sequences of identities which are to be eventually satisfied. Profinite semigroups give an adequate framework to treat this correspondence with pseudoidentities between pseudowords replacing identities between words. This perspective is extensively studied in the books Almeida (1994) and Rhodes and Steinberg (2009).

To make things more concrete, the variety of finite aperiodic semigroups (i.e., finite semigroups without nontrivial groups as subsemigroups) corresponds to a special class of languages, called star free languages, which has an essentially combinatorial description. A finite semigroup S is aperiodic if there exists $n \geqslant 1$ such that $s^{n+1} = s^n$ for all $s \in S$. These semigroups are thus defined by eventually satisfying a sequence of identities. In a profinite semigroup, one has an operator $s \mapsto s^\omega$ and the family of aperiodic profinite semigroups is defined by only one pseudoidentity, namely $s^{\omega+1} = s^\omega$.

Later, the first author has initiated the study of the connection between free profinite semigroups and symbolic dynamics (see Almeida (2002a)). He has shown (Almeida 2005b) that minimal subshifts correspond to maximal non-singleton regular $\mathcal{J}$-classes in the natural $\mathcal{J}$-order of the free profinite monoid. Moreover, the Schützenberger group of such a $\mathcal{J}$-class is a dynamical invariant of the subshift (Costa 2006; Almeida and Costa 2016). Finally, it is shown in Almeida and Costa (2016) that if a minimal system satisfies the tree condition (as defined in Berthé et al. (2015)), the corresponding group is a free profinite group.

© Springer Nature Switzerland AG 2020
J. Almeida et al., *Profinite Semigroups and Symbolic Dynamics*, Lecture Notes in Mathematics 2274, https://doi.org/10.1007/978-3-030-55215-2_1

In this book, we give a gentle introduction to the notions used in profinite algebra and develop the link with minimal sets and symbolic dynamics. This link seems natural once we realize that both semigroup theory and symbolic dynamics have developed with strong connections with automata theory and the theory of formal languages. Indeed, a symbolic dynamical system is completely determined by a language of finite words, consisting of the finite blocks appearing in elements of the dynamical system. The introduction of the sofic systems by Weiss (1973) (the systems whose language of blocks is recognized by a finite automaton, equivalently, by a finite semigroup) made such connections rather clear. The connection with automata theory plays a major role in the book Lind and Marcus (1995), mostly concerned with sofic systems, and which we give as a supporting reference for the introduction to symbolic dynamics.

Minimal subshifts form another major important class of dynamical systems, quite different from sofic systems. They also present connections with automata theory in the more studied case of systems defined by primitive substitutions, as one may see in the books Fogg (2002) and Lothaire (2002), which are our guiding references concerning minimal systems.

In the context of this book, our motivation to explore the profinite world is the following. We are interested in the situation where we fix a uniformly recurrent set F on an alphabet A (in general a non rational set, like the set of factors of the Fibonacci word). We want to study sets of the form $F \cap L$ where L is the inverse image through a morphism $\varphi : A^* \to M$ of a subset of a finite monoid M. The aim is thus to develop a theory of automata observing through the filter of a non rational set.

We are particularly interested in sets L of the form

1. $L = wA^*w$ for some word $w \in F$ (linked to the complete return words to w).
2. $L = h^{-1}(H)$ where $h : A^* \to G$ is a morphism onto a finite group G and H is a subgroup of G.

This problem has been studied successfully in a number of cases starting with a Sturmian set F in Berstel et al. (2012) and progressively generalizing to sets called tree sets in Berthé et al. (2015), nowadays also known as dendric sets.

The framework of profinite semigroups allows one to work simultaneously with all rational sets L. This is handled, as we shall see, both through the definition of an inverse limit and through a topology on words. This topology is defined by introducing a distance on words: two words are close if one needs a morphism φ on a large monoid M to distinguish them. For example, for any word x, the powers $x^{m!}$ of x with $m \geqslant n$ will not be distinguished by a monoid with less than n elements. Thus, one can consider a limit, which is not a word but is called a pseudoword, and is denoted x^ω, which has the same image by all continuous morphisms φ onto a finite monoid M.

The moment of inspiration that led to this work was the derivation in Almeida and Costa (2016) of a result on profinite subgroups of free profinite monoids (the Schützenberger group of a dendric set is free) whose proof was obtained using the so called Return Theorem, which is the main result of Berthé et al. (2015). The Return

Theorem is a result about the sets of finite blocks of an important class of symbolic dynamical systems (the dendric systems, which include the Sturmian systems), and it gives a global version of properties like the Finite Index Basis Property, as defined in Berthé et al. (2015). One of the goals of the present work is to develop this connection between symbolic dynamical systems and free profinite monoids.

We begin in Chap. 2 with two motivating examples: the first one concerns profinite integers (closely related to p-adic integers) and the second one the profinite natural integers. These will appear later as the free profinite group and the free profinite semigroup on one generator respectively. We develop the proof of a result concerning linear recurrent sequences (the Skolem-Mahler-Lech Theorem) and of the topological dynamical systems built on p-adic integers called odometers. Chapter 2 is a prelude to the following chapters, and has a somewhat informal presentation. Several well-known definitions appearing there are formalized only in the following chapters, which are essentially self-contained.

In Chap. 3, we recall some basic definitions concerning topological groups and semigroups. We refer to Willard (2004) for a general introduction to the notions concerning topological spaces. As we only consider Hausdorff spaces, we include the Hausdorff property in the definition of compact space. Most of the topological spaces that we shall consider in this work are metrizable. This allows to use the characterization of several topological notions (such as continuity or compactness) in terms of sequences, instead of nets. We give in Sect. 3.9 an introduction to the basic notions concerning profinite semigroups. We have chosen a simplified presentation which uses the class of all semigroups instead of working inside a pseudovariety of semigroups. It simplifies the statements but the proofs work essentially in the same way.

In Chap. 4 we specialize to free profinite groups and semigroups. We first recall some basic notions on free semigroups and formal languages. We define recognizable and rational languages and state without proof Kleene's Theorem (Theorem 4.2.8). In Sect. 4.3, we recall some elementary notions concerning free groups. We define the Stallings automaton of a finitely generated subgroup and use it to prove Hall's Theorem asserting that every finitely generated subgroup of a free group is a free factor of a subgroup of finite index (Theorem 4.3.4). In Sect. 4.4 we introduce free profinite semigroups, and in Sect. 4.5 we see their elements, the pseudowords, as operations. Free profinite groups are introduced in Sect. 4.6. In Sect. 4.7, we describe the presentations of profinite semigroups and groups. In Sect. 4.8 we describe results concerning codes in profinite semigroups. The main result, from Margolis et al. (1998), is that every finite code is a profinite code (Theorem 4.8.3). In Sect. 4.9, we come to an important point although not developed systematically in this book, concerning the relativization of the notion of profinite semigroup to particular classes of varieties of finite semigroups (also called pseudovarieties).

In Chap. 5, we come to the basic notions of symbolic dynamics with shift spaces. We are essentially interested in looking at the language of blocks of a shift space, namely recurrent languages (corresponding to transitive shift spaces) and specially uniformly recurrent languages (corresponding to minimal shift spaces). We give in

Sect. 5.5 some emphasis to primitive substitutive shifts, which are minimal shift spaces obtained by iterating a primitive substitution. In Sect. 5.6 we introduce uniformly recurrent pseudowords, which are closely connected with minimal shift spaces. We prove an important result (Theorem 5.6.7) showing that the uniformly recurrent pseudowords can be characterized in algebraic terms, as the $\mathcal{J}$-maximal infinite elements of the free profinite monoid.

In Chap. 6, we introduce important particular cases of shift spaces, namely the Sturmian shifts and their generalization, the tree shifts, also called dendric shifts, introduced in Berthé et al. (2015). We also introduce return words and the notion of limit return sets. We show that the limit return sets, which are sets of pseudowords, generate maximal subgroups of the free profinite monoid.

In Chap. 7, we prove several results concerning the presentation of the Schützenberger group $G(F)$ of a uniformly recurrent set F. As a main result, we prove (Theorem 7.3.4) that the group $G(F)$ is a free profinite group whenever F is a tree set (Almeida and Costa 2016).

Finally, in Chap. 8, we study the groups associated with bifix codes. We are especially interested in bifix codes included in a recurrent set F and maximal for this property, called F-maximal bifix codes. One of the main results in this chapter relates F-complete bifix codes with bases of subgroups of finite index of the free group (Theorem 8.4.2). We mention another important result shown in Chap. 8, motivated by the natural assignment (extensively studied in the theory of codes, as one may see in the reference book Berstel et al. (2009)) of a finite group G_Z to each recognizable code Z. The group G_Z is any of the maximal subgroups of the minimum ideal of the syntactic monoid of Z^*. If the bifix code Z is maximal in A^* and $F \subseteq A^*$ is uniformly recurrent, then $X = Z \cap F$ is a finite F-maximal bifix code to which one naturally associates a finite group $G_X(F)$. This group is any of the maximal subgroups of the minimum $\mathcal{J}$-class of the syntactic monoid of X^* intersecting the image of F. In Chap. 8 we give necessary and sufficient conditions (that hold in particular for uniformly recurrent tree sets) for G_Z and $G_X(F)$ to be equivalent permutation groups, in what we call the Charged Code Theorem. The proof makes use of the profinite group $G(F)$. The Charged Code Theorem is from Kyriakoglou and Perrin (2017) and Almeida et al. (2020), with the special case where F is Sturmian first shown in Berstel et al. (2012) without any explicit use of pseudowords.

Each chapter ends with a section of exercises followed by a section of solutions and a section of notes in which we gather the bibliographic references of the results of the chapter.

This book is intended to be readable by a student with a basic mathematical background. Most general notions used (such as the basic notions of topology) are defined and most results used are proved.

Acknowledgements This manuscript was written in connection with a workshop held in Marne la Vallée in January 2016 and gathering Marie-Pierre Béal, Valérie Berthé, Francesco Dolce, Pavel Heller, Julien Leroy, Jean-Eric Pin and the authors. We wish to thank the colleagues who have read parts of the manuscript and helped us to improve the presentation, in particular, Christian Choffrut and Jean-Éric Pin. We also thank Hendrik Lenstra who has shown interest in our presentation of profinite integers and Pierre de La Harpe who has provided advice on the presentation of profinite spaces on a preliminary version of this book. Finally, we thank the anonymous referees who helped us to improve the manuscript in several places.

Jorge Almeida gratefully acknowledges partial support by CMUP (UID/MAT/ 00144/2019), which is funded by FCT (Portugal) with national (MATT'S) and European structural funds (FEDER) under the partnership agreement PT2020.

The work of Alfredo Costa was carried out in part at City College of New York, CUNY, whose hospitality is gratefully acknowledged, with the support of the FCT sabbatical scholarship SFRH/BSAB/150401/2019, and it was partially supported by the Centre for Mathematics of the University of Coimbra—UIDB/00324/2020, funded by the Portuguese Government through FCT/MCTES.

Chapter 2
Prelude: Profinite Integers

2.1 Introduction

We place at the beginning of this book a presentation of a subject which is part of number theory and may give the reader an intuition to follow the rest of the chapters.

Actually, the notion of profinite integers (and the related notion of profinite natural integers) is connected with the classical field of p-adic analysis. It allows one to work in the completion of integers under a topology which is different from the topology induced by the usual Euclidean distance.

We begin by introducing profinite integers and p-adic integers (Sect. 2.2), before defining profinite natural integers (Sect. 2.3). The group of profinite integers will turn out in Chap. 4 to be the free profinite group with one generator while the profinite natural integers will appear as the free profinite monoid on one generator.

In Sect. 2.4 we use the notions presented on p-adic numbers to give a proof of a statement (the Skolem-Mahler-Lech Theorem on the set of zeroes of a recognizable series) which does not use p-adic numbers in its formulation. This gives an early example of an application of profinite algebra in the domain of ordinary integers. It is also an example of the interaction between profinite algebra and automata theory since the notion of recognizable series is closely related with the notion of recognizable language which is one of the main objects of study in this book.

In the last section (Sect. 2.5), we introduce odometers, which are examples of topological dynamical systems. Interestingly, the rings of p-adic integers and of profinite integers are obtained as particular cases. This indicates a link between profinite groups and topological dynamical systems, a point which will be the main focus of further development of this book.

© Springer Nature Switzerland AG 2020
J. Almeida et al., *Profinite Semigroups and Symbolic Dynamics*, Lecture Notes
in Mathematics 2274, https://doi.org/10.1007/978-3-030-55215-2_2

2.2 Profinite Integers

Consider the map assigning to each $x \in \mathbb{Z}$ the rational number

$$|x|_\pi = \begin{cases} 1/r(x) & \text{if } x \neq 0 \\ 0 & \text{otherwise} \end{cases}$$

where $r(x) = \min\{n \geqslant 1 \mid x \not\equiv 0 \bmod n\}$ if $x \neq 0$. We may as well assume that $r(0) = \infty$, with the usual convention that $1/\infty = 0$.

The values of $r(x)$ for $1 \leqslant x \leqslant 15$ are displayed in Table 2.1.

We note that $r(x)$ is always a power of a prime. Indeed, for any integer x and prime p, let $\mathrm{ord}_p(x)$ be the exponent of the highest power of p diving x. Then $r(x)$ and $\mathrm{ord}_p(x)$ are related by

$$r(x) = \min\{p^{\mathrm{ord}_p(x)+1} \mid p \geqslant 2 \text{ prime}\}.$$

The map $x \mapsto |x|_\pi$ is a norm on the additive group $\mathbb{Z}$, called the *profinite norm*. Recall that, by definition, a norm on an additive group G is a map $x \mapsto |x|$ from G into the nonnegative real numbers satisfying the two following conditions:

(i) $|x| = 0$ if and only if $x = 0$,
(ii) the triangular inequality $|x + y| \leqslant |x| + |y|$.

Condition (i) is clearly satisfied. The triangle inequality even holds in the stronger form (called the *ultrametric inequality*)

$$|x + y|_\pi \leqslant \max\{|x|_\pi, |y|_\pi\}. \tag{2.1}$$

The ultrametric inequality follows from the inequality

$$r(x + y) \geqslant \min\{r(x), r(y)\}$$

which obviously holds since $x \equiv 0 \bmod n$ and $y \equiv 0 \bmod n$ implies $x + y \equiv 0 \bmod n$. We are adopting the convention that $\infty > n$ for all $n \in \mathbb{Z}$.

As for any norm, the profinite norm defines a distance on $\mathbb{Z}$ called the *profinite distance* and defined by

$$d(x, y) = |x - y|_\pi.$$

Table 2.1 The first values of $r(x)$

x	1	2	3	4	5	6	7	8	9	10	11	12	13	14	15
$r(x)$	2	3	2	3	2	4	2	3	2	3	2	5	2	3	2

It satisfies the ultrametric inequality

$$d(x, z) \leqslant \max\{d(x, y), d(y, z)\}.$$

A distance d satisfying the ultrametric inequality is called a *ultrametric distance*. The use of a ultrametric distance, that we shall meet all along this book, has surprising features compared to the usual Euclidean distance. As an illustration, let us verify that in a set E with respect to an ultrametric distance, all triangles are isosceles. Consider indeed $x, y, z \in E$ and assume without loss of generality that

$$d(x, y) \leqslant d(x, z) \leqslant d(y, z).$$

By the ultrametric inequality, we have

$$d(y, z) \leqslant \max\{d(y, x), d(x, z)\} = \max\{d(x, y), d(x, z)\} = d(x, z).$$

Thus $d(y, z) = d(x, z)$ so that the triangle (x, y, z) is isosceles.

The profinite norm is related to the *p-adic norm* defined for every prime number p by

$$|x|_p = \begin{cases} p^{-\operatorname{ord}_p(x)} & \text{if } x \neq 0 \\ 0 & \text{otherwise} \end{cases}$$

The p-adic norm is not only a norm of Abelian groups, as the profinite norm, but a ring norm satisfying the additional inequality

$$|xy| \leqslant |x||y|$$

Indeed, we have $\operatorname{ord}_p(xy) = \operatorname{ord}_p(x)\operatorname{ord}_p(y)$ and thus $|xy|_p = |x|_p|y|_p$. Note that the profinite norm is not a ring norm since for example $r(12) = 5$ while $r(3)r(4) = 2 \cdot 3 = 6$ and thus $|12|_\pi > |3|_\pi|4|_\pi$. On the other hand, it is true that $|xy|_\pi^2 \leqslant |x|_\pi|y|_\pi$ (Exercise 2.1).

The completion of $\mathbb{Z}$ under the profinite distance is called the set of *profinite integers*, denoted $\widehat{\mathbb{Z}}$.

Let us briefly review the construction of the completion of $\mathbb{Z}$ with respect to the profinite distance. A sequence (x_n) of integers is a *Cauchy sequence* if for every $\varepsilon > 0$ there is an $n \geqslant 1$ such that for all $i, j \geqslant n$ one has $d(x_i, x_j) \leqslant \varepsilon$. We call two Cauchy sequences $(x_n), (y_n)$ equivalent if $d(x_n, y_n) \to 0$ when $n \to \infty$. The completion $\widehat{\mathbb{Z}}$ of $\mathbb{Z}$ under the distance d is the set of equivalence classes of Cauchy sequences in $\mathbb{Z}$. The set $\mathbb{Z}$ is included in $\widehat{\mathbb{Z}}$ because a constant sequence with value x can be identified with x (and that for $x \neq y$ the constant sequences with values x and y are not equivalent).

An alternative way to define profinite integers is the following. Consider the set of sequences of integers $(x_n)_{n \geqslant 1}$ such that

(i) $0 \leqslant x_n \leqslant n - 1$
(ii) one has $x_n \equiv x_m \mod n$ whenever $n \mid m$.

Each element x_n in the sequence is considered as an element of the ring $\mathbb{Z}/n\mathbb{Z}$. In this way, the set of these sequences is a subset of the direct product of all rings $\mathbb{Z}/n\mathbb{Z}$ called the *inverse limit* of the cyclic groups $\mathbb{Z}/n\mathbb{Z}$, denoted $\varprojlim \mathbb{Z}/n\mathbb{Z}$ (we shall describe inverse limits in general in Chap. 3). We consider the product topology on the set of these sequences and we observe that it contains $\mathbb{Z}$ by considering eventually constant sequences. Actually, we have

$$\widehat{\mathbb{Z}} = \varprojlim \mathbb{Z}/n\mathbb{Z}.$$

This results from the fact that the product topology on $\varprojlim \mathbb{Z}/n\mathbb{Z}$ is the topology defined by the profinite norm and that $\mathbb{Z}$ is dense in both spaces.

The profinite norm (and the profinite distance) can be extended to $\widehat{\mathbb{Z}}$. Indeed if (x_n) is a Cauchy sequence, the sequence $|x_n|_\pi$ is also a Cauchy sequence (this time in $\mathbb{R}$) because if $|x_i - x_j|_\pi \leqslant \varepsilon$, then

$$|x_i|_\pi - |x_j|_\pi \leqslant |x_i - x_j|_\pi \leqslant \varepsilon.$$

This is because $|x_i|_\pi = |x_i - x_j + x_j|_\pi \leqslant |x_i - x_j|_\pi + |x_j|_\pi$. Thus the norm of the limit can be defined as $\lim |x_n|_\pi$. One can also define the norm of an element $x = (x_n)$ of the inverse limit $\varprojlim \mathbb{Z}/n\mathbb{Z}$ by $r(x) = \min\{n \geqslant 1 \mid x_n \neq 0\}$ and $|x|_\pi = 1/r(x)$.

The operations on $\mathbb{Z}$ can be extended to the set $\widehat{\mathbb{Z}}$ by using the ring structure of each $\mathbb{Z}/n\mathbb{Z}$ in the inverse limit $\varprojlim \mathbb{Z}/n\mathbb{Z}$ (this can also be done using Cauchy sequences, see Exercise 2.2). In this way, the set of profinite integers becomes a ring called the *ring of profinite integers*. This ring is actually a *topological ring*, in the sense that the ring operations are continuous.

Euclidean division of a profinite integer $x \in \widehat{\mathbb{Z}}$ by an ordinary integer $n \geqslant 1$ is well defined. Indeed if $x = \lim x_i$, with $x_i \in \mathbb{Z}$, set

$$x_i = nq_i + r_i$$

with $0 \leqslant r_i < n$ and $q_i \in \mathbb{Z}$. Taking a subsequence of the sequence of pairs (q_i, r_i), we may assume that all r_i are equal and that the sequence q_i converges to some $q \in \widehat{\mathbb{Z}}$. By continuity of the ring operations, we get $x = nq + r$, as required. Moreover, the quotient q and the residue r are unique (Exercise 2.5).

Therefore, similarly to the relationship between $\mathbb{Z}$ and $\mathbb{Z}/n\mathbb{Z}$, we have the associated map $x \mapsto x \mod n$ from $\widehat{\mathbb{Z}}$ onto the $\mathbb{Z}/n\mathbb{Z}$, which is a group morphism. It is the canonical projection from $\widehat{\mathbb{Z}} = \varprojlim \mathbb{Z}/n\mathbb{Z}$ to $\mathbb{Z}/n\mathbb{Z}$. This framework gives a natural meaning for the notation $x \equiv y \mod n$ when $x, y \in \widehat{\mathbb{Z}}$.

Likewise, the completion of $\mathbb{Z}$ with respect to the p-adic norm is a ring denoted $\mathbb{Z}_p$ and called the *ring of p-adic integers*. Here too, we have an alternative description via an inverse limit:

$$\mathbb{Z}_p = \varprojlim \mathbb{Z}/p^n \mathbb{Z}.$$

In particular, we may also write $x \equiv y \bmod p^n$ when $x, y \in \mathbb{Z}_p$ have the same image in $\mathbb{Z}/p^n \mathbb{Z}$.

The following result gives the right way to handle profinite integers through an expansion similar to the expansion of real numbers with respect to an integer basis.

Proposition 2.2.1 *Any profinite integer has a unique expansion*

$$x = c_1 + c_2 2! + c_3 3! + \cdots \tag{2.2}$$

with digits $0 \leqslant c_i \leqslant i$.

Proof For $x \in \widehat{\mathbb{Z}}$, set $r_n = x \bmod (n+1)!$. The sequence (r_n) satisfies

(i) $0 \leqslant r_n < (n+1)!$,
(ii) $r_n \equiv r_{n+1} \bmod (n+1)!$

Thus $0 \leqslant r_{n+1} - r_n < (n+2)!$. Consequently we have $r_{n+1} = r_n + c_{n+1}(n+1)!$ with $0 \leqslant c_{n+1} \leqslant n+1$. Then (r_n) is a Cauchy sequence converging to x and x has the expansion (2.2). ∎

The expansion (2.2) is denoted $x = (\ldots c_3 c_2 c_1)_!$. It forms the *factorial number system*.

Note that we have the equality

$$-1 = (\ldots 321)_!$$

which holds because $1 + 2.2! + \cdots + n.n! = (n+1)! - 1$, as one may verify by induction on n.

In the ring $\mathbb{Z}_p$, every element x has a unique expansion

$$x = a_0 + a_1 p + a_2 p^2 + \cdots \tag{2.3}$$

with $0 \leqslant a_i < p$. One has actually $x \equiv a_0 + \cdots + a_n p^n \bmod p^{n+1}$. Moreover $\mathrm{ord}_p(x)$ is the least n such that $a_n \neq 0$. For example, we have

$$-1 = 1 + 2 + 2^2 + \cdots$$

in $\mathbb{Z}_2$ and

$$\frac{1}{2} = 2 + 3 + 3^2 + \cdots$$

in $\mathbb{Z}_3$.

Proposition 2.2.2 *The ring* $\widehat{\mathbb{Z}}$ *is the direct product of the rings* $\mathbb{Z}_p$ *of p-adic integers, that is,*

$$\widehat{\mathbb{Z}} = \prod_p \mathbb{Z}_p.$$

Proof Since $\mathbb{Z}$ is embedded in each ring $\mathbb{Z}_p$, the ring $\mathbb{Z}$ embeds in the direct product $\prod_p \mathbb{Z}_p$. This embedding is continuous since each embedding of $\mathbb{Z}$ in $\mathbb{Z}_p$ is continuous. For every finite set S of primes, the Chinese Remainder Theorem shows that the image of $\mathbb{Z}$ in $\prod_{p \in S} \mathbb{Z}_p$ is dense. Since S can be chosen arbitrarily large, this shows that the image of $\mathbb{Z}$ is dense in $\prod_p \mathbb{Z}_p$, and thus the equality. ∎

The ring $\mathbb{Z}_p$ is an integral domain (see Exercise 2.6). Its field of fractions is the field of *p-adic numbers*, denoted $\mathbb{Q}_p$. The p-adic norm can be extended to $\mathbb{Q}$ as follows. For a, b relatively prime integers, set $\mathrm{ord}_p(a/b) = \mathrm{ord}_p(a) - \mathrm{ord}_p(b)$. Then $|x|_p = p^{-\mathrm{ord}_p x}$ is a norm on $\mathbb{Q}$ and $\mathbb{Q}_p$ is the completion of $\mathbb{Q}$ with respect to this norm.

Any element x of $\mathbb{Q}_p$ has a unique expansion

$$x = a_{-r} p^{-r} + \cdots + a_0 + a_1 p + \cdots$$

which can be obtained as follows. If $x \in \mathbb{Z}_p$, or equivalently if $|x|_p \leqslant 1$, then the expansion is given by Eq. (2.3). Otherwise, let $r = |x|_p$. Then $y = p^{-r} x$ is in $\mathbb{Z}_p$ and thus has an expansion $y = b_0 + b_1 p + \cdots$. We obtain $x = b_0 p^{-r} + \cdots + b_r + b_{r+1} p + \cdots$ which has the required form.

On the contrary, $\widehat{\mathbb{Z}}$ is not an integral domain. Let indeed $x \in \widehat{\mathbb{Z}}$ be equal to 1 in $\mathbb{Z}_2$ and 0 in any other $\mathbb{Z}_p$ and let $y \in \widehat{\mathbb{Z}}$ be 1 in $\mathbb{Z}_3$ and 0 in any other $\mathbb{Z}_p$. Then $xy = 0$ although $x, y \neq 0$.

It is possible to define profinite Fibonacci numbers. Recall that the Fibonacci numbers are defined by $F_0 = 0$, $F_1 = 1$ and $F_n = F_{n-1} + F_{n-2}$. The definition can be extended to negative n by letting $F_n = (-1)^{n-1} F_{-n}$ (Table 2.2).

The function $n \mapsto F_n$ is continuous. Indeed, for any n, there is m such that $i \equiv j \bmod m!$ implies $F_i \equiv F_j \bmod n!$. To see this, consider the matrix

$$R = \begin{bmatrix} 1 & 1 \\ 1 & 0 \end{bmatrix}. \tag{2.4}$$

We have for all $n \geqslant 1$, $[F_{n-1}\ F_{n-2}]R = [F_n\ F_{n-1}]$ and thus for every $k, n \geqslant 0$,

$$[F_{n+k}\ F_{n+k-1}] = [F_n\ F_{n-1}]R^k.$$

Table 2.2 The Fibonacci numbers F_n for $-5 \leqslant n \leqslant 5$

n	-5	-4	-3	-2	-1	0	1	2	3	4	5
F_n	-5	3	-2	1	-1	0	1	1	2	3	5

Since R has determinant 1, it is invertible in $\mathbb{Z}$ and thus there is for every $n \geqslant 1$ an integer m such that $R^m \equiv I$ mod n. Then if $i \equiv j$ mod $m!$, we have $F_i \equiv F_j$ mod $n!$.

One can actually show a stronger statement: for $n \geqslant 4$, if $i \equiv j$ mod $n!$ then $F_i \equiv F_j$ mod $n!$ (see Exercise 2.9). This implies that for $k \geqslant 3$, the first k digits of the factorial expansion of F_i only depend on the first k digits of i.

2.3   Profinite Natural Integers

We now consider the set $\mathbb{N}$ of natural integers. It is a *semiring*, which means that it is a semigroup for addition and that the multiplication distributes over addition (but it is not a group for addition since there are no negative elements).

Given $i \geqslant 0$ and $p \geqslant 1$, we define on $\mathbb{N}$ an equivalence relation by $x \equiv y$ mod (i, p) if $x = y$ or if $x, y \geqslant i$ and $x \equiv y$ mod p. The quotient of $\mathbb{N}$ by this equivalence is a semiring denoted $N_{i,p}$. The semiring $N_{i,p}$ has $i + p$ elements since $N_{i,p} = \{0, 1, \ldots, i, \ldots, i + p - 1\}$ (see Fig. 2.1).

As an additive semigroup $N_{i,p}$ has a neutral element, namely 0, and thus it is a monoid. For $i \geqslant 1$, it has another element x which is idempotent, that is, such that $2x = x$. Let indeed $\omega_{i,p}$ be the least multiple of p which is at least equal to i. Then $2\omega_{i,p} = \omega_{i,p}$. This is the only nonzero idempotent. Indeed, in $N_{i,p}$, the set of elements of the form $\omega_{i,p} + x$ is a group isomorphic to $\mathbb{Z}/p\mathbb{Z}$, which contains exactly one idempotent.

We define a distance on the natural integers by

$$
d_+(x, y) = \begin{cases} 0 & \text{if } x = y \\ 1/r_+(x, y) & \text{otherwise} \end{cases}
$$

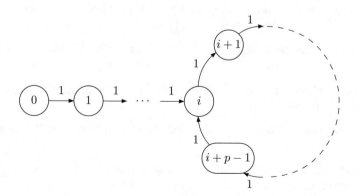

Fig. 2.1 The semigroup $N_{i,p}$

where

$$r_+(x, y) = \min\{i + p \mid x \not\equiv y \bmod (i, p)\}.$$

It can be easily verified that d_+ is an ultrametric distance (see Exercise 2.10). The completion of $\mathbb{N}$ with respect to this distance is the set of *profinite natural integers*, denoted $\widehat{\mathbb{N}}$.

As an important example, the sequence $n!$ converges in $\widehat{\mathbb{N}}$. Indeed, for $n < m$, $n! \equiv m! \bmod (i, p)$ holds if $i \leqslant n!$ and $p \leqslant n$. We denote $\omega = \lim n!$, where the limit is taken in $\widehat{\mathbb{N}}$. Note that $n!$ converges to 0 in $\widehat{\mathbb{Z}}$ but that $\omega \neq 0$ since $n! \not\equiv 0 \bmod (1, 1)$ and thus $d_+(n!, 0) \geqslant 1/2$ for $n \geqslant 1$.

As for profinite integers, an alternative way to define profinite natural integers is to consider the inverse limit $\varprojlim N_{i,p}$ of the family $N_{i,p}$ formed of the sequences $x_{i,p}$ of elements of $\mathbb{N}$ such that

(i) $x_{i,p} \leqslant i + p - 1$,
(ii) one has $x_{i,p} \equiv x_{j,q} \bmod (i, p)$ whenever $i \leqslant j$ and $p \mid q$.

We have $\widehat{\mathbb{N}} = \varprojlim N_{i,p}$. Indeed, $\mathbb{N}$ is included in $\varprojlim N_{i,p}$ by considering eventually constant sequences and the topology on $\mathbb{N}$ defined by the distance d_+ is the same as the product topology on $\varprojlim N_{i,p}$.

This allows us to define a semiring structure on $\widehat{\mathbb{N}}$. We prove now some properties of this semiring.

We first show that $\widehat{\mathbb{Z}}$ is a morphic image of $\widehat{\mathbb{N}}$. We denote by $\sim$ the equivalence on $\widehat{\mathbb{N}}$ defined by $x \sim y$ if $x \equiv y \bmod p$ for every $p \geqslant 1$.

Proposition 2.3.1 *The quotient of $\widehat{\mathbb{N}}$ by the equivalence $\sim$ is a group isomorphic to $\widehat{\mathbb{Z}}$.*

Proof The map $x \mapsto x \bmod p$ from $\mathbb{N}$ into $\mathbb{Z}/p\mathbb{Z}$ is defined and surjective for every $p \geqslant 1$. It defines a continuous map from $\mathbb{N}$ into $\varprojlim \mathbb{Z}/p\mathbb{Z}$ which extends to a surjective map from $\widehat{\mathbb{N}}$ onto $\widehat{\mathbb{Z}}$.   ∎

The map from $\widehat{\mathbb{N}}$ onto $\widehat{\mathbb{Z}}$ sending each element on its equivalence class modulo $\sim$ is called the *canonical projection* from $\widehat{\mathbb{N}}$ onto $\widehat{\mathbb{Z}}$.

Proposition 2.3.2 *The additive semigroup $\widehat{\mathbb{N}}$ has two idempotents, 0 and $\omega = \lim n!$.*

Proof Let $x \neq 0$ be an idempotent in $\widehat{\mathbb{N}}$. Then for every $i, p \geqslant 1$, the image of x in $N_{i,p}$ is $\omega_{i,p}$ and thus $x = \omega$.   ∎

Note that, in particular, $\omega \sim 0$. The next proposition shows that $\widehat{\mathbb{Z}}$ is embedded in $\widehat{\mathbb{N}}$.

Proposition 2.3.3 *The set K of elements $\widehat{\mathbb{N}}$ of the form $\omega + x$ for some $x \in \widehat{\mathbb{N}}$ is a group isomorphic to $\widehat{\mathbb{Z}}$. Moreover $\widehat{\mathbb{N}} \setminus K = \mathbb{N}$.*

Proof If $\omega + x \sim \omega + y$, we have $x \equiv y \bmod p$ for every $p \geq 1$ and thus $x \sim y$. This shows that the restriction of the canonical projection to $\omega + \widehat{\mathbb{N}}$ is injective. This proves the first assertion. Next, an $x \in \widehat{\mathbb{N}}$ is in $\widehat{\mathbb{N}} \setminus K$ if and only if there are i, p such that the image of x in $N_{i,p}$ belongs to $\{0, 1, \ldots, i-1\}$. This proves the second assertion.   ∎

2.4   Zero Set of a Recognizable Series

A power series $\sigma = \sum_{n \geq 0} s_n z^n$ with coefficients in $\mathbb{Q}$ is said to be *recognizable* if there is an integer $d \geq 1$ and a triple (λ, M, γ) consisting of a row vector $\lambda \in \mathbb{Q}^d$, a matrix $M \in \mathbb{Q}^{d \times d}$ and a column vector $\gamma \in \mathbb{Q}^d$ such that

$$s_n = \lambda M^n \gamma$$

for every $n \geq 1$. We say that the series σ is *recognized* by the triple (λ, M, γ). It is equivalent to saying that there exists a finite dimensional vector space V, a vector $\lambda \in V$, an endomorphism μ of V and a linear form γ on V such that $s_n = \langle \lambda \mu^n, \gamma \rangle$.

Example 2.4.1 Let $\sigma = \sum_{n \geq 0} F_n z^n$ where F_n is the n-th Fibonacci number. Then σ is recognized by the triple

$$\lambda = \begin{bmatrix} 1 & 0 \end{bmatrix}, \quad M = \begin{bmatrix} 1 & 1 \\ 1 & 0 \end{bmatrix}, \quad \gamma = \begin{bmatrix} 0 \\ 1 \end{bmatrix}.$$

The *zero set* of a series $\sigma = \sum_{n \geq 0} s_n z^n$ is the set of indices n such that $s_n = 0$. We will show how p-adic numbers allow to prove the following result, known as the Skolem-Mahler-Lech Theorem (see the Notes Section for references).

Theorem 2.4.2 *The zero set of a recognizable series is eventually periodic.*

We first prove the following lemma. A recognizable series is said to be *regular* if it can be recognized by a triple (λ, M, γ) with the matrix M invertible.

Lemma 2.4.3 *For any recognizable series $\sigma = \sum_{n \geq 0} s_n z^n$, there is an integer k such that the series $\sum_{n \geq 0} s_{n+k} z^n$ is regular.*

Proof Set $V = \mathbb{Q}^d$ and consider the sequence of subspaces $V \supseteq VM \supseteq VM^2 \supseteq \cdots$. The dimensions of these subspaces cannot decrease strictly more than d times and thus there is an integer $k \leq d$ such that $VM^{k+1} = VM^k$. Let $W = VM^k$ and let μ be the multiplication by M on W. Then $s_{n+k} = \langle (\lambda M^k) \mu^n, \gamma \rangle$ for all $n \geq 0$. Since μ is an isomorphism, this shows that the series $\sum_{n \geq 0} s_{n+k} z^n$ is regular.   ∎

The proof of Theorem 2.4.2 gives us a taste of the power of p-adic analysis. One of the first examples that is usually given to illustrate the differences between p-adic analysis and classical real analysis, is the fact that in $\mathbb{Q}_p$ a series $\sum a_n$ converges if and only if the sequence (a_n) converges to 0. It is also under mild conditions that one can commute the order in a double series in $\mathbb{Q}_p$, as next stated (see the Notes for a reference) (for a proof, see Gouvêa (1997, Proposition 4.1.4)).

Proposition 2.4.4 *For each $j, k \in \mathbb{N}$, let $b_{jk} \in \mathbb{Q}_p$, and suppose that we have*

1. $\lim\limits_{k \to \infty} b_{kj} = 0$ *for every j;*
2. $\lim\limits_{j \to \infty} b_{kj} = 0$ *uniformly in k, that is, for every $\varepsilon > 0$, there is N not depending on k such that $j \geqslant N \implies |b_{kj}| < \varepsilon$.*

Then, in $\mathbb{Q}_p$, both double series $\sum\limits_{j \geqslant 0} \sum\limits_{k \geqslant 0} b_{ij}$ and $\sum\limits_{k \geqslant 0} \sum\limits_{j \geqslant 0} b_{ij}$ converge and have equal sums.

Notice that $\lim\limits_{j \to \infty} b_{kj} = 0$ uniformly in k if b_{kj} is always of the form $a_{kj} p^j$ for some $a_{kj} \in \mathbb{Z}_p$, and that is the situation under which we shall next apply Proposition 2.4.4 in the proof of Theorem 2.4.2.

Proof of Theorem 2.4.2 Since the zero sets of the series $\sum_{n \geqslant 0} s_n z^n$ and $\sum_{n \geqslant k} s_n z^n$ differ by a finite set, we may assume that $\sigma = \sum_{n \geqslant 0} s_n z^n$ is regular. Let (λ, M, γ) be a triple recognizing σ with M invertible. We may also assume that λ, M, γ have integer coefficients. Let p be an odd prime not dividing $\det(M)$. Then M is invertible in $\mathbb{Z}/p\mathbb{Z}$ and, thus, there is an integer $m \geqslant 1$ such that $M^m \equiv I \bmod p$.

We are going to show that the zero set of σ is eventually periodic of period m. For this, we will show that, for every $r = 0, \ldots, m - 1$, the set

$$Z(r) = \{n \geqslant 0 \mid s_{nm+r} = 0\}$$

is either finite or equal to $\mathbb{N}$.

By the choice of m there is an integer matrix N such that $M^m = I + pN$. Consider the function P defined on $\mathbb{Z}$ by

$$P(n) = \lambda M^r (I + pN)^n \gamma$$

in such a way that $s_{nm+r} = P(n)$. We claim that

$$P(n) = \sum_{j \geqslant 0} p^j P_j(n) \tag{2.5}$$

where the P_j are polynomials in n with integer coefficients and the sum is taken in the ring $\mathbb{Z}_p$ of p-adic integers. Indeed, we have by the Binomial Formula

$$(I + pN)^n = \sum_{k \geqslant 0} \binom{n}{k} p^k N^k.$$

But (see Exercise 2.13)

$$\operatorname{ord}_p(k!) = \lfloor \frac{k}{p} \rfloor + \lfloor \frac{k}{p^2} \rfloor + \cdots \leqslant \frac{k}{p} + \frac{k}{p^2} + \cdots = \frac{k}{p-1}$$

and so we have

$$\operatorname{ord}_p(p^k/k!) = k - \operatorname{ord}_p(k!) \geqslant k\frac{p-2}{p-1}.$$

Since $p > 2$, we have $\lim_{k \to \infty} \operatorname{ord}_p(p^k/k!) = \infty$. Therefore, we may consider a natural number $K(j)$ such that $\operatorname{ord}_p(p^k/k!) > j$ whenever $k > K(j)$. This shows that $p^k/k! \in \mathbb{Z}_p$ and that we can write

$$p^k/k! = \sum_{j \geqslant 0} a_{kj} p^j$$

with $0 \leqslant a_{kj} < p$ and $a_{kj} = 0$ whenever $k > K(j)$. Then the polynomials

$$P_j(n) = \sum_{k=0}^{K(j)} a_{kj}(\lambda M^r N^k \gamma)n(n-1)\cdots(n-k+1)$$

satisfy the following chain of equalities, where in the second one we use Proposition 2.4.4.

$$\sum_{j \geqslant 0} P_j(n)p^j = \sum_{j \geqslant 0}\sum_{k \geqslant 0} a_{kj}(\lambda M^r N^k \gamma)n(n-1)\cdots(n-k+1)p^j$$

$$= \sum_{k \geqslant 0}\left(\sum_{j \geqslant 0} a_{kj} p^j\right)(\lambda M^r N^k \gamma)n(n-1)\cdots(n-k+1)$$

$$= \sum_{k \geqslant 0}\frac{p^k}{k!}(\lambda M^r N^k \gamma)n(n-1)\cdots(n-k+1)$$

$$= \sum_{k \geqslant 0} p^k\binom{n}{k}(\lambda M^r N^k \gamma)$$

$$= \lambda M^r(I + pN)^n \gamma = P(n).$$

This proves our claim.

Assume that $Z(r)$ is infinite. Then, for every $j \geqslant 0$, $P(n)$ modulo p^j is a polynomial in n which has an infinity of roots and thus is zero. Thus $P_j(n) = 0$ for every $j \geqslant 0$ and thus $P = 0$, which implies $Z(r) = \mathbb{N}$. ∎

2.5   Odometers

Let (p_n) be a strictly increasing sequence of natural integers such that p_n divides p_{n+1} for all n. We consider the set $\mathbb{Z}_{(p_n)}$ of infinite expansions

$$x = x_0 p_0 + x_1 p_1 + x_2 p_2 + \cdots$$

with $0 \leqslant x_n < p_n$ called (p_n)-*adic integers*. Formally, this amounts to consider the direct product $\prod_{n \geqslant 0} \mathbb{Z}/p_n\mathbb{Z}$ with the product topology of the discrete topologies. It is a compact topological ring containing $\mathbb{Z}$ as a dense subring.

When $p_n = p^n$ for all n, we find again the ring of p-adic integers. When $p_n = n!$ for all n, we find the ring of profinite integers.

A *topological dynamical system* is a pair (X, T) consisting of a nonempty compact topological space and a homeomorphism $T : X \to X$. We shall see many of these systems later in this book.

Let $T : \mathbb{Z}_{(p_n)} \to \mathbb{Z}_{(p_n)}$ be the transformation $T(x) = x + 1$. The pair $(\mathbb{Z}_{(p_n)}, T)$ is a topological dynamical system called an *odometer*.

Let (X, T) be a topological dynamical system. The *orbit* of a point $x \in X$ is the set $\{T^n(x) \mid x \in X\}$.

A topological dynamical system (X, T) is said to be *minimal* if the orbit of every point $x \in X$ is dense in X.

Proposition 2.5.1 *Every odometer is a minimal topological dynamical system.*

Proof Let $x \in X$. Set $x = x_0 p_0 + x_1 p_1 + \cdots$ with $0 \leqslant x_n < p_n$ and $y_n = x_0 p_0 + \cdots + x_n p_n$. Let $z \in \mathbb{Z}$ and set $z_n = z \bmod p_n$. Then

$$\lim T^{z_n - y_n}(x) = \lim(z_n + x_{n+1} p_{n+1} + \cdots) = z$$

showing that the closure of the orbit of x contains z. Since $\mathbb{Z}$ is dense in $\mathbb{Z}_{(p_n)}$, this gives the conclusion. ∎

2.6   Exercises

2.6.1   Section 2.2

2.1 Let $x, y \in \mathbb{Z}$. Show that $|xy|_\pi \leqslant \sqrt{|x|_\pi |y|_\pi}$.

2.2 Show that if (x_n) and (y_n) are Cauchy sequences in $\mathbb{Z}$, then $(x_n + y_n)$ and $x_n y_n$ are Cauchy sequences and that the corresponding result only depends on the equivalence classes of (x_n), (y_n).

Solve Exercises 2.3 and 2.4 using the definition of $\widehat{\mathbb{Z}}$ as the completion of $\mathbb{Z}$ under the profinite distance.

2.3 Show that if n and m are elements of $\mathbb{Z}$, then n divides m in $\mathbb{Z}$ if and only if n divides m in $\widehat{\mathbb{Z}}$.

2.4 Show that if $n \in \mathbb{Z}$ and $x \in \widehat{\mathbb{Z}}$ are such that $nx = 0$, then $n = 0$ or $x = 0$.

2.5 Establish the uniqueness of the quotient and of the remainder in the Euclidean division in $\widehat{\mathbb{Z}}$ of a profinite integer x by an integer $n \geqslant 1$.

2.6 Prove that $\mathbb{Z}_p$ is an integral domain.

2.7 Show that for every $k \geqslant 0$,

$$F_k = 1/\sqrt{5}(\varphi^k - \hat{\varphi}^k)$$

where $\varphi = (1 + \sqrt{5})/2$ and $\hat{\varphi} = (1 - \sqrt{5})/2$. This closed form of the Fibonacci numbers is known as *Binet's Formula*.

2.8 The nth *Pisano period*, denoted $\pi(n)$ is the least period of the sequence $F_k \bmod n$.

(i) Verify the values of $\pi(n)$ for $n \leqslant 15$ given in Table 2.3.
(ii) Show that if p is prime and $p \equiv \pm 1 \bmod 10$, then $\pi(p) \mid p - 1$. (Hint: use the fact that if $p \equiv \pm 1 \bmod 10$, then 5 is a nonzero square modulo p and use the analogue of Binet's Formula in the field $\mathbb{Z}/p\mathbb{Z}$.)
(iii) Show that if p is prime and $p \equiv \pm 3 \bmod 10$, then $\pi(p) \leqslant 2(p + 1)$. (Hint: use the fact that 5 is not a square modulo p and that Binet's Formula holds in the field GF_{p^2}.)
(iv) Show that if m, n are relatively prime, $\pi(mn)$ is the least common multiple of $\pi(m)$ and $\pi(n)$.
(v) Show that $\pi(p^k) \mid p^{k-1}\pi(p)$. (Hint: for $p \neq 2$ prove using Binet's Formula that the sequences $u_n = F_n$ and $v_n = F_{n-1} + F_{n+1}$ satisfy for $n, a \geqslant 0$

$$u_{an} = 2^{1-a}u_n(Ku_n^2 + pv_n^{a-1}) \tag{2.6}$$

$$u_{an+1} = 2^{-a}(Ku_n^2 + au_nv_n^{a-1} + v_n^a) \tag{2.7}$$

for some integer K. For $p = 2$, use the formulas

$$u_{2n} = u_n(u_{n-1} + u_{n+1}) \tag{2.8}$$

$$u_{2n+1} = u_{n+1}^2 + u_n^2 \tag{2.9}$$

which can be verified by induction.)

Table 2.3 The Pisano periods of $n \leqslant 15$

n	1	2	3	4	5	6	7	8	9	10	11	12	13	14	15
$\pi(n)$	1	3	8	6	20	24	16	12	24	60	10	24	28	48	40

2.9 Show that for $n \geqslant 4$, if $i \equiv j \bmod n!$ then $F_i \equiv F_j \bmod n!$. (Hint: using Exercise 2.8, show that, for prime p and $k \geqslant 1$, if $p^k \mid n!$ then $\pi(p^k) \mid n!$.)

2.6.2   Section 2.3

2.10 Verify that d_+ is a ultrametric distance.

2.6.3   Section 2.4

2.11 Show that $\mathrm{ord}_p(p^n!) = 1 + p + \cdots + p^{n-1}$.

2.12 Show that if a is not divisible by p, then $\mathrm{ord}_p((ap^n)!) = a(1+p+\cdots+p^{n-1})$.

2.13 Show that $\mathrm{ord}_p(k!) = \lfloor \frac{k}{p} \rfloor + \lfloor \frac{k}{p^2} \rfloor + \cdots$.

2.7   Solutions

2.7.1   Section 2.2

2.1 It suffices to consider the case where $xy \neq 0$. Let q be a prime such that $r(xy) = q^{\mathrm{ord}_q(xy)+1}$. Then we have $r(xy) \geqslant q$ and

$$r(xy) = q^{\mathrm{ord}_q(x)+1} \cdot q^{\mathrm{ord}_q(y)+1} \cdot q^{-1} \geqslant r(x) \cdot r(y) \cdot q^{-1} \geqslant \frac{r(x) \cdot r(y)}{r(xy)},$$

thus $r(xy)^2 \geqslant r(x) \cdot r(y)$.

2.2 We have

$$d(x_i + y_i, x_j + y_j) \leqslant \max\{d(x_i + y_i, x_j + y_i), d(x_j + y_i, x_j + y_j)\}$$
$$\leqslant \max\{|x_i - x_j|_\pi, |y_i - y_j|_\pi\} = \max\{d(x_i, x_j), d(y_i, y_j)\}$$

and thus $(x_n + y_n)$ is a Cauchy sequence. If we start with equivalent sequences (x_n') and (y_n'), then

$$d(x_i' + y_i', x_i + y_i) \leqslant \max\{d(x_i' + y_i', x_i' + y_i), d(x_i' + y_i, x_i + y_i)\}$$
$$\leqslant \max\{|y_i' - y_i|_\pi, |x_i' - x_i|_\pi\} = \max\{d(y_i', y_i), d(x_i', x_i)\}.$$

This settles the case concerning the sequence $(x_n + y_n)$. For the one concerning the sequence $(x_n y_n)$, one follows small variations of these reasonings, after noting that

$$d(x_i y_i, x_j y_j) \leqslant \max\{d(x_i y_i, x_j y_i), d(x_j y_i, x_j y_j)\}$$

$$\leqslant \max\{|y_i|_\pi^{\frac{1}{2}} \cdot |x_i - x_j|_\pi^{\frac{1}{2}}, |x_j|^{\frac{1}{2}} \cdot |y_i - y_j|^{\frac{1}{2}}\},$$

the last inequality holding by Exercise 2.1.

2.3 The "only if" is obvious. Conversely, suppose that there is $x \in \widehat{\mathbb{Z}}$ such that $m = nx$. Let (x_i) be a sequence of integers converging to x. Then we have $\lim_{i \to \infty} d(nx_i, m) = 0$. By the definition of the distance d, it follows that for every integer $k \geqslant 1$ there is j_k such that $nx_i \equiv m \bmod k$ for all $i > j_k$, where this the congruence is taken in $\mathbb{Z}$. Choosing $k = n$, we conclude that n divides m in $\mathbb{Z}$.

2.4 Suppose that $n \neq 0$. Let (x_i) be a sequence of integers converging to x. Then we have $\lim_{i \to \infty} |nx_i|_\pi = 0$. Therefore, for every integer $k \geqslant 1$ there is j_k such that $nx_i \equiv 0 \bmod nk$ for all $i > j_k$. But $nx_i \equiv 0 \bmod nk$ implies $x_i \equiv 0 \bmod k$. Hence $|x_i|_\pi < \frac{1}{k}$ if $i > \max\{j_1, j_2, \ldots, j_k\}$. Taking limits, we get $|x|_\pi \leqslant \frac{1}{k}$ for every $k \geqslant 1$, which means that $|x|_\pi = 0$, that is, $x = 0$.

2.5 Let $q_1, q_2 \in \widehat{\mathbb{Z}}$ and $r_1, r_2 \in \mathbb{Z}$, with $0 \leqslant r_1, r_2 < n$, be such that $nq_1 + r_1 = nq_2 + r_2$. First notice that $n | (r_1 - r_2)$ (cf. Exercise 2.3), whence $r_1 = r_2$. Therefore, we have $n(q_1 - q_2) = 0$. In view of Exercise 2.4, we get $q_1 = q_2$.

2.6 Let x, y be two nonzero elements of $\mathbb{Z}_p$. Then there exist $m, n \geqslant 0$ such that $x \not\equiv 0 \bmod p^n$ and $y \not\equiv 0 \bmod p^m$. Since $xy \not\equiv 0 \bmod p^{n+m}$, we conclude that $xy \neq 0$.

2.7 Since φ and $\hat{\varphi}$ are the roots of $x^2 - x - 1$, the sequences φ^k and $\hat{\varphi}^k$ follow the same recurrence as the Fibonacci numbers. Since the values of both sides coincide for $k = 0, 1$, the result follows.

2.8

(i) The residues modulo 2 of the Fibonacci numbers are $0, 1, 1, 0, 1, 1, \ldots$ showing that $\pi(2) = 3$. The other cases are treated similarly.

(ii) Since the roots of $x^2 - x - 1$ exist and are distinct in $\mathbb{Z}/p\mathbb{Z}$, we have the analogue of Binet's Formula in $\mathbb{Z}/p\mathbb{Z}$. Since the multiplicative group of $\mathbb{Z}/p\mathbb{Z}$ is cyclic of order $p - 1$, we have $\pi(p) \mid p - 1$.

(iii) The roots α, β of $x^2 - x - 1$ in F_{p^2} are distinct and satisfy $\alpha^p = \beta$ and $-1 = \alpha\beta$. Thus $\alpha^{2(p+1)} = (\alpha\beta)^2 = 1$.

(v) By Binet's Formula, we have $u_n = (\varphi^n - \hat{\varphi}^n)/\sqrt{5}$. Similarly since $v_0 = 2, v_1 = 1$, we have $v_n = \varphi^n + \hat{\varphi}^n$. Thus, by the Binomial Formula,

$$u_{an} = (\varphi^{an} - \hat{\varphi}^{an})/\sqrt{5}$$

$$= (2^{-a}(\sqrt{5}u_n + v_n)^p - 2^{-a}(-\sqrt{5}u_n + v_n)^a)/\sqrt{5}$$

$$= 2^{1-a} \sum_{j \text{ odd}} \binom{a}{j} 5^{(j-1)/2} u_n^j v_n^{a-j}$$

$$= 2^{1-a} u_n (K u_n^2 + p v_n^{a-1}).$$

The derivation of the second formula is analogous. Assume now that $t \geqslant 1$ is the largest integer such that $\pi(p^t) = \pi(p)$. We show by induction on $e \geqslant t$ that $\pi(p^e) = p^{e-t} \pi(p)$. Indeed, taking into account the fact that $\gcd(u_n, v_n)$ is 1 or 2, Eq. (2.6) shows that if n is the least integer such that $u_n \equiv 0 \bmod p^e$ but $u_n \not\equiv 0 \bmod p^{e+1}$, then pn is the least integer such that $u_{pn} \equiv 0 \bmod p^{e+1}$ but $u_{pn} \not\equiv 0 \bmod p^{e+2}$. Let $x \geqslant 1$ be the least integer such that $u_{pnx+1} \equiv 1 \bmod p^{e+1}$ (so that $\pi(p^{e+1}) = pnx$). Then Eq. (2.7) with nx instead of n and $a = p$ shows that $(v_{nx}/2)^p \equiv 1 \bmod p^{e+1}$ and thus $v_{nx} \equiv 2 \bmod p^e$ and finally $u_{nx} \equiv 1 \bmod p^e$ showing that $\pi(p^e) = nx$. The derivation of the case $p = 2$ using Eqs. (2.8) and (2.9) is similar.

2.9 For $p = 2$, by Exercise 2.8 (v), $\pi(p^k)$ divides $2^{k-1} \cdot 3$ which divides $n!$ since 2^k divides $n!$.

For $p = 5$, $\pi(p^k)$ divides $4 \cdot 5^k$ which divides $n!$ since $n > 3$.

For $p \equiv \pm 1 \bmod 10$, by Exercise 2.8 (ii), $\pi(p^k)$ divides $p^{k-1}(p - 1)$ which divides $n!$ because $p^k \mid n!$ implies $p \mid n!$ which in turn implies $p - 1 < n$.

Finally, for $p \equiv \pm 3 \bmod 10$, by Exercise 2.8 (iii), $\pi(p^k)$ divides $2(p + 1)p^{k-1}$ which divides $n!$ because $2(p + 1)$ divides $n!$ (indeed, $p \mid p^k \mid n!$ implies $n \geqslant p$ so $2(p + 1)$ is the product of 4 and $(p + 1)/2$ which divide $n!$ and are the same factor only for $p = 7$ in which case $2(p + 1) = 4$ which divides $n!$) and $2(p + 1)$ and p^{k-1} are coprime.

2.7.2   Section 2.3

2.10 The inequality $d_+(x, z) \leqslant \max\{d_+(x, y), d_+(y, z)\}$ follows from $r_+(x, z) \geqslant \min\{r_+(x, y), r_+(y, z)\}$.

2.7.3   Section 2.4

2.11 For $1 \leqslant t \leqslant n$, there are p^{n-t} multiples of p^t in $\{1, 2, \ldots, p^n\}$ which are the ap^t for $1 \leqslant a \leqslant p^{n-t}$.

2.12 There are ap^{n-t} multiples of p^t in $\{1, 2, \ldots, ap^n\}$ which are the bp^t for $1 \leqslant b \leqslant ap^{n-t}$.

2.13 Let $k! = a_0 + a_1 p + \cdots + a_n p^n$ be the expansion of $k!$ in base p. For $1 \leqslant t \leqslant n$ there are $a_t + \cdots + a_n p^{n-t} = \lfloor \frac{k}{p^t} \rfloor$ multiples of p^t in $\{1, 2, \ldots, k!\}$ which are the bp^t for $1 \leqslant b \leqslant a_t + \cdots + a_n p^{n-t}$.

2.8 Notes

The group of profinite integers is introduced in Fried and Jarden (2008) under the name of *Prüfer group*. The factorial number system is described in Knuth (1998, page 175). Profinite Fibonacci numbers are described in Lenstra (2005). Exercise 2.8 is from Wall (1960). Exercise 2.9 is due to Hendrik Lenstra (personal communication).

Recognizable series as defined here are a particular case of a more general notion (see Berstel and Reutenauer (2011)). We shall come back to the use of the term recognizable in this case in Chap. 4.

The Skolem-Mahler-Lech Theorem (Theorem 2.4.2) is a well-known example of a statement on ordinary integers proved using p-adic analysis although p-adic numbers are not part of the statement. A few remarks are in order about this theorem. First the theorem is usually stated for linear recurrent sequences and not for recognizable series. The equivalence of the two notions is classical. See Berstel and Reutenauer (2011) for example. Next, the complete form of the theorem is on an arbitrary field of characteristic 0 and not only on $\mathbb{Q}$ (the statement on $\mathbb{Q}$ was actually first proved by Skolem alone). Next, a proof which does not explicitly use p-adic analysis was proposed by Hansel (1986). It is however very close to the one presented here. Finally, it is interesting to note that the result is nonconstructive in the sense that the proof gives no indication on how to effectively compute the zero set as pointed out in Tao (2007). Actually the question whether the zero set of a recognizable series is empty is not known to be decidable.

For a proof of Proposition 2.4.4 see for example Gouvêa (1997, Proposition 4.1.4).

A general reference concerning topological dynamical systems is Walters (1982). Odometers are described in Berthé and Rigo (2010).

Chapter 3
Profinite Groups and Semigroups

3.1 Introduction

In this chapter, we recall the basic definitions of topological groups and semigroups (and monoids, as well) and, in particular, compact ones. A reader familiar with finite semigroups will find in compact semigroups a familiar feature: the closure of the semigroup generated by any element contains an idempotent. For a reader with a previous familiarity with topological groups, one of the first main differences is that in a compact semigroup there can be more than one idempotent.

Section 3.2 is intended to be used as a memento of the basic notions of topology used in the book. We have made the choice to use general topological spaces instead of metric spaces because it clarifies some arguments, although all topological spaces which we shall use are metrizable. The main drawback is that sequences have to be replaced by nets (and subsequences by the more delicate notion of subnet).

In Sects. 3.3 and 3.4, we recall the basic definitions and properties of topological semigroups and groups. We continue with the classical presentation of Green's relations in a topological semigroup (Sect. 3.6). Profinite semigroups are introduced in Sect. 3.9 using the notion of projective limit introduced in Sect. 3.8.

3.2 Topological and Metric Spaces

We give a condensed introduction to the basic notions of topology which may serve as a reminder for readers already with some knowledge of this field and as a first approach to others.

© Springer Nature Switzerland AG 2020
J. Almeida et al., *Profinite Semigroups and Symbolic Dynamics*, Lecture Notes in Mathematics 2274, https://doi.org/10.1007/978-3-030-55215-2_3

3.2.1 Topological Spaces

Definition and First Examples

Recall that a *topological space* is a set X together with a family of subsets of X, called *open sets*, which is such that

(i) X and $\emptyset$ are open,
(ii) any finite intersection of open sets is open,
(iii) an arbitrary union of open sets is open.

The complements of open sets are the *closed sets*. A set which is both closed and open is called *clopen*.

Example 3.2.1 The *discrete topology* on a set X corresponds to the choice of all subsets being open (and thus also closed). One then also says that X is a *discrete space*.

Example 3.2.2 The *trivial topology* on a set X corresponds to the choice of $\emptyset$ and X being the unique open sets.

Any subset S of a topological space X inherits the topology of X by considering as open sets the intersections of the form $U \cap S$ with U open in X. This topology is called the *induced topology* and S is a *subspace* of X.

A *basis* for the topology of X is a collection $\mathcal{B}$ of open sets such that any open set is a union of elements of $\mathcal{B}$. A *subbasis* is a collection $\mathcal{C}$ of open sets such that the finite intersections of elements of $\mathcal{C}$ form a basis of the topology.

Example 3.2.3 The usual topology of the set $\mathbb{R}$ of real numbers is the one where the intervals of the form (a, b) form a basis of the topology. The collection of intervals of the form $(a, +\infty)$ or $(-\infty, b)$ is a subbasis for the same topology. The intervals of the form $[a, b]$ are examples of closed sets for this topology.

A *neighborhood* of a point $x \in X$ is a subset of X containing an open set which contains x.

Example 3.2.4 For the topology in Example 3.2.3, the set $]0, 2]$ is not open (as it contains no open interval containing 2), but it is a neighborhood of 1.

The *topological closure*, or just *closure*, of a set U contained in the topological space X, usually denoted by $\overline{U}$, is the intersection of all closed sets containing U. When the equality $\overline{U} = X$ holds one says that U is *dense* in X. It turns out that U is dense in X if and only if U intersects every nonempty open set of X (cf. Exercise 3.1).

Example 3.2.5 The set of rational numbers is dense in the usual topology of $\mathbb{R}$.

A *Hausdorff space* is a topological space such that any two distinct points have disjoint neighborhoods. The spaces in Examples 3.2.1 and 3.2.3 are Hausdorff.

In contrast, the trivial space in Example 3.2.2 is not Hausdorff, except when $\text{Card}(X) \leqslant 1$. In this book we are only interested in Hausdorff spaces.

A direct product $X = \prod_{i \in I} X_i$ of a family of topological spaces X_i is a topological space for the *product topology* defined as follows. For $i \in I$, denote by $\pi_i : X \to X_i$ the projection on X_i. The open sets of the product topology are the unions of finite intersections of sets of the form $\pi_i^{-1}(U)$ where $U \subseteq X_i$ is open. Thus a subbasis of the product topology is formed of the sets $\prod_{i \in I} U_i$ where the $U_i \subseteq S_i$ are open and $U_i = S_i$ for all i except one. A basis of the product topology is formed of the sets $\prod_{i \in I} U_i$ where the $U_i \subseteq X_i$ are open and $U_i = X_i$ for all but a finite number of indices i.

Example 3.2.6 We view the product $\prod_{n \in \mathbb{N}} \mathbb{Z}/n\mathbb{Z}$ as a topological space for the product topology, by endowing each quotient $\mathbb{Z}/n\mathbb{Z}$ with the discrete topology. The sets of the form $\{(x_n)_{n \in \mathbb{N}} \in \prod_{n \in \mathbb{N}} \mathbb{Z}/n\mathbb{Z} \mid x_k = r\}$, for $k \geqslant 0$ and $r \in \mathbb{Z}/k\mathbb{Z}$, form a subbasis of the space.

Consider a topological space Y together with a set I, and the set Y^I of functions $I \to Y$. The elements of Y^I may be seen as the elements of the product of copies of X indexed by I, with the i-th component of $f \in Y^I$ being $f(i)$. In this particular context, the product topology with which we endow Y^I is frequently called the *topology of pointwise convergence*, or simply *pointwise topology*.

Example 3.2.7 Let A be a discrete space. The set $X = A^{\mathbb{N}}$ of right infinite sequences of elements of A is a topological space for the pointwise topology (we see X as the product space $X = \prod_{i \in \mathbb{N}} X_i$, with $X_i = A$ for all i). The sets of the form $\{x_0 x_1 x_2 \ldots \mid x_i = a\}$, where $a \in A$ and $i \in \mathbb{N}$, form a subbasis of X.

Nets

A *directed set* is a partially ordered set such that any two elements have a common upper bound. Every chain (that is, totally ordered set) is a directed set.

Example 3.2.8 The set $\mathbb{N}$ of natural integers with the usual order is a chain. The same set with the order $x \leqslant y$ if x divides y is also directed, but not a chain.

We now introduce nets, which are a generalization of the usual notion of a sequence indexed by the natural integers. A *net* $(x_i)_{i \in I}$ in a topological space X is a sequence of elements $x_i \in X$ indexed by a directed set I. We occasionally use the notation (x_i) instead of $(x_i)_{i \in I}$.

The term *sequence* will from hereon be reserved to nets of the form $(x_n)_{n \geqslant p}$ for some integer p, that is, nets indexed by sets having the same order as that of the natural integers. This is precisely the familiar use of the term sequence.

The net (x_i) *converges* to a limit $x \in X$ if it is eventually in every neighborhood of x, that is, for every neighborhood U of x there is an $i \in I$ such that U contains all x_j for $i \leqslant j$. We extended the usual notation for sequences, writing $x_i \to x$, or

$\lim x_i = x$, if (x_i) converges to x. When X is Hausdorff, the limit is unique (see Exercise 3.2).

Example 3.2.9 If X is a discrete space, then $(x_i) \rightarrow x$ if and only if there is i_0 such that $x_i = x$ whenever $i_0 \leqslant i$.

A set is closed if and only if it contains the limits of all nets of its elements (see Exercise 3.3).

In the next lines we introduce the notion of subnet which replaces, for nets, the notion of subsequence. A subset J of a directed set I, with partial order $\leqslant$, is called *cofinal* if for every $i \in I$ there is $j \in J$ such that $i \leqslant j$. If J is cofinal, it is directed (see Exercise 3.4). A map $\varphi : J \rightarrow I$ between directed sets is *increasing* when $j_1 \leqslant j_2$ implies $\varphi(j_1) \leqslant \varphi(j_2)$ for every $j_1, j_2 \in J$. A *subnet* of a net $(x_i)_{i \in I}$ is a net of the form $(x_{\varphi(j)})_{j \in J}$ where J is a directed set and $\varphi : J \rightarrow I$ is an increasing map such that $\varphi(J)$ is cofinal in I. Frequently, the subnet $(x_{\varphi(j)})_{j \in J}$ is denoted by $(x_{i_j})_{j \in J}$, that is, $\varphi(j)$ is denoted i_j.

Formally, a *subsequence* of a sequence (x_n) is then just a subnet (x_{n_m}) of (x_n) that turns out to be also a sequence.[1]

A *cluster point* (or *accumulation point*) of a net $(x_i)_{i \in I}$ in a topological space X is a point $x \in X$ such that for every neighborhood U of x and every $i \in I$, there is some $j \geqslant i$ such that $x_j \in U$. It turns out that x is a cluster point of $(x_i)_{i \in I}$ if and only if some subnet of $(x_i)_{i \in I}$ converges to x, see Exercise 3.8.

Continuity

A map $f : X \rightarrow Y$ between topological spaces X, Y is *continuous* if for every open set $U \subseteq Y$, the set $f^{-1}(U)$ is open in X. Note that the composition of continuous functions is still continuous. The identity on X is a trivial example of a continuous function. Let us see another couple of simple but important examples.

Example 3.2.10 If X is endowed with the discrete topology and Y is any topological space, then every map $f : X \rightarrow Y$ is continuous.

Example 3.2.11 A map $f : \mathbb{R} \rightarrow \mathbb{R}$ is continuous for the usual topology of $\mathbb{R}$ if and only if it is continuous in the familiar sense of Calculus.

When we have a continuous bijection $f : X \rightarrow Y$ for which the inverse $f^{-1} : Y \rightarrow X$ is also continuous, f is said to be a *homeomorphism*, and X are Y are *homeomorphic* spaces. Homeomorphic spaces are considered "equal" topological spaces, like isomorphic groups are "equal" groups.

---

[1] In what is perhaps a more familiar setting, in the definition of a subsequence $(x_{n_m})_{m \geqslant 1}$ of a sequence $(x_n)_{n \geqslant 1}$, one often also requires the mapping $m \mapsto n_m$ to be strictly increasing, not merely increasing. But this subtle difference is of no consequence for the sequel, as we require the set $\{n_m \mid m \geqslant 1\}$ to be cofinal in the set of natural integers.

We have the following useful characterization of continuity in terms of nets: a map f is continuous if and only if the net $f(x_i)$ converges to $f(x)$ for every convergent net x_i with limit x (see Exercise 3.11).

Example 3.2.12 Given a direct product $X = \prod_{i \in I} X_i$ of a family of topological spaces X_i, it follows directly from the definition of the product topology that the projections $\pi_i : X \rightarrow X_i$ are continuous. Quite often, one uses the description of this fact in terms of nets: a net $(x_k)_{k \in K}$ of elements of X converges to x if and only if, for every $i \in I$, the net $\pi_i(x_k)$ converges to $\pi_i(x)$.

We now look at a particular instance of the previous example.

Example 3.2.13 For the pointwise convergence topology of Y^X, one has $f_k \rightarrow f$ if and only if $f_k(x) \rightarrow f(x)$ for every $x \in X$, which explains the terminology.

3.2.2 Metric Spaces

A *metric space* is a set S with a *distance*, that is, with a map $d : S \times S \rightarrow \mathbb{R}_+$ such that

 (i) $d(x, y) = 0$ if and only if $x = y$,
 (ii) $d(x, y) = d(y, x)$,
 (iii) $d(x, y) \leqslant d(x, z) + d(z, y)$.

Such a metric space is sometimes denoted as the pair (S, d).

Example 3.2.14 The Euclidean distance on $\mathbb{R}^n$ is defined by the equality $d(x, y)^2 = \sum_{i=1}^{n} (x_n - y_n)^2$.

The *open ball* centered at $x \in S$ with radius ε is the set

$$B_d(x, \varepsilon) = \{y \in S \mid d(x, y) < \varepsilon\}.$$

In the absence of ambiguity, we drop the subscript d in this notation, writing $B(x, \varepsilon)$ instead.

A metric space is a topological space for which the open sets are the unions of open balls. A topological space is *metrizable* if its topology can be defined by a distance. There can be however several distances defining the same topology (see Exercise 3.14 for example).

Example 3.2.15 The topology of the real numbers is metrizable, as it may be defined by the Euclidian distance.

Example 3.2.16 The discrete topology of a set A is metrizable: it is defined by the metric d such that $d(x, y) = 1$ if and only if $x \neq y$, whenever $x, y \in A$.

Metric spaces have many properties which make them easier to handle than general topological spaces.

First, any metric space is Hausdorff (see Exercise 3.13).

Next, a subset U of a metric space X is closed if and only if it contains the limits of all sequences of its elements (Exercise 3.15).

Finally, a map $f : X \to Y$ between topological spaces with X metrizable is continuous if and only if $(f(x_n))_{n \geqslant 1} \to f(x)$ whenever $(x_n)_{n \geqslant 1} \to x$ (see Exercise 3.16).

Note that a map $f : X \to Y$ from a metric space (X, d_X) to a metric space (Y, d_Y) is continuous if, for every $x \in X$, and every $\varepsilon > 0$, there is $\delta_{x,\varepsilon} > 0$ such that $d_X(x, y) \leqslant \delta_{x,\varepsilon}$ implies $d_Y(f(x), f(y)) \leqslant \varepsilon$. The map f satisfies the stronger property of being *uniformly continuous* if for every $\varepsilon > 0$ there is $\delta_\varepsilon > 0$ such that $d_X(x, y) \leqslant \delta_\varepsilon$ implies $d_Y(f(x), f(y)) \leqslant \varepsilon$.

The concrete topological spaces which play a role in this book are metrizable spaces. However, we do not make the implicit hypothesis that all topological spaces that we consider from hereon are supposed to be metrizable because that would hide the precise hypotheses needed for each statement.

Proposition 3.2.17 *A denumerable product of metrizable spaces is metrizable.*

The proof is left as an exercise (see Exercise 3.19). For example, the space $A^{\mathbb{N}}$ from Example 3.2.7 is metrizable. A non denumerable product of metrizable spaces need not be metrizable (see Exercises 3.21 and 3.22).

A *Cauchy sequence* (already met in Chap. 2) in a metric space (X, d) is a sequence (x_n) such that for every $\varepsilon > 0$ there is an integer $n \geqslant 1$ such that, for all $i, j \geqslant n$, one has $d(x_i, x_j) \leqslant \varepsilon$.

A metric space is *complete* if every Cauchy sequence converges.

The *completion* of a metric space X is the complete metric space X' containing X as a dense subspace with the following universal property: if Y is a complete metric space and $f : X \to Y$ is uniformly continuous, then there is a unique uniformly continuous function from X' to Y which extends f (see Exercise 3.23).

3.2.3 Compact Spaces

An *open cover* of a space X is a family $(U_i)_{i \in I}$ consisting of open sets U_i and such that $X = \bigcup_{i \in I} U_i$. A *subcover* of the open cover $(U_i)_{i \in I}$ is an open cover of the form $(U_i)_{i \in J}$, with $J \subseteq I$. To say that the subcover is finite means that J is finite.

A topological space X is *compact* if it is Hausdorff and if from any open cover of X, one can extract a finite subcover of X.

Example 3.2.18 Every finite set, when endowed with the discrete topology, is a compact topological space.

Next is an infinite example, obtained by adjoining to the set $\mathbb{N}$ of natural numbers a point at infinity (appropriately denoted ∞).

Example 3.2.19 Consider the topological space $\mathbb{N}_\infty = \mathbb{N} \cup \{\infty\}$, where ∞ is an extra element added to the discrete space $\mathbb{N}$ of natural numbers, in such a way that the open sets of $\mathbb{N}_\infty$ are the subsets of $\mathbb{N}$ and the complements in $\mathbb{N}_\infty$ of the finite subsets of $\mathbb{N}$ (compare with Exercise 3.21). Then $\mathbb{N}_\infty$ is a compact space. In this topology, for any sequence (x_n) of natural integers, the expression $(x_n)_n \to \infty$ has the familiar meaning used in Calculus.

A closed subset of a compact space is compact (Exercise 3.25).

Example 3.2.20 As a natural companion to Example 3.2.19, we have the compact space $(\mathbb{Z}_+)_\infty = \mathbb{N}_\infty \setminus \{0\}$, a closed subspace of $\mathbb{N}_\infty$.

A family (U_i) of subsets of X has the *finite intersection property* if any intersection of finitely many of the U_i is nonempty. A family of open sets of the space X is an open covering of X if and only if the family of its complements has an empty intersection. Hence, we have the following useful equivalent formulation of the definition of compact space: a Hausdorff topological space X is compact if and only if any family of closed subsets of X having the finite intersection property has a nonempty intersection.

Example 3.2.21 If the family of nonempty closed sets $(U_i)_{i \in I}$ of a compact space X is a chain (that is, $U_i \subseteq U_j$ or $U_j \subseteq U_i$ for each $i, j \in I$), then $\bigcap_{i \in I} U_i \neq \emptyset$.

A space X is compact if and only if every net of elements of X has a convergent subnet (Exercise 3.27). Since for metric spaces it suffices to consider sequences and subsequences (see Exercise 3.17 for more precision on this), it follows easily that a compact metric space is complete (Exercise 3.32).

Example 3.2.22 A familiar Calculus exercise consists in showing that every sequence of elements of the interval $[a, b]$ has a convergent subsequence. It follows from this property that $[a, b]$ is a compact space.

Next we state without proof the famous *Tychonoff Theorem* (see again the Notes for reference).

Theorem 3.2.23 *Any product of compact spaces is compact.*

Example 3.2.24 Let A be a finite set, endowed with the discrete topology. The set $X = A^{\mathbb{Z}}$ of biinfinite sequences of elements of A is a topological space for the product topology. Since A is finite, X is compact. Likewise, the space $A^{\mathbb{N}}$ from Example 3.2.7 is compact, under the assumption that A is finite.

Concerning continuous functions, compact spaces have the following useful properties. First, if $f : X \to Y$ is a continuous bijection between compact spaces X and Y, then the inverse $f^{-1} : Y \to X$ is also continuous (see Exercises 3.30 and 3.31). Second, all continuous functions between compact metric spaces are uniformly continuous.

3.3   Topological Semigroups

We begin this section by formally recalling the notions of semigroup and monoid, which, despite being so simple and well known, are perhaps sometimes not so easily remembered as groups are.

3.3.1   Semigroups and Monoids

A *semigroup* is a nonempty set with an associative operation. More formally, a semigroup is a pair (S, m) formed by a nonempty set S and a map $m : S \times S \to S$, denoted $m(s, t) = s \cdot t$, for which the associative law $(s \cdot t) \cdot r = s \cdot (t \cdot r)$ holds for all $s, t, r \in S$. The function m is the semigroup operation.

Example 3.3.1 The set $\mathbb{Z}_+$ of positive integers is a semigroup under the operation of addition. It is also a semigroup under the operation of multiplication. To distinguish these two semigroup structures, one may denote the first by $(\mathbb{Z}_+, +)$, and the second by $(\mathbb{Z}_+, \cdot)$.

In general, we refer to a semigroup just via the underlying set, if the semigroup operation is understood from the context. So, for example, when we talk about the *additive* semigroup $\mathbb{Z}_+$, we mean the structure $(\mathbb{Z}_+, +)$.

A *monoid* is a semigroup with a neutral element (a *neutral element* of a semigroup S, also called an *identity*, is some, necessarily unique, element $n \in S$ for which the equalities $s \cdot n = n \cdot s = s$ hold for all $s \in S$). Therefore, groups are examples of monoids.

As is common for groups, the semigroup operation is usually denoted multiplicatively. In the multiplicative notation, the neutral element of a monoid is usually denoted by 1. However, we should also have in mind the corresponding additive notation in some examples, like the additive monoid $\mathbb{N}$ of natural numbers (whose neutral element is 0).

Monoids are algebraic structures pretty much as other classic structures (like groups or rings), with the neutral element being a distinguished element. In fact, we may see monoids as abstract algebras in the sense of universal algebra. For these reasons, we restrain ourselves of explicitly defining the notions of submonoid, product of monoids, monoid morphism, endomorphism, and others (similar to the familiar correspondent notions for groups and rings). The same goes for semigroups.

Example 3.3.2 Take any set X. The set X^X of functions $X \to X$, with respect to the composition of functions, is a monoid, where we are considering functions in the classical sense, acting on the left, with composition $(f \circ g)(x) = f(g(x))$. The identity on X is the neutral element of X^X. The monoid X^X is a submonoid of the monoid of partial functions on X, and these two monoids are submonoids of the larger monoid of binary relations on X, for the usual composition of binary relations.

An element u in a monoid M is said to be *invertible*, or a *unit of M*, when there is $v \in M$ such that $uv = vu = 1$. The element v is unique, and is called the *inverse* of u. The set of units of M is a submonoid of M which in fact is a group, the *group of units of M*.

Example 3.3.3 The group of units of the monoid X^X, met in Example 3.3.2 is the group of bijections $X \to X$.

An *anti-isomorphism* between the semigroups S, T is a bijection $\varphi : S \to T$ such that $\varphi(x \cdot y) = \varphi(y) \cdot \varphi(x)$ for all $x, y \in S$. We then say that T is *anti-isomorphic* to S, and we may also say that T is the *dual* of S (note that all duals of S are isomorphic, that is, the dual is unique up to isomorphism). Informally speaking, two anti-isomorphic semigroups may be seen as mirrors of each other. Note these notions carry on immediately for monoids.

Example 3.3.4 Take any set X. If, instead of functions $X \to X$, we consider *right transformations on X* (that is, binary relations f on X such that f^{-1} is a function $X \to X$, so that it makes sense to use the notation $x \cdot f = y$ when $(y, x) \in f$), then we get another monoid, denoted T_X, anti-isomorphic to the monoid X^X in Example 3.3.2. In this context, the elements of X^X may be called *left transformations on X*. The group of units of T_X is S_X, the group of permutations $\sigma : i \in X \to i \cdot \sigma \in X$, acting on the right of X.

In the same way, the monoid of partial functions on X, which we may call *left partial transformations on X*, has its dual companion, the anti-isomorphic monoid of *right partial transformations on X*, denoted by PT_X.

When $\mathrm{Card}(X) = n$, the monoids T_X, S_X and PT_X may be represented by T_n, S_n and PT_n, respectively, as their isomorphism class is indeed determined by $\mathrm{Card}(X)$. Next is a more concrete set of examples.

Example 3.3.5 On the left of Fig. 3.1 we see the four elements of T_2, and in the middle we see the nine elements of PT_2. In both cases, at the top we see the group of units, S_2.

Finally, on the right of Fig. 3.1 we see the the 27 elements of T_3, and the six elements at the top form the group of units S_3.

We reassure the reader that the notation $u = [i, j, k]$ for the right transformation $u \in T_3$ is the expected one: it means that $1 \cdot u = i$, $2 \cdot u = j$ and $3 \cdot u = k$. Likewise for the notation $v = [i, j]$ adopted for an element $v \in PT_2$, with the additional convention that $k \cdot v = \_$ means that k does not belong to the domain of v. The elements u such that $u^2 = u$ are marked in Fig. 3.1 with a star.

An element u of a semigroup S such that $u^2 = u$ is said to be an *idempotent*. We shall revisit certain features of Fig. 3.1. For now, we take immediate advantage of Fig. 3.1 to introduce some special types of idempotents. In a semigroup S, a *right-zero* is an element x such that $sx = x$ for every $s \in S$. In Fig. 3.1, the right-zeros are the elements in the lower row. Dually, a *left-zero* is an element x such that $xs = x$ for every $s \in S$. Finally, a *zero* is an element that is simultaneously a right-zero and left-zero. A zero, if it exists, is unique. The idempotent $[\_, \_]$, the empty relation

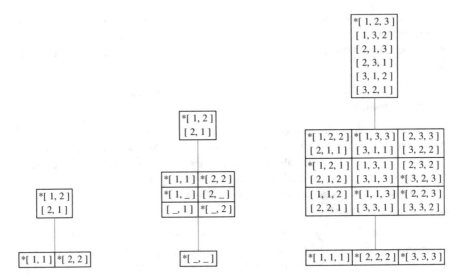

Fig. 3.1 The monoids T_2, PT_2 and T_3

on $\{1, 2\}$, is the zero of PT_2, also depicted in Fig. 3.1. Indeed, the empty relation is the zero of the monoid of binary relations on X, for any set X. A zero is also said to be an *absorbing* element. The terminology "zero" is meant for the multiplicative notation, in which case the absorbing element may be denoted by 0.

Given a semigroup S that is not a monoid, it is quite often convenient to consider the monoid S^1 consisting of S together with an extra element 1, by extending to $S^1 = S \cup \{1\}$ the semigroup operation of S, with 1 being the neutral element of S^1. When S is a monoid, one defines $S^1 = S$, consistent with the use of the multiplicative notation 1 for the neutral element of S.

Example 3.3.6 If S is the additive semigroup $\mathbb{Z}_+$ of positive integers, then S^1 is (isomorphic to) the monoid $\mathbb{N} = \mathbb{Z}_+ \cup \{0\}$ of natural integers.

3.3.2 Interplay Between Algebra and Topology

A *topological semigroup* is a semigroup S endowed with a Hausdorff topology such that the semigroup operation $S \times S \to S$ is continuous. The continuity of the operation means that if $(s_i, t_i) \to (s, t)$ in the product topology of $S \times S$, then $s_i t_i \to st$ in S. Therefore, since limits are unique in Hausdorff spaces, if the nets $(s_i)_{i \in I}$ and $(t_i)_{i \in I}$ converge, then the equality $\lim s_i \cdot \lim t_i = \lim s_i t_i$ holds.

Example 3.3.7 For the usual topology, the set $\mathbb{R}_+$ of positive real numbers is a topological semigroup under addition.

A *compact semigroup* is a topological semigroup which is compact as a topological space.

Example 3.3.8 A finite semigroup can always be viewed as a compact semigroup under the discrete topology.

Along this book, we systematically view finite semigroups as compact semigroups, with no need of reminding that we are endowing them with the discrete topology. We next present two natural examples of infinite compact semigroups, of different cardinals.

Example 3.3.9 For the topology of real numbers, the interval $[0, 1]$ is a compact semigroup under multiplication.

Example 3.3.10 Let the compact space $(\mathbb{Z}_+)_\infty$ from Example 3.2.20 be endowed with the semigroup structure resulting from the extension to $(\mathbb{Z}_+)_\infty$ of the additive semigroup structure of $\mathbb{Z}_+$, done via the equalities $x + \infty = \infty + x = \infty$ for all $x \in (\mathbb{Z}_+)_\infty$. Then $(\mathbb{Z}_+)_\infty$ is a compact semigroup.

Every subsemigroup of a topological semigroup is a topological semigroup for the induced topology. Similarly, every product of topological semigroups is a topological semigroup for the product topology. In particular, by Tychonoff's Theorem, every product of compact semigroups is a compact semigroup.

Recall that an *idempotent* in a semigroup S is an element $e \in S$ such that $e^2 = e$. A semigroup may not have idempotents (e.g., the additive semigroup $\mathbb{Z}_+$), but every compact semigroup contains at least one idempotent, a property that we shall revisit in Sect. 3.7, and that we prove in the next few lines.

Proposition 3.3.11 *Any compact semigroup contains an idempotent.*

Proof Let S be a compact semigroup. With respect to the inclusion relation, every chain of closed subsemigroups of S has nonempty intersection, since S is compact. Clearly, such intersection is a closed subsemigroup of S. Therefore, by Zorn's Lemma, S contains a minimal closed subsemigroup T. For any $t \in T$, the sets tT and Tt are closed subsemigroups contained in T, thus $tT = Tt = T$ by minimality. In particular, given $t \in T$, there is $e \in T$ such that $te = t$, and there is $s \in T$ such that $e = st$. Then $e^2 = ste = st = e$, thus e is an idempotent. ∎

The multiplicative compact semigroup $[0, 1]$ has two idempotents, 0 and 1, while the additive compact semigroup $(\mathbb{Z}_+)_\infty$ has only one idempotent (the point ∞). Idempotents may be very abundant in a compact semigroup.

Example 3.3.12 Every nonempty compact space X is a compact semigroup for the unique semigroup operation in which every element of X is a left-zero.

A *topological (compact) monoid* is a topological (compact) semigroup with a neutral element.

Example 3.3.13 By letting $x + \infty = \infty + x = \infty$ for all $x \in \mathbb{N}_\infty$, the space $\mathbb{N}_\infty$ from Example 3.2.19 becomes a compact monoid, with 0 as a neutral element.

An *isolated point* in a topological space X is an element $x \in X$ such that $\{x\}$ is open. If S is a topological (compact) semigroup which is not a monoid, then S^1 is viewed as a topological (compact) monoid, extending the topology of S to S^1 by adding 1 as an isolated point.

Example 3.3.14 Similarly to Example 3.3.6, if $S = (\mathbb{Z}_+)_\infty$ then the compact monoid S^1 is (isomorphic) to the compact monoid $\mathbb{N}_\infty = (\mathbb{Z}_+)_\infty \cup \{0\}$.

In a semigroup S, an element u of S is a *factor* of an element v of S if $v \in S^1 u S^1$. If $v \in u S^1$, then u is said to be a *prefix* of v. Dually, if $v \in S^1 u$, then u is a *suffix* of v.

Proposition 3.3.15 *The set of factors, the set of prefixes, and the set of suffixes of an element of a compact semigroup, are all closed sets.*

Proof It suffices to show that the set of factors is closed, as the other two cases may be similarly treated. Let S be a compact semigroup and let $(u_i)_{i \in I}$ be a net of factors of $x \in S$ converging to $u \in S$. For each $i \in I$, let p_i, q_i be elements of S^1 such that $x = p_i u_i q_i$ for all $i \in I$. As S^1 is compact, the product $S^1 \times S^1$ is compact, and so the net $(p_i, q_i)_{i \in I}$ has some subnet converging in $S^1 \times S^1$. If (p, q) is the limit of this subnet, then, because the semigroup operation is continuous, we have $x = puq$, and thus u is a factor of x. ∎

3.3.3   Generating Sets

If X is a nonempty subset of a semigroup S, then we denote by X^+ the intersection of all subsemigroups of S containing X. Since X is nonempty, so is X^+. In fact, X^+ is a subsemigroup of S, called the *subsemigroup of S generated by X*, and the following equality holds:

$$X^+ = \bigcup_{n \geqslant 1} \{x_1 \cdots x_n \mid x_1, \ldots, x_n \in X\}.$$

If $S = X^+$, then one says that X is *a generating set* of S, or that X *generates* S. Of course, if X generates S, then so does every subset of S containing X.

Example 3.3.16 The *additive* semigroup $(\mathbb{Z}_+, +)$ of positive integers is generated by the set $\{1\}$, and the *multiplicative* semigroup $(\mathbb{Z}_+, \cdot)$ of positive integers is generated by the set of prime numbers together with 1.

Example 3.3.17 Consider the partial transformations $a = [2, \_]$ and $b = [\_, 1]$, represented by the labeled directed graph in Fig. 3.2. The subsemigroup of PT_2 generated by $\{a, b\}$ has five elements: besides a and b, we also have $ab = [1, \_]$, $ba = [\_, 2]$ and the zero $a^2 = b^2 = [\_, \_]$.

Fig. 3.2 Generators of B_2

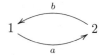

A map $\alpha : A \to S$ such that $\alpha(A)$ generates the semigroup S is said to be a *generating map* of S. We may also say that S is *A-generated*. For instance, when $A = \{a, b\}$, the semigroup B_2 met in Example 3.3.17 is A-generated, since it is generated by the mapping $\alpha : A \to B_2$ such that $\alpha(a) = [2, \_]$ and $\alpha(b) = [\_, 1]$. Note that if a semigroup S is A-generated for some set A, then it is X-generated whenever $\text{Card}(X) \geqslant \text{Card}(A)$: for example, if $X = \{x, y, z\}$, then B_2 is X-generated, with a generating map being $\beta : X \to B_2$ such that $\beta(x) = \alpha(a)$ and $\beta(y) = \beta(z) = \alpha(b)$.

We use Example 3.3.17 to help introduce the following notion, which we shall revisit later. The *transition semigroup* of a labeled directed multigraph is the semigroup of binary relations (on the vertex set) generated by the set of binary relations defined by some label—where the binary relation δ_x defined by the label x is such that $(p, q) \in \delta_x$ if and only if there is an edge from p to q labeled by x. In Example 3.3.17, we have two labels, a and b, the binary relations δ_a and δ_b are the partial transformations $[2, \_]$ and $[\_, 1]$, respectively, and the transition semigroup is B_2.

We now look at topological semigroups. If X is a nonempty subset of a topological semigroup S, then $\overline{X^+}$ is a closed subsemigroup of S (Exercise 3.36). Clearly, $\overline{X^+}$ is the intersection of all closed subsemigroups of S containing X, named the *closed subsemigroup of X generated by S*. If X^+ is dense in S, we may say that X generates S (as a topological semigroup), but one should be aware of being in the context of *topological* semigroups, as one may have $\overline{X^+} = S$, but not $X^+ = S$. Likewise, if $\alpha : A \to S$ is such that $\alpha(A)^+$ is dense in the topological semigroup S, then we say that α is a generating map of S and that S is A-generated.

Example 3.3.18 As topological (additive) semigroups, $(\mathbb{Z}_+)_\infty$ is generated by $\{1\}$, and $\mathbb{N}_\infty$ is generated by $\{0, 1\}$.

Given any subset X of a monoid M, we denote by X^* the intersection of all submonoids of M containing X. Note that X^* is a submonoid of M, which is called the *submonoid of M generated by X*. Observe that $\emptyset^* = \{1\}$, and that if $X \neq \emptyset$ then $X^* = \{1\} \cup X^+$. Other related definitions carry on immediately for (topological) monoids, *mutatis mutandis*.

Example 3.3.19 The additive monoid $\mathbb{N}$ of natural numbers is generated (as a monoid) by the set $\{1\}$. But viewing $\mathbb{N}$ as an additive semigroup, $\{1\}$ is not one of its generator sets: the subsemigroup generated by $\{1\}$ is $\mathbb{Z}_+$.

Example 3.3.20 As a topological monoid, the compact additive monoid $\mathbb{N}_\infty$ is generated by $\{1\}$.

3.4 Topological Groups

A *topological group* is a group with a Hausdorff topology such that multiplication and taking the inverse are continuous operations, and so topological groups are special examples of topological semigroups.

Example 3.4.1 The set $\mathbb{R}$ of real numbers with the usual topology is a topological group under addition.

In the following lines, we use the fact that every subgroup of a topological group is a topological group for the induced topology.

In any topological group G, and for every $g \in G$, the translation $x \mapsto xg$ is a homeomorphism from G onto G. Hence, every coset Hg of an open (resp. closed) subgroup H of G is open (resp. closed).

Every open subgroup H of a topological group G is also closed since its complement is the union of all cosets Hg for $g \in G \setminus H$, which are open sets.

Example 3.4.2 The set $\mathbb{Z}$ is a subgroup of the additive group $\mathbb{R}$. It is closed, but not open, since its complement is dense in $\mathbb{R}$.

Next we mention a property of compact groups, that is, topological groups whose topology is compact.

Proposition 3.4.3 *In a compact group, a subgroup is open if and only if it is closed and of finite index.*

Proof Assume that H is an open subgroup of G. We have already seen that H is also closed. The union of the cosets of H form a covering of G by open sets. Since G is compact, there is a finite subfamily covering G and thus H has finite index.

Conversely, if H is a closed subgroup of finite index, then the complement of H is the finite union of the cosets Hg for $g \notin H$, each one a closed set, and thus H is open. ∎

For groups, the expressions "compact semigroup" and "compact group" describe the same thing, as next seen in a more precise manner.

Proposition 3.4.4 *Let S be a compact semigroup. If S is a group, then it is a compact group with respect to the topology of S as a compact semigroup.*

Proof It remains to show that the map $s \mapsto s^{-1}$ is continuous in S. Suppose that (s_i) is a net of elements of S converging to s. It follows from the continuity of the semigroup operation of S, and from the equality $s_i^{-1} s_i = 1$, that every convergent subnet of (s_i^{-1}) must converge to s^{-1}. Since S is compact, we conclude that (s_i^{-1}) converges to s^{-1} (cf. Exercise 3.33). ∎

A subsemigroup G of a semigroup S is said to be a *subgroup of S* if the semigroup G is actually a group. If S is a compact semigroup, then we use the terminology *closed subgroup* for a subgroup of S which is topologically closed:

this terminology is adequate, because a closed subgroup of a compact semigroup is indeed a compact group, by Proposition 3.4.4.

3.5 Quotients of Compact Semigroups

The goal of this section is to present a companion for compact semigroups of the familiar isomorphism theorem for groups. In particular, we need to build a proper notion of quotient of a compact semigroup.

Given an equivalence relation ρ on a set X, let $[x]_\rho$ denote the equivalence class of $x \in X$ with respect to ρ, that is, $[x]_\rho = \{y \in X \mid (x, y) \in \rho\}$. The *quotient set* X/ρ is the set of the ρ-classes $[x]_\rho$. The *quotient map* $q_\rho : X \to X/\rho$ is defined by $q_\rho(x) = [x]_\rho$.

The *kernel* of a map $f : X \to Y$ is the set, denoted $\operatorname{Ker} f$, of all pairs (u, v) in $X \times X$ such that $f(u) = f(v)$. Note that the equivalence relation ρ is the kernel of q_ρ, and that $\operatorname{Ker} f$ is always an equivalence relation. A fact one should also keep in mind is that the associated canonical map $X/\operatorname{Ker} f \to f(X)$, defined by the correspondence $[x]_{\operatorname{Ker} f} \mapsto f(x)$, is a well-defined bijection. We shall often use the alternative notation $\operatorname{Im} f$ for the image $f(X)$.

Remark 3.5.1 In group theory, the kernel of a group morphism $\varphi : G \to H$ is not defined as a subset of $G \times G$, but instead as the normal subgroup $N = \varphi^{-1}(1)$ of G. However, there is a simple and close relationship with the semigroup concept. In fact, the equivalence classes for the relation $\operatorname{Ker} \varphi$ are precisely the elements of G/N.

A *congruence* ρ of a semigroup S is an equivalence relation on S which is compatible with the operation of S. This means that for every pair $(u, v) \in \rho$ and element $w \in S$, one has $(wu, wv) \in \rho$ and $(uw, vw) \in \rho$. We may use the notation $u \equiv v \bmod \rho$ for $(u, v) \in \rho$. For such a congruence ρ, the quotient S/ρ is endowed with a natural structure of semigroup, given by $[s]_\rho[t]_\rho = [st]_\rho$, making the quotient map $q_\rho : S \to S/\rho$ an onto morphism of semigroups with kernel ρ. Conversely, the kernel of every morphism of semigroups is a congruence.

Theorem 3.5.2 *If $\varphi : S \to T$ is a morphism of semigroups, then $S/\operatorname{Ker} \varphi$ and $\operatorname{Im} \varphi$ are isomorphic semigroups.*

Proof The mapping $[s]_{\operatorname{Ker} \varphi} \mapsto \varphi(s)$ is an isomorphism $S/\operatorname{Ker} \varphi \to \operatorname{Im} \varphi$. ∎

The next lines prepare the analog of Theorem 3.5.2 for compact semigroups. An equivalence relation on a topological space X is said to be *closed*, if it is a closed subset of $X \times X$. If ρ is a closed equivalence, its classes are closed (Exercise 3.38). The converse is not true (see Exercise 3.39).

Example 3.5.3 If $f : X \to Y$ is a continuous map and Y is Hausdorff, then the equivalence relation $\operatorname{Ker} f$ is closed: indeed, whenever (x_i, y_i) is a net of elements of Ker f converging to (x, y), one has $f(x) = \lim f(x_i) = \lim f(y_i) = f(y)$ (note that we are using the uniqueness of limits in Hausdorff spaces).

Given a topological space X and an equivalence relation ρ on X, we endow X/ρ with the *quotient topology*, the topology where a subset U of X/ρ is open if and only if $q_\rho^{-1}(U)$ is an open set with respect to the topology of X. Note that $q_\rho^{-1}(U)$ is just the union of the ρ-classes of X belonging to U. If X is compact, then X/ρ is compact if and only if ρ is closed (Exercise 3.42).

Example 3.5.4 If $f : X \to Y$ is a continuous map between compact spaces, then $X/\mathrm{Ker}\,f$ is compact, and the canonical map $X/\mathrm{Ker}\,f \to \mathrm{Im}\,f$ is a homeomorphism (in Exercise 3.43 we ask to fill the details justifying this assertion).

With the data we now have, we are ready to state the next proposition, giving an adequate definition of quotient of a compact semigroup.

Proposition 3.5.5 *Let S be a compact semigroup. Suppose that ρ is a closed congruence of S. Then the quotient semigroup S/ρ is a compact semigroup for the quotient topology.*

Proof We already saw that ρ being closed and S being compact entails S/ρ being a compact space. It only remains to show that the semigroup operation of S/ρ is continuous. The following diagram commutes, where m_S is the semigroup operation of S and $m_{S/\rho}$ is the semigroup operation of S/ρ:

$$
\begin{array}{ccc}
S \times S & \xrightarrow{\ q_\rho \times q_\rho\ } & S/\rho \times S/\rho \\
\downarrow{\scriptstyle m_S} & & \downarrow{\scriptstyle m_{S/\rho}} \\
S & \xrightarrow{\ \ q_\rho\ \ } & S/\rho
\end{array}
$$

Since the equality $q_\rho \circ m_S = m_{S/\rho} \circ (q_\rho \times q_\rho)$ holds, and as we know already that q_ρ, m_S and $q_\rho \times q_\rho$ are continuous, it follows from the compactness of $S \times S$ and $S/\rho \times S/\rho$ that $m_{S/\rho}$ is also continuous (cf. Exercise 3.30). ∎

We may now formulate the isomorphism theorem for compact semigroups, which, in view of the proof of Theorem 3.5.2, one immediately recognizes as a special case of Example 3.5.4.

Theorem 3.5.6 *If $\varphi : S \to T$ is a continuous morphism of compact semigroups, then $S/\mathrm{Ker}\,\varphi$ and $\mathrm{Im}\,\varphi$ are isomorphic compact semigroups.*

We wish to expand a little more on the comparison of the group and semigroup contexts, both in the non-topological and in the compact cases. If the (compact) semigroup S has a neutral element 1, then for any congruence ρ the quotient S/ρ is a (compact) monoid with neutral element $[1]_\rho$, so that the canonical morphism $q_\rho : S \to S/\rho$ becomes a (continuous) morphism of (compact) monoids. If, moreover, S is a group, then S/ρ is a (compact) group, with inversion given by $[u]_\rho^{-1} = [u^{-1}]_\rho$, and q_ρ is then a (continuous) morphism of (compact) groups. It is easy to verify that if ρ is a (closed) congruence on the (compact) group G, then the ρ-class of the identity is a (closed) normal subgroup and that, conversely, for every (closed) normal subgroup H of the group G, the equivalence ρ formed of the pairs (u, v) such that $uv^{-1} \in H$ is a (closed) congruence ρ_H such that $H = [1]_{\rho_H}$. In particular, if $\varphi : G \to H$ is a (continuous) group morphism and $N = \varphi^{-1}(1)$, then the associated congruence ρ_N is the kernel congruence $\operatorname{Ker}\varphi$. In view of this discussion, one sees that Theorem 3.5.2 indeed generalizes for semigroups the classical isomorphism theorem for groups.

3.6 Ideals and Green's Relations

In this section, we give some definitions for (topological) monoids that clearly carry on to (topological) semigroups via the construction $S \rightsquigarrow S^1$. A first introduction within the monoid context makes it possible to avoid the somewhat unpleasant notation S^1.

Consider a monoid M. An *ideal* (also called a *two-sided ideal*) in M is a nonempty subset J of M such that $MJM \subseteq M$. The ideal generated by $x \in M$ is MxM, the intersection of all ideals containing x.

A *right ideal* (resp. *left ideal*) in M is a nonempty subset J of M such that $JM \subseteq M$ (resp. $MJ \subseteq M$). The right ideal (resp. left ideal) generated by $x \in M$ is xM (resp. Mx), the intersection of all right-ideals (resp. left ideals) containing x.

Remark 3.6.1 Note that xM, Mx and MxM are closed when M is a compact monoid (cf. Proposition 3.3.15).

3.6.1 Green's Relations

The *Green relations* $\mathcal{J}$, $\mathcal{R}$ and $\mathcal{L}$ in a monoid M are equivalence relations that capture information about the ideal structure of M. They are defined by:

- $x \mathrel{\mathcal{J}} y$ if and only if $MxM = MyM$;
- $x \mathrel{\mathcal{R}} y$ if and only if $xM = yM$;
- $x \mathrel{\mathcal{L}} y$ if and only if $Mx = My$.

The relation $\mathcal{R}$ is a *left congruence*, that is $x \mathrel{\mathcal{R}} y$ implies $ux \mathrel{\mathcal{R}} uy$. Symmetrically $\mathcal{L}$ is a *right congruence*.

One defines the $\mathcal{J}$-*order* by letting $x \leqslant_{\mathcal{J}} y$ if and only if y is a factor of x, that is, if and only if $MxM \subseteq MyM$. Hence, two elements x, y are $\mathcal{J}$-equivalent if and only if each one is a factor of the other. We write $x <_{\mathcal{J}} y$ if $x \leqslant_{\mathcal{J}} y$ but x and y are not $\mathcal{J}$-equivalent.

Replacing the notion of factor by prefix, one obtains the $\mathcal{R}$-*order* (resp. $\mathcal{L}$-*order*). Thus, $x \leqslant_{\mathcal{R}} y$ if and only if y is a prefix of x, that is, if and only if $xM \subseteq yM$. Similarly, $x \leqslant_{\mathcal{L}} y$ if and only if y is a suffix of x, that is, if and only if $Mx \subseteq My$.

Each of the relations $\leqslant_{\mathcal{J}}$, $\leqslant_{\mathcal{R}}$ and $\leqslant_{\mathcal{L}}$ is transitive and reflexive, but not anti-symmetric in general. The reader should also have in mind the trivial fact that when $\mathcal{K}$ is $\mathcal{J}$, $\mathcal{R}$ or $\mathcal{L}$, then $x \, \mathcal{K} \, y$ if and only if $x \leqslant_{\mathcal{K}} y$ and $y \leqslant_{\mathcal{K}} x$.

Example 3.6.2 Let f and g be elements of the monoid of functions $X \to X$. Then $f \, \mathcal{L} \, g$ if and only if Ker $f =$ Ker g, and $f \, \mathcal{R} \, g$ if and only if Im $f =$ Im g. A dual result holds for T_X, which extends naturally to PT_X (see Exercise 3.47).

The notion of ideal extends to all semigroups by defining an ideal of a semigroup S which is not a monoid as being an ideal of S^1 not containing 1. The same for right and left ideals. That is, a nonempty subset I of any semigroup S is an ideal if and only if $S^1 I S^1 \subseteq S$; it is a right ideal if and only if $I S^1 \subseteq S$; and its a left ideal if and only if $S^1 I \subseteq S$.

The Green relations on the semigroup S are the restriction to S of the Green relations on S^1. For example, $x, y \in S$ are $\mathcal{J}$-equivalent if and only if $S^1 x S^1 = S^1 y S^1$. The ideals of the form $S^1 x S^1$ are the *principal ideals* of S.

In this paragraph we introduce two other important equivalence relations derived from the relations $\mathcal{R}$ and $\mathcal{L}$. The first one is the intersection of $\mathcal{R}$ and $\mathcal{L}$, denoted $\mathcal{H}$. In any semigroup, one has $\mathcal{RL} = \mathcal{LR}$ (see Exercise 3.49) and one denotes $\mathcal{D}$ the equivalence $\mathcal{RL} = \mathcal{LR}$, which is the supremum of $\mathcal{R}$ and $\mathcal{L}$ in the lattice of all equivalence relations on the semigroup. The relations $\mathcal{H}$ and $\mathcal{D}$ complete the classic set of Green's relations.

For each Green relation $\mathcal{K}$, we denote by $K(x)$ the $\mathcal{K}$-class of an element x.

Example 3.6.3 Let us return to Fig. 3.1. Each inner rectangle represents an $\mathcal{H}$-class, and inside each external big rectangle, each row of inner rectangles represents an $\mathcal{R}$-class, and each column represents an $\mathcal{L}$-class (in agreement with the characterization of the Green relations on T_X and PT_X given by Example 3.6.2). For instance, in PT_2 one has

$$H([1, 1]) = \{[1, 1]\}, \quad R([1, 1]) = \{[1, 1], [2, 2]\}, \quad L([1, 1]) = \{[1, 1], [1, \_], [\_, 1]\},$$

and

$$H([1, 2]) = R([1, 2]) = L([1, 2]) = D([1, 2]) = \{[1, 2], [2, 1]\}.$$

The $\mathcal{D}$-classes are represented by the bigger external rectangles, and so both T_3 and PT_2 have three $\mathcal{D}$-classes, while T_2 has two $\mathcal{D}$-classes. We shall see that $\mathcal{D} = \mathcal{J}$ in every finite monoid (Proposition 3.6.5).

The representation of the Green relations of a finite semigroup as done in Fig. 3.1 is called an *eggbox diagram* (cf. Example 3.6.3). The idempotents are marked with an asterisk. The spatial disposition of the $\mathcal{J}$-classes is that of the Hasse diagram of the partial order on the set of $\mathcal{J}$-classes, induced by $\leqslant_{\mathcal{J}}$: two $\mathcal{J}$-classes K_1 and

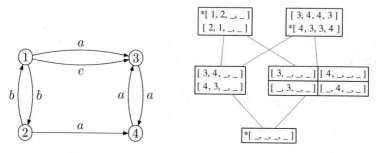

Fig. 3.3 Labeled multigraph and its transition semigroup

K_2 are connected by a line, with K_1 in some level below K_2, if and only if K_1 and K_2 are consecutive with respect to the $\mathcal{J}$-order. In the three semigroups depicted in Fig. 3.1, the $\mathcal{J}$-order induces a total order on the set of $\mathcal{J}$-classes.

Example 3.6.4 In Fig. 3.3 we find a labeled directed multigraph with its transition semigroup S at the right. In this example, the $\mathcal{J}$-classes are not linearly ordered.

There are several "symmetries" in each $\mathcal{D}$-class of a semigroup. For example, inside a $\mathcal{D}$-class, any two $\mathcal{R}$-classes have the same cardinal, and a dual fact holds for $\mathcal{L}$-classes: see Exercise 3.52 (and also the couple of exercises following it).

3.6.2   Stable Semigroups

A semigroup S is said to be *stable* when the following holds: $x \leqslant_\mathcal{R} y$ and $x \mathrel{\mathcal{J}} y$ implies $x \mathrel{\mathcal{R}} y$, for every $x, y \in S$, and the dual property concerning $\mathcal{L}$ also holds.

Proposition 3.6.5 *Every compact semigroup is stable.*

Proof Suppose that S is a compact semigroup. Let $x \in S$ and $u \in S$ be such that $xu \mathrel{\mathcal{J}} x$. We want to show that $xu \mathrel{\mathcal{R}} x$. The hypothesis $xu \mathrel{\mathcal{J}} x$ entails that $x = rxus$, for some $r, s \in S^1$. It follows that $x = r^n x(us)^n$ for every $n \geqslant 1$. Since S is compact, the closure of the semigroup generated by us contains an idempotent e by Proposition 3.3.11. Let $(us)^{n_i}$ be a net converging to e. Thanks to the compactness of S, by taking subnets we may assume that $(us)^{n_i-1}$ and r^{n_i} are also convergent, to t and r_0 respectively. Then $e = ust$ and $x = r_0xe$ whence

$$(xu)st = x(ust) = xe = r_0xe = x,$$

showing that $x \in xuS$ and thus that $x \mathrel{\mathcal{R}} xu$.   ∎

On stable semigroups, we have the following simplification.

Proposition 3.6.6 *In a stable semigroup, one has $\mathcal{D} = \mathcal{J}$.*

Proof Clearly $x \mathcal{D} y$ implies $x \mathcal{J} y$ in any semigroup. Assume conversely that $x \mathcal{J} y$, for x, y in a stable semigroup S. Then we have $x = uyv$ for some u, $v \in S^1$. As x is a factor of y, we get that uy and yv are in the $\mathcal{J}$-class of y. Since S is stable, we have $uy \mathcal{L} y$ and $yv \mathcal{R} y$. And since $\mathcal{R}$ is a left congruence, $uyv \mathcal{R} uy$ also holds. Then $y \mathcal{L} uy \mathcal{R} uyv = x$, showing that $x \mathcal{D} y$.   ∎

Therefore, according to Propositions 3.6.5 and 3.6.6, one has $\mathcal{D} = \mathcal{J}$ in every compact semigroup. Next we present another important consequence of assuming that a semigroup is compact, concerning its ideal structure.

Proposition 3.6.7 *A compact semigroup has a unique minimal ideal. This minimal ideal is a principal ideal, and it is also a $\mathcal{J}$-class.*

Proof Consider the family of closed ideals of a compact semigroup S. This family is nonempty (S is itself a closed ideal) and every chain of elements of the family has nonempty intersection, since S is compact. Note that the intersection of such a chain is also a closed ideal. Therefore, by Zorn's Lemma, there is a minimal closed ideal I of S. Let u be an element of I. Then $S^1 u S^1 \subseteq I$, by the definition of ideal in a semigroup. Since $S^1 u S^1$ is itself a closed ideal (by Exercise 3.35), we actually have $S^1 u S^1 = I$, by the minimality of I. This also shows that I is the $\mathcal{J}$-class of the elements of I.

If K is any ideal of S, then the intersection $I \cap K$ is nonempty (it contains IK) and is an ideal contained in the minimal ideal I, whence $I \subseteq K$ by the minimality of I. This shows that I is the unique minimal ideal of S.   ∎

We denote by $K(S)$ the unique minimal ideal of a compact semigroup S. Because of its uniqueness, we may say that $K(S)$ is the *minimum ideal* of S. One should have in mind that $K(S)$ is topologically closed, since it is a principal ideal. Note also that the elements of $K(S)$ are those that are minimal for the relation $\leqslant_{\mathcal{J}}$, that is $u \in K(S)$ if and only if $u \leqslant_{\mathcal{J}} s$ for every $s \in S$. Therefore, in the eggbox diagram of a finite semigroup, the minimum ideal is the $\mathcal{J}$-class at the bottom.

See Exercise 3.58 for properties concerning right and left ideals, related to Proposition 3.6.7.

3.6.3 The Schützenberger Group

It is easy to verify that an $\mathcal{H}$-class of a semigroup is a group if and only if it contains an idempotent (Exercise 3.59). Clearly, every subgroup of a semigroup is contained in an $\mathcal{H}$-class containing an idempotent. Therefore, the $\mathcal{H}$-classes containing idempotents are precisely the maximal subgroups of the semigroup.

Example 3.6.8 The group of units of a monoid is a maximal subgroup.

In a compact semigroup, all Green relations are closed (see Exercise 3.57). In particular, since a maximal subgroup of a semigroup is an $\mathcal{H}$-class, every maximal subgroup of a compact semigroup is a compact group.

An element s of the semigroup S is *regular* if there is some $x \in S$ such that $sxs = s$. In any semigroup, if $\mathcal{K}$ is one of the Green relations $\mathcal{D}$, $\mathcal{R}$ or $\mathcal{L}$ (recall that $\mathcal{D} = \mathcal{J}$ in a compact semigroup, and, more generally, in a stable semigroup), then a $\mathcal{K}$-class contains a regular element if and only if all its elements are regular, if and only if it contains an idempotent (Exercise 3.56); and for that reason one says that a $\mathcal{K}$-class is *regular* when some of (equivalently, all of) its elements are regular.

Example 3.6.9 Returning to the semigroup in Example 3.6.4, and looking at Fig. 3.3, one sees that it has two non-regular $\mathcal{D}$-classes: that of $[3, 4, \_, \_]$ and that of $[3, \_, \_, \_]$ (recall that the idempotents are marked with an asterisk).

Example 3.6.10 The minimum ideal $K(S)$ of a compact semigroup S is a regular $\mathcal{J}$-class, because $K(S)$ is a compact semigroup and every compact semigroup contains idempotents.

In a semigroup S, two idempotents e, f are $\mathcal{D}$-equivalent if and only if $e = xy$ and $f = yx$ for some $x, y \in S$ (this is part of Exercise 3.51). We use this elementary fact in the proof of the following property, which takes a relevant role in later chapters of this book.

Proposition 3.6.11 *Let D be a regular $\mathcal{D}$-class of a semigroup S. If H and K are maximal subgroups of D, then H and K are isomorphic groups. Moreover, if S is a compact semigroup, then H and K are isomorphic compact groups.*

Proof Let e and f be the idempotents respectively belonging to H and K. We may take elements $x, y \in S$ with $xy = e$ and $yx = f$ (Exercise 3.51). Consider the mapping $\varphi : H \to S$ defined by $\varphi(u) = yux$. For all $u, v \in H$, one has $\varphi(uv) = yuvx = yuevx = yux \cdot yvx = \varphi(u) \cdot \varphi(v)$. Therefore, φ is a semigroup morphism. As $\varphi(e) = f$, and the morphic image of a group by a semigroup morphism is a group, we deduce that $\varphi(H) \subseteq K$. Symmetrically, we may consider the morphism $\psi : K \to S$ given by $\psi(u) = xuy$. Then we have $\psi(\varphi(u)) = xyuxy = eue = u$ for all $u \in H$, and, similarly, $\varphi(\psi(u)) = u$ for all $u \in K$, establishing that φ and ψ are mutually inverse[2] isomorphisms between H and K. If S is a compact semigroup, then it follows immediately from the continuity of the multiplication of S that φ and ψ are continuous. ∎

Thanks to Proposition 3.6.11, we may identify, up to isomorphism (of compact) groups, the maximal subgroups of a regular $\mathcal{D}$-class D of a (compact) semigroup. The isomorphism class of the maximal subgroups of D is called the *Schützenberger group* of D, and is denoted $G(D)$. If S is compact, we view $G(D)$ as a compact group. Note that $G(D)$ is a profinite group if S is profinite.

Example 3.6.12 The Schützenberger group of the $\mathcal{D}$-class of $[2, 2, 1]$ in T_3 is the cyclic group of order two: see Fig. 3.1, where we see that this $\mathcal{D}$-class indeed has six maximal subgroups isomorphic to $\mathbb{Z}/2\mathbb{Z}$.

---

[2] Alternatively, to show that $\varphi : H \to K$ is bijective, one could have used Green's Lemma (cf. Exercise 3.52).

In fact, every $\mathcal{D}$-class (even if non-regular) is associated with a group via a more general construction of Marcel-Paul Schützenberger (whence the terminology *Schützenberger group*). Albeit not necessary to the sequel, the reader may wish to look at the outline of this construction, made in the following paragraphs.

Let H be an $\mathcal{H}$-class of a semigroup S. Set

$$T(H) = \{x \in S^1 \mid Hx = H\}.$$

Note that $T(H)$ is a submonoid of S^1. One also has

$$T(H) = \{x \in S^1 \mid Hx \cap H \neq \emptyset\},$$

a useful equality which is easy to show (Exercise 3.67). Each $x \in T(H)$ defines a mapping $\rho_x : H \to H$ described by the correspondence $h \mapsto hx$. The set of right translations of the form ρ_x, with $x \in T(H)$, is a group of permutations on H, which is denoted $\Gamma(H)$ and has the same cardinal of H (see Exercise 3.68).

The groups of the form $\Gamma(H)$, corresponding to different $\mathcal{H}$-classes H contained in the same $\mathcal{D}$-class D, are isomorphic (Exercise 3.69). For every $\mathcal{D}$-class, we say that the isomorphism class $G(D)$ of the groups of the form $\Gamma(H)$, with H an $\mathcal{H}$-class of D, is the *Schützenberger group of D*—a definition consistent with the one first given in the special case where D is regular, as every maximal subgroup H is isomorphic to $\Gamma(H)$ (Exercise 3.70).

Example 3.6.13 Looking in Fig. 3.3 at the eggbox diagram of the semigroup from Example 3.6.4, one sees that the Schützenberger group of the $\mathcal{D}$-class of $[3, 4, \_, \_]$, which is not regular, is the group $\mathbb{Z}/2\mathbb{Z}$, since this $\mathcal{D}$-class is an $\mathcal{H}$-class with two elements. Similarly, the Schützenberger group of the non-regular $\mathcal{D}$-class of $[3, \_, \_, \_]$ is the trivial group.

For compact semigroups, there are also natural topological versions of the Schützenberger group and of its properties (see Exercise 3.69).

3.7   The ω-Power

A semigroup, or a monoid, generated by one element is called *cyclic* or *monogenic*. Given an element s of a semigroup (resp. monoid), we may use the notation s^+ (resp. s^*) for the monogenic semigroup (resp. monoid) generated by $\{s\}$.

Let M be a finite monoid generated by an element s. The *index* of s (or M) is the least integer $i \geqslant 0$ such that $s^{i+k} = s^i$ for some $k \geqslant 1$, and its *least period* is the least integer $p \geqslant 1$ such that $s^{i+p} = s^i$. Up to isomorphism, such a monoid depends on i and p only, and for that reason we denote it by $M_{i,p}$. It can be represented as in Fig. 3.4.

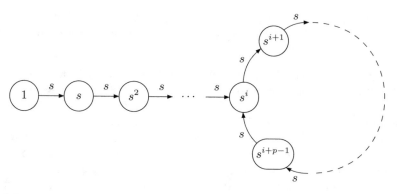

Fig. 3.4 The monoid $M_{i,p}$ generated by s

Note in particular that the monoid $M_{i,p}$ has $i + p$ elements, and that it is a group if and only if $i = 0$. The monoid $M_{i,p}$ is actually the same, up to the isomorphism $n \mapsto s^n$, as the monoid $N_{i,p}$ met in Chap. 2.

Finally, if $i > 0$, we denote by $C_{i,p}$ the subsemigroup $M_{i,p} \setminus \{1\}$ of $M_{i,p}$. Every finite cyclic semigroup is isomorphic to $C_{i,p}$, for a unique pair (i, p).

Monogenic semigroups and monoids are clearly commutative, and so all their Green relations coincide (cf. Exercise 3.50), a fact that the reader should have in mind when reading the following proposition.

Proposition 3.7.1 *The semigroup $C_{i,p}$, generated by s, contains a unique idempotent e. The $\mathcal{J}$-class of e is a cyclic group of order p. The idempotent e equals s^k if and only if k is a multiple of p greater or equal than i.*

Proof Consider the set $H = \{s^i, s^{i+1}, \ldots, s^{i+p-1}\}$, corresponding to the circular part in Fig. 3.4. Clearly, if s^m is an idempotent of $C_{i,p}$, where $m > 0$, then $s^m \in H$, that is, $m \geqslant i$. Let k be the least multiple of p greater than or equal to i. Since $s^{j+p} = s^j$ for every $j \geqslant i$, we know that s^k is an idempotent belonging to H, that H is an $\mathcal{H}$-class (and a $\mathcal{J}$-class), and that the map $n \mapsto s^{k+n}$ is a well defined isomorphism from $\mathbb{Z}/p\mathbb{Z}$ onto H. In particular, $e = s^k$ is the unique idempotent of $C_{i,p}$ and, whenever t is a positive integer, $s^t = e$ if and only if t is a multiple of p greater than or equal to i. ∎

Recall that we see finite semigroups as special cases of compact semigroups, by endowing them with the discrete topology. Proposition 3.7.1 has a sort of generalization for compact semigroups. Indeed, in any compact semigroup S, just as in a finite semigroup, the closure of the semigroup generated by an element $s \in S$ contains a unique idempotent, denoted s^ω. More precisely, one has the following proposition, where, by a *monogenic topological semigroup (generated by s)*, we mean a topological semigroup S of the form $S = \overline{s^+}$, with $s \in S$.

Proposition 3.7.2 *Let S be a monogenic compact semigroup. Then S contains a unique idempotent. The $\mathcal{J}$-class of S containing this idempotent is a group and it is the minimum ideal $K(S)$.*

Proof By Proposition 3.3.11, S contains an idempotent e. By hypothesis, there is $s \in S$ such that S is the closure of $s^+ = \{s^k \mid k \in \mathbb{Z}_+\}$. Hence, we may take a net $(s^{n_i})_{i \in I}$ converging to e, with $n_i \in \mathbb{Z}_+$ for every $i \in I$. Fix $k \in \mathbb{Z}_+$. Because e is idempotent, we have $\lim_{i \in I} s^{kn_i} = (\lim_{i \in I} s^{n_i})^k = e$. Since s^k is a factor of s^{kn_i}, we conclude that $e \leqslant_{\mathcal{J}} s^k$, since $S^1 s^k S^1$ is closed (see Exercise 3.35). For the same reason, it follows that $e \leqslant_{\mathcal{J}} t$ for every $t \in S$. In particular, e belongs to the $\mathcal{J}$-class $K(S)$. Since S is commutative (cf. Exercise 3.37), its Green relations coincide, and so $K(S)$ is an $\mathcal{H}$-class of S with an idempotent, thus a group. It follows that e is the unique idempotent of S.   ∎

In the special case of finite semigroups we also have the following.

Proposition 3.7.3 *Let S be a finite semigroup. For every $s \in S$, the sequence $(s^{n!})$ converges to s^ω. More precisely, for every $n \geqslant \mathrm{Card}(S)$, one has $s^\omega = s^{n!}$.*

Proof Let (i, p) be such that the subsemigroup of S generated by s is (isomorphic to) to $C_{i,p}$. There is a multiple k of p such that $i \leqslant k < i + p$. By Proposition 3.7.1, we have $s^\omega = s^k$. Since $k \leqslant \mathrm{Card}(C_{i,p}) \leqslant \mathrm{Card}(S)$, if $n \geqslant \mathrm{Card}(S)$, then $n!/k$ is an integer, thus $s^{n!} = (s^k)^{n!/k} = (s^\omega)^{n!/k} = s^\omega$.   ∎

Let S be a monogenic compact semigroup, and let $s \in S$. For every $k \geqslant 1$, we denote by $s^{\omega+k}$ the product $s^\omega \cdot s^k$. As s^ω belongs to the group $K(S)$, so does $s^{\omega+k}$, and so we may consider the inverse of $s^{\omega+k}$ in $K(S)$, denoted $s^{\omega-k}$. Since $s^{\omega-k} = s^{\omega-k} \cdot s^\omega$, we actually have $s^{\omega-k} \cdot s^k = s^\omega$ for every $k \geqslant 1$. Finally, by letting $s^{\omega+0} = s^\omega$, the power $s^{\omega+k}$ gets defined for every integer k.

Proposition 3.7.4 *Let M be a compact monoid. An element u of M is invertible if and only if $u^\omega = 1$. If u is invertible, then $u^{\omega-1}$ is its inverse.*

Proof It follows from the equality $u^\omega = u \cdot u^{\omega-1}$ that u is invertible, with inverse $u^{\omega-1}$, if $u^\omega = 1$. Conversely, suppose that u is invertible. Then 1 and u are $\mathcal{H}$-equivalent. As the $\mathcal{H}$-class H of the idempotent 1 is a group, one has $u^+ \subseteq H$. Every $\mathcal{H}$-class is topologically closed, thus $\overline{u^+} \subseteq H$ and $u^\omega \in H$. Since H contains only one idempotent, we must have $u^\omega = 1$.   ∎

Corollary 3.7.5 *In a compact group, the closed subsemigroups coincide with the closed subgroups.*

Proof Let T be a closed subsemigroup of a compact group S, and let $s \in T$. From $s \in \overline{s^+}$ we get $s^{\omega-1} \in T$. But $s^\omega = 1$, as S is a group, thus $s^{\omega-1} = s^{-1}$ by Proposition 3.7.4.   ∎

For a subset X of a group G, let $\langle X \rangle$ be the subgroup of G generated by X. If G is a topological group, then $\overline{\langle X \rangle}$ is the intersection of all closed subgroups containing X, that is, the closed subgroup generated by X. By Corollary 3.7.5, if G is compact, then $\overline{X^+} = \overline{\langle X \rangle}$ for every nonempty subset X of G.

3.8 Projective Limits

Profinite semigroups constitute a special class of compact semigroups. They are the projective limits of finite semigroups, for which reason we begin by formally introducing the notion of projective limit. This notion holds not only for semigroups, but also for monoids, groups, rings, sets, indeed, for every algebraic structure (in the sense given in universal algebra).

We want to define profinite semigroups as some kind of limit of finite semigroups in such a way that properties true in all finite semigroups will remain true in profinite semigroups. For this we need the notion of projective limit of topological semigroups.

A *projective system* (or *inverse system*) of topological semigroups is given by

 (i) a directed set I, that is a partially ordered set in which any two elements have a common upper bound;
 (ii) for each $i \in I$, a topological semigroup S_i;
 (iii) for each pair $i, j \in I$ with $i \geqslant j$, a continuous morphism $\psi_{i,j} : S_i \to S_j$, said to be a *connecting morphism*, for which $\psi_{i,i}$ is the identity on S_i and such that, whenever $i \geqslant j \geqslant k$, one has $\psi_{i,k} = \psi_{j,k} \circ \psi_{i,j}$, that is, the following diagram is always commutative:

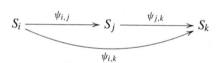

Such a projective system is briefly denoted by $(S_i, \psi_{i,j}, I)$.

When reading the following example, recall that we view finite semigroups as topological semigroups for the discrete topology.

Example 3.8.1 The set $\mathbb{Z}_+$ of positive integers, ordered by divisibility (that is, $m \leqslant n$ if and only if $m \mid n$), is a directed set. The family of cyclic groups $(\mathbb{Z}/n\mathbb{Z})_{n \in \mathbb{Z}_+}$ forms a projective system for the morphisms $\psi_{n,m}$ defined by $\psi_{n,m}(x) = x \bmod m$.

In the same way, the family of cyclic groups $(\mathbb{Z}/n!\mathbb{Z})_{n \in \mathbb{Z}_+}$, indexed by the set $\mathbb{Z}_+$ with the natural total order on integers, is a projective system.

Given a projective system $(S_i, \psi_{i,j}, I)$, a family $\psi_i : S \to S_i$ of continuous morphisms, from a fixed topological semigroup S into all the semigroups S_i, is said to be *compatible* if $\psi_{i,j} \circ \psi_i = \psi_j$ whenever $i \geqslant j$, that is, if the following diagram is always commutative:

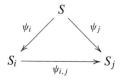

A *projective limit* (or *inverse limit*) of the projective system $(S_i, \psi_{i,j}, I)$ is a topological semigroup S together with a compatible family of continuous morphisms $\psi_i : S \rightarrow S_i$ such that the following universal property holds (see Fig. 3.5): for any topological semigroup T and any compatible family of continuous morphisms $\tau_i : T \rightarrow S_i$, there exists a unique continuous morphism $\theta : T \rightarrow S$ such that $\psi_i \circ \theta = \tau_i$ for all $i \in I$. We may denote by (S, ψ_i) such a projective limit.

Proposition 3.8.2 *The projective limit of a projective system $(S_i, \psi_{i,j}, I)$ is unique, in the following sense: if (S, ψ_i) and (S', ψ_i') are two projective limits of the projective system $(S_i, \psi_{i,j}, I)$, there is a continuous isomorphism $\theta : S \rightarrow S'$ such that $\psi_i' \circ \theta = \psi_i$ for each $i \in I$.*

Proof Suppose that (S, ψ_i) and (S', ψ_i') are two projective limits of the projective system $(S_i, \psi_{i,j}, I)$. By the universal property of the projective limit, there exist continuous morphisms $\theta : S \rightarrow S'$ and $\theta' : S' \rightarrow S$ such that $\psi_i' \circ \theta = \psi_i$ and $\psi_i \circ \theta' = \psi_i'$. The morphism $\rho = \theta' \circ \theta$ is such that $\psi_i \circ \rho = \psi_i$ (see Fig. 3.6). By the uniqueness of the morphism stipulated in the universal property, this forces $\theta' \circ \theta$ to be the identity on S. One proves in the same way that $\theta \circ \theta'$ is the identity on S'. Thus θ is a continuous isomorphism.   ∎

Assuming the projective system is known, it is customary to use the notation $\varprojlim_{i \in I} S_i$, or even $\varprojlim S_i$, for the unique (up to isomorphism of topological semigroups) projective limit of a projective system $(S_i, \psi_{i,j}, I)$, if such a limit exists. The following proposition addresses the question of the existence of the projective limit $\varprojlim S_i$, which it turns out to be related to our choice of not allowing the empty set to be a semigroup.

Fig. 3.5 The projective limit.

Fig. 3.6 The morphisms θ and θ'

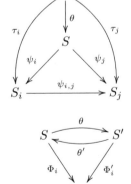

Proposition 3.8.3 *Consider a projective system* $(S_i, \psi_{i,j}, I)$. *Let S be the subset of* $\prod_{i \in I} S_i$ *consisting of all* $(s_i)_{i \in I}$ *such that, for all* $i, j \in I$ *with* $i \geqslant j$, *one has*

$$\psi_{i,j}(s_i) = s_j. \tag{3.1}$$

The projective system has a projective limit if and only if $S \neq \emptyset$. *Moreover, if* $S \neq \emptyset$, *then S is a topological subsemigroup of* $\prod_{i \in I} S_i$, *endowed with the product topology, and* $S = \varprojlim_{i \in I} S_i$, *where, for each k, the associated morphism* $\psi_k : S \to S_k$ *is the restriction to S of the projection from* $\prod_{i \in I} S_i$ *to* S_k.

Proof If $s = (s_i)_{i \in I}$ and $t = (t_i)_{i \in I}$ belong to S, then $st = (s_i t_i)_{i \in I}$ also belongs to S because $\psi_{i,j}$ are morphisms of semigroups. Therefore, S is a topological subsemigroup of $\prod_{i \in I} S_i$ if $S \neq \emptyset$.

Let T be a topological semigroup together with a compatible family of continuous morphisms $\tau_i : T \to S_i$ (note that such data exists if the projective system has a projective limit). Since we defined semigroups as being nonempty, we may take some $t \in T$. For such a t, one has $(\tau_i(t))_{i \in I} \in S$, by the definition of compatible family of morphisms. Therefore, we have $S \neq \emptyset$, and a continuous morphism $\theta : T \to S$, given by $\theta(t) = (\tau_i(t))_{i \in I}$. Moreover, if we have a continuous morphism $\theta' : T \to S$ such that $\psi_i \circ \theta' = \tau_i$, then $\theta'(t) = (\tau_i(t))_{i \in I} = \theta(t)$ for every $t \in T$, thus $\theta = \theta'$. This shows that $S \neq \emptyset$ if and only if $\varprojlim S_i$ exists, with $S = \varprojlim S_i$ if $S \neq \emptyset$. ∎

The projective limit $\varprojlim_{i \in I} S_i$ may be empty, that is, the semigroup $\varprojlim_{i \in I} S_i$ may not exist (we excluded empty sets from being semigroups).

Example 3.8.4 Let I be the set of positive integers with the usual order, let each S_i be the additive semigroup $\mathbb{Z}_+$, endowed with the discrete topology, and let the connecting morphisms $\psi_{n,m} : \mathbb{Z}_+ \to \mathbb{Z}_+$ be defined by $\psi_{n,m}(x) = 2^{n-m}x$. Then $\varprojlim_{i \in I} S_i$ is empty. Indeed, if (x_n) is such that $\psi_{n,m}(x_n) = x_m$ for $n \geqslant m$, then x_1 is divisible by arbitrary high powers of 2.

We will in general handle the projective limit $\varprojlim S_i$ as being the subsemigroup S of $\prod_{i \in I} S_i$ described in Proposition 3.8.3.

Next, we present some more useful remarks regarding the topology of the projective limit $\varprojlim_{i \in I} S_i$.

Remark 3.8.5 If the set I is uncountable, then the projective limit may be non metrizable. However, if I is at most countable and the topology of each semigroup S_i is metrizable, then $\prod_{i \in I} S_i$ is metrizable by Proposition 3.2.17, and therefore $\varprojlim_{i \in I} S_i$ is metrizable.

Remark 3.8.6 A net $(s_n)_{n \in \mathcal{N}}$ converges in $\varprojlim_{i \in I} S_i$ if and only if, for every $i \in I$, the net $(\psi_i(s_n))_{n \in \mathcal{N}}$ converges in S_i (cf. Example 3.2.12). This implies that $\varprojlim_{i \in I} S_i$ is a closed subset of the product $\prod_{i \in I} S_i$.

In contrast with Example 3.8.4, we have the following statement.

Proposition 3.8.7 *Consider a projective system* $(S_i, \psi_{i,j}, I)$ *of compact semigroups. Then the projective limit* $S = \varprojlim S_i$ *exists and it is a compact semigroup. If, moreover, the connecting morphisms* $\psi_{i,j} : S_i \to S_j$ *are onto, then each component projection* $\psi_i : S \to S_i$ *is onto.*

Proof Consider the subspace S as in the statement of Proposition 3.8.3. Since S is a closed subspace of the compact semigroup $\prod_{i \in I} S_i$, it is compact. To see that it is nonempty, consider for each $j \in I$ the subsemigroup T_j of $\prod_{i \in I} S_i$ consisting of the (s_i) for which one has $\psi_{j,k}(s_j) = s_k$ for all k such that $k \leqslant j$. Observe that $S = \bigcap_{j \in I} T_j$. Note also that $j \leqslant j'$ implies $T_j \supseteq T_{j'}$. We shall also use the fact that T_j is closed, justified by the same arguments sustaining that S is closed. Moreover, for each $x \in S_j$ there is $(s_i) \in T_j$ such that $s_j = x$: just let s_k be $\psi_{j,k}(x)$ if $k \leqslant j$, and let it be some arbitrary element of S_k otherwise. Therefore, T_j is nonempty. Let us see that the family of sets of the form T_j has the finite intersection property, i.e., that any intersection of finitely many T_j is nonempty. Indeed, if $j_1, \ldots, j_n \in I$, let j be a common upper bound of the j_ℓ. Then $\bigcap_{\ell=1}^{n} T_{j_\ell}$ contains the nonempty set T_j. Because $\prod_{i \in I} S_i$ is compact, the intersection $\bigcap_{j \in I} T_j$ of closed sets is nonempty. In view of Proposition 3.8.3, this shows that the projective limit $S = \varprojlim S_i$ exists, and it is a compact semigroup.

We proceed to show that the projections $\psi_j : S \to S_j$ are onto. The corresponding extension to the component projection $\prod_{i \in I} S_i \to S_j$ will be denoted by $\bar{\psi}_j$ in the next few lines. Fix $j \in I$, and let $y \in S_j$. We want to show that, for every finite subset F of I, the set $\bigcap_{i \in F}(T_i \cap \bar{\psi}_j^{-1}(y))$ is nonempty. Consider a common upper bound ℓ of the elements of $F \cup \{j\}$. Since $T_\ell \subseteq T_i$ for every $i \in F$, we are reduced to show that $T_\ell \cap \bar{\psi}_j^{-1}(y)$ is nonempty. Because $\psi_{\ell,j}$ is onto, there is $x \in \psi_{\ell,j}^{-1}(y)$. As observed in the previous paragraph, there is $s = (s_i)$ in T_ℓ with $s_\ell = x$ and $s_k = \psi_{\ell,k}(x)$ whenever $k \leqslant \ell$. In particular, we have $s_j = \psi_{\ell,j}(x) = y$, thus $s \in T_\ell \cap \bar{\psi}_j^{-1}(y)$ and $\bigcap_{i \in F}(T_i \cap \bar{\psi}_j^{-1}(y)) \neq \emptyset$. Since $S = \bigcap_{i \in I} T_i$ is compact, it follows that the set $S \cap \psi_j^{-1}(y) = \bigcap_{i \in I}(T_i \cap \bar{\psi}_j^{-1}(y))$ is nonempty, establishing that $\psi_j : S \to S_j$ is onto. ∎

Therefore, in a sense, no information is lost when passing to the projective limit, provided the projective system is formed by onto continuous morphisms between compact semigroups. Every projective limit of compact semigroups is the projective limit of a projective system with onto connecting morphisms (Exercise 3.76).

Just as for semigroups, one defines in the same way a projective system and a projective limit of topological monoids or groups, and indeed, of any kind of topological abstract algebra (including in particular topological spaces, which are viewed as topological algebras without operations).

Example 3.8.8 The projective limit of the family of cyclic groups $\mathbb{Z}/n\mathbb{Z}$, that we met in Example 3.8.1, is the group part of the ring $\widehat{\mathbb{Z}} = \varprojlim_{n \geqslant 1} \mathbb{Z}/n\mathbb{Z}$, where the latter projective limit is in the category of compact rings.

The group $\widehat{\mathbb{Z}}$ already appeared in Chap. 2, before our formal introduction of projective limits. Its monoid counterpart $\widehat{\mathbb{N}}$ also appeared in Chap. 2, as well as the equality $\hat{\mathbb{N}} = \varprojlim N_{i,p}$. Concerning the corresponding semigroup counterpart, we see that the projective limit $\varprojlim C_{i,p}$ is a compact semigroup isomorphic to $\hat{\mathbb{N}} \setminus \{1\}$.

In Example 3.8.8, we could as well have used the family $\mathbb{Z}/n!\mathbb{Z}$, since in fact we have a natural isomorphism between $\varprojlim_{n \geqslant 1} \mathbb{Z}/n\mathbb{Z}$ and $\varprojlim_{n \geqslant 1} \mathbb{Z}/n!\mathbb{Z}$. Indeed, if $J \subseteq I$ is cofinal, then $\varprojlim_{j \in J} S_j = \varprojlim_{i \in I} S_i$ (see Exercise 3.74).

3.9   Profinite Semigroups: Definition and Characterizations

A *profinite semigroup* is a projective limit of a projective system of finite semigroups, endowed with the discrete topology. Note that every profinite semigroup is a compact semigroup (cf. Proposition 3.8.7).

Example 3.9.1 Let us recall Example 3.8.8. The additive group $\hat{\mathbb{Z}}$ of profinite integers is a profinite semigroup (which is a group). Likewise, $\hat{\mathbb{N}}$ and $\hat{\mathbb{N}} \setminus \{1\}$ are profinite semigroups.

Every finite semigroup S is profinite: indeed S is the projective limit of the projective system consisting solely of S, with the identity on S as the sole connecting morphism.

The next generalization of Proposition 3.7.3 is an example of how features of finite semigroups may be preserved when a projective limit of them is taken.

Proposition 3.9.2 *Let S be a profinite semigroup. For every $s \in S$, the sequence $(s^{n!})$ converges to s^ω.*

Proof Let $S = \varprojlim_{i \in I} S_i$, for some projective system $(S_i, \psi_{j,i}, I)$ of finite semigroups. Thanks to Proposition 3.7.3, for each projection $\psi_i : S \to S_i$ we have

$$\lim_{n \to \infty} \psi_i(s^{n!}) = \lim_{n \to \infty} \psi_i(s)^{n!} = \psi_i(s)^\omega$$

and so, as we are dealing with the pointwise topology, one gets

$$s^{n!} = (\psi_i(s^{n!}))_{i \in I} \to (\psi_i(s)^\omega)_{i \in I}.$$

Note that $(\psi_i(s)^\omega)_{i \in I}$ is idempotent, since each component is idempotent. Therefore, $\lim s^{n!}$ equals s^ω, the unique idempotent in the closed subsemigroup of S generated by s.   ∎

The next lines prepare a more informative characterization of profinite semigroups, to be given afterwards, in Theorem 3.9.3.

A topological space is *connected* if it is not the union of two disjoint nonempty open sets. A subset of a topological space is connected if it is connected as a subspace, that is to say it cannot be covered by the union of two disjoint open sets.

A continuous image of a connected set is connected (Exercise 3.78). A union of connected sets with nonempty intersection is connected (Exercise 3.79).

Every topological space decomposes as a union of disjoint connected subsets, called its *connected components*. The connected component of x is the union of all connected sets containing x.

A topological space is *totally disconnected* if its connected components are singletons. Any product of totally disconnected spaces is totally disconnected (Exercise 3.80) and every subspace of a totally disconnected space is totally disconnected.

As an important particular case, a *Cantor space* is a compact metric space without isolated points and which is totally disconnected. All Cantor spaces are homeomorphic (see the Notes for a reference).

A topological space is *zero-dimensional* if it admits a basis consisting of clopen sets.

A compact space is zero-dimensional if and only if it is totally disconnected (see Exercise 3.82).

The following result gives several alternative direct definitions of profinite semigroup without using projective limits. A topological semigroup is *residually finite* if for any distinct $u, v \in S$ there exists a continuous semigroup morphism $\varphi : S \to T$ into a finite semigroup T such that $\varphi(u) \neq \varphi(v)$.

Theorem 3.9.3 *The following conditions are equivalent for a compact semigroup S.*

  (i) *S is profinite,*
 (ii) *S is residually finite as a topological semigroup,*
(iii) *S is a closed subsemigroup of a direct product of finite semigroups,*
(iv) *S is totally disconnected,*
 (v) *S is zero-dimensional.*

We first prove the following lemma. The *syntactic conruence ρ_K* of a subset K in a semigroup S is defined by $s \equiv t \bmod \rho_K$ if one has for all $u, v \in S^1$, $usv \in K \Leftrightarrow utv \in K$. An equivalence on a topological space X is *open* if it is an open subset of $X \times X$. Actually, an equivalence is open if and only it is clopen (see Exercise 3.40).

Lemma 3.9.4 (Hunter) *If S is a compact semigroup and K is a clopen subset of S, the syntactic congruence of K is open (and thus has a finite number of classes).*

Proof Consider a net (x_i, y_i) of elements of $S \times S$ such that $x_i \not\equiv y_i \bmod \rho_K$ and converging to (x, y). By definition of ρ_K, for each n there exist $u_i, v_i \in S$ such that $u_i x_i v_i \in K$ and $u_i y_i v_i \notin K$, or vice versa. Up to taking a subnet, we may assume that the same case occurs for all i (cf. Exercise 3.5), and without loss of generality we assume that it is the first case. Because S is compact, we may also assume that $u_i \to u$ and $v_i \to v$. Since K is closed, $uxv \in K$ and since the complement of K is closed, $uyv \in S \setminus K$. Thus x, y are not equivalent and we conclude that the complement of ρ_K is closed. ∎

Proof of Theorem 3.9.3 The explicit construction of the projective limit shows that
(i) $\Rightarrow$ (ii) and (ii) $\Rightarrow$ (iii) is immediate from the definitions.

For (iii) $\Rightarrow$ (i), suppose that $\Phi : S \to \prod_{i \in I} S_i$ is an injective continuous
morphism from a compact semigroup S into a product of finite semigroups. We may
assume that for each $i \in I$ the map $\pi_i \circ \Phi : S \to S_i$ is onto. For each subset J of I,
let $\pi_J : \prod_{i \in I} S_i \to \prod_{i \in J} S_i$ be the natural projection. We build a projective system
of finite semigroups by considering all onto mappings of the form $\Phi_F : S \to S_F$
where F is a finite subset of I and $\Phi_F = \pi_F \circ \Phi$. The indexing set of the projective
system is the directed set of all finite subsets of I under the inclusion ordering. If
$F \subseteq F'$, then the connecting morphism $S_{F'} \to S_F$ is the natural projection from
$S_{F'}$ onto S_F. Then S is the projective limit of this system of finite semigroups.

Since a product of totally disconnected spaces is totally disconnected, we have
(iii) $\Rightarrow$ (iv). The equivalence (iv) $\Leftrightarrow$ (v) holds for any compact space. Finally, the
implication (v) $\Rightarrow$ (ii) results from Lemma 3.9.4.   ∎

We illustrate Theorem 3.9.3 with the following example.

Example 3.9.5 Consider the Cantor space $S = A^{\mathbb{N}}$ of infinite words on a finite
alphabet A. Considered as a semigroup with the product $xy = x$ for every $x, y \in S$,
it is clearly a topological semigroup and thus it is profinite. We may obtain S as
the projective limit of a directed system of finite semigroups by considering the
finite semigroups $S_n = A^n$ for $n \geqslant 1$ with the product $xy = x$. The connecting
morphisms $\psi_{i,j}$ are defined for $i \geqslant j$ by $\psi_{i,j}(x) = y$ if y is the prefix of length j
of x.

Corollary 3.9.6 *The class of profinite semigroups is closed under taking closed
subsemigroups and direct products.*

Proof We already remarked that the class of totally disconnected spaces is closed
under taking direct products and subspaces. Since the class of compact spaces is also
closed under taking direct products and closed subspaces, the conclusion follows
from Theorem 3.9.3.   ∎

The following example shows that the class of profinite semigroups is not closed
under taking continuous closed morphic images.

Example 3.9.7 Let $A = \{0, 1\}$. Every $x \in A^{\mathbb{N}}$ can be considered as the expansion
in base 2 of a real number in $[0, 1]$. This defines a continuous map π from $A^{\mathbb{N}}$ onto
$[0, 1]$. Both spaces can be turned into compact semigroups by making all elements
left-zeros. The first space is totally disconnected (making $A^{\mathbb{N}}$ a profinite semigroup),
but not the second. Hence π is a continuous morphism from a profinite semigroup
onto a compact semigroup which is not profinite.

A subset K of a semigroup S is *recognized* by a morphism $\varphi : S \to T$ if $K = \varphi^{-1}\varphi(K)$.

Proposition 3.9.8 *Let S be a profinite semigroup. A subset $K \subseteq S$ is clopen if
and only if it is recognized by a continuous morphism $\varphi : S \to T$ into a finite
semigroup T.*

Proof If K is clopen then, by the definition of the congruence ρ_K, it is the union of the ρ_K-classes intersected by K, and so K is recognized by the continuous projection from S onto the finite semigroup S/ρ_K, according to Hunter's Lemma (Proposition 3.9.4). The converse is immediate.   ∎

3.10   Profinite Groups and Profinite Monoids

A *profinite group* (respectively, *profinite monoid*) is a projective limit of a projective system of finite groups in the category of topological groups (respectively, topological monoids), that is, all morphisms considered—connecting morphisms, projections—must be morphisms of topological groups (respectively, topological monoids). Therefore, profinite groups and profinite monoids are special examples of profinite semigroups.

Proposition 3.10.1 *If a profinite semigroup is a group, then it is a profinite group.*

Proof Indeed, let S be the projective limit of a family S_i of finite semigroups. We may assume that the projection morphisms $\psi_i : S \to S_i$ are surjective. If S is a group, since the image of a group by a semigroup morphism is a group, each S_i is a finite group and thus S is a profinite group.   ∎

It thus follows from Theorem 3.9.3 that a compact group is profinite if and only if it is totally disconnected. Likewise, a compact group is profinite if and only if it is residually finite (a topological group G is *residually finite* if for every distinct $u, v \in G$ there is a continuous morphism $\varphi : G \to H$ into a finite group such that $\varphi(u) \neq \varphi(v)$ holds). Therefore, a closed subgroup of a profinite group is profinite. Indeed, every subgroup of G is residually finite for the induced topology, thus any compact subgroup of G is profinite.

The preceding remarks about profinite groups have analog versions in the monoid realm.

The following result contrasts with the fact that a quotient of a profinite semigroup needs not be profinite (Example 3.9.7).

Proposition 3.10.2 *The quotient of a profinite group by a closed normal subgroup is profinite.*

Proof Let H be a closed normal subgroup of a profinite group G. Let $(G_i)_{i\in I}$ be a projective system of finite groups such that $G = \varprojlim_{i \in I} G_i$, with compatible morphisms $\varphi_i : G \to G_i$ and connecting morphisms $\varphi_{i,j} : G_i \to G_j$. Set $H_i = \varphi_i(H)$. Then H_i is a normal subgroup of $\operatorname{Im} \varphi_i$. The direct product $K = \prod_{i \in I} \operatorname{Im} \varphi_i / H_i$ is a profinite group, by Corollary 3.9.6. Consider the continuous morphism $\zeta : G \to K$ naturally associated to this product. The group $\operatorname{Im} \zeta$ is profinite, again by Corollary 3.9.6. Therefore, by Theorem 3.5.6, it suffices to show that $H = \zeta^{-1}(1)$. That is, we want to show the equality

$$\bigcap_{i \in I} \varphi_i^{-1}(H_i) = H. \tag{3.2}$$

It is trivial that $H \subseteq \varphi_i^{-1}(H_i)$. Conversely, let $x \in \bigcap_{i \in I} \varphi_i^{-1}(H_i)$. For each $i \in I$, take $h_i \in H$ with $\varphi_i(x) = \varphi_i(h_i)$. Let h be an accumulation point of the net $(h_i)_{i \in I}$. If $k \geqslant i$ then we have $\varphi_i(x) = \varphi_{k,i}(\varphi_k(x)) = \varphi_{k,i}(\varphi_k(h_k)) = \varphi_i(h_k)$. As h is an accumulation point of $(h_k)_{k \geqslant i}$, it follows that $\varphi_i(x) = \varphi_i(h)$ for every $i \in I$, thus $x = h$. Since $h \in H$ because H is closed, this establishes (3.2).   ■

Note the specific property of compact groups that was crucial in the proof of Proposition 3.10.2: the kernel of a continuous morphism of compact groups is determined by the inverse image of the neutral element.

A simple example of a compact group which is not profinite is the circle, that is, the multiplicative group of complex numbers of modulus 1, which has no nontrivial image which is a finite group.

3.11   The Profinite Distance

The *profinite distance* or *profinite metric* on a profinite semigroup S is the map $d_S : S \times S \to \mathbb{R}_+$ defined by

$$d_S(u, v) = \begin{cases} 1/r_S(u, v) & \text{if } u \neq v \\ 0 & \text{otherwise} \end{cases}$$

where $r_S(u, v)$ is the minimal cardinality of a finite semigroup F for which there is a continuous morphism $\varphi : S \to F$ such that $\varphi(u) \neq \varphi(v)$. We also use the notations d and r instead of d_S and r_S, respectively. The distance d is actually *ultrametric* since it satisfies the condition

$$d(u, w) \leqslant \max\{d(u, v), d(v, w)\},$$

which is stronger than the triangle inequality. When S is the monoid of natural integers, we find again the distance d_+ introduced in Chap. 2.

A topological semigroup S is said to be *finitely generated* if $S = \overline{X^+}$ for some nonempty finite subset X of S.

Proposition 3.11.1 *For a finitely generated profinite semigroup, the topology is induced by the natural metric.*

Proof Let S be a finitely generated profinite semigroup. Denote by $B_\varepsilon(u)$ the open ball $\{v \in S \mid d(u, v) < \varepsilon\}$. Let K be a clopen set in S. By Proposition 3.9.8, there is a continuous morphism $\varphi : S \to F$ into a finite monoid M which recognizes K. Let $n = \text{Card}(F)$ and set $\varepsilon = 1/n$. For $s \in K$ and $t \in B_\varepsilon(s)$, we have $d(s, t) < \varepsilon$ and thus $r(s, t) > n$. It follows that $\varphi(s) = \varphi(t)$. Since φ recognizes K, we conclude that $t \in K$. Therefore the ball $B_\varepsilon(s)$ is contained in K and so K is a union of open balls. Thus, since the clopen sets form a basis of the topology, any open set is a union of open balls.

Conversely, consider the open ball $B = B_{1/n}(u)$. Since there is a finite number of isomorphism types of semigroups with at most n elements, and since S is finitely generated, there are finitely many kernels $\rho_1, \dots, \rho_k$ of continuous morphisms from S into such semigroups and so their intersection $\rho = \bigcap_{i=1}^{k} \rho_i$ is a clopen congruence on S. The continuous canonical morphism $\varphi : S \to S/\rho$ onto the finite semigroup S/ρ is such that $\varphi(u) = \varphi(v)$ if and only if $[u]_{\rho_i} = [v]_{\rho_i}$ for all i (cf. Exercise 3.45), if and only if $r(u, v) > n$ (cf. Theorem 3.5.6). Hence $B = \varphi^{-1}\varphi(B)$ so that B is open. ∎

The following elementary statement is useful.

Proposition 3.11.2 *Let S and T be a profinite semigroups. Then every continuous morphism $f : S \to T$ is contracting, that is, for every $s, t \in S$ we have*

$$d_T(f(s), f(t)) \leqslant d_S(s, t).$$

Proof We may assume that $f(s) \neq f(t)$. Then, there is a continuous morphism $\varphi : T \to F$ from T into a finite semigroup F such that $\varphi(f(s)) \neq \varphi(f(t))$. Choose φ with F having the smallest possible cardinal, so that the equality $r_T(f(s), f(t)) = \text{Card}(F)$ holds. As the composition $\varphi \circ f$ is again a continuous morphism, we conclude that $r_S(s, t) \leqslant r_T(f(s), f(t))$. ∎

3.12 Profinite Monoids of Continuous Endomorphisms

For a topological semigroup S, the set of its continuous endomorphisms is a monoid for the composition of functions, denoted $\text{End}(S)$. By default, we consider $\text{End}(S)$ as a topological submonoid of S^S, for the pointwise topology of S^S.

The following theorem gives an appropriate framework concerning several connections between profinite semigroups and symbolic dynamics, that we shall see along this book.

Theorem 3.12.1 *Let S be a finitely generated profinite semigroup. Then $\text{End}(S)$ is a profinite monoid.*

Before proving Theorem 3.12.1, we pause for some comments. Suppose that S is a finitely generated profinite semigroup. The set $\text{Aut}(S)$ of continuous automorphisms of S is a maximal subgroup of $\text{End}(S)$. Since $\text{End}(S)$ is a profinite monoid by Theorem 3.12.1, $\text{Aut}(S)$ is itself a profinite group. In particular, the group of continuous automorphisms of a finitely generated profinite group is also a profinite group. On the other hand, an infinitely generated profinite group may be such that its group of continuous automorphisms is not a profinite group, with respect to the pointwise topology (cf. Exercise 3.88). Therefore, the condition "finitely generated" in Theorem 3.12.1 is not superfluous.

To prove Theorem 3.12.1, we need a series of preliminary results.

Lemma 3.12.2 *If S is a finitely generated profinite semigroup, then* $\mathrm{End}(S)$ *is compact.*

Proof For fixed $s, t \in S$, the set of $f \in S^S$ such that $f(st) = f(s)f(t)$ is closed for the product topology since the multiplication is continuous. Thus the monoid of all (not necessarily continuous) endomorphisms of S is closed.

Consider a net (f_i) in of continuous endomorphisms of S converging to some $f \in S^S$. Given $s, t \in S$, by continuity of the profinite metric, we have $d(f(s), f(t)) = \lim_{i \in I} d(f_i(s), f_i(t))$ while $d(f_i(s), f_i(t)) \leqslant d(s, t)$ by Proposition 3.11.2. Hence $d(f(s), f(t)) \leqslant d(s, t)$, which shows that f is continuous in view of Proposition 3.11.1. Thus $\mathrm{End}(S)$ is a closed subset of the compact space S^S and therefore it is compact. ∎

Given a topological semigroup S the *evaluation map* $\varepsilon : \mathrm{End}(S) \times S \to S$ is defined by

$$\varepsilon : (f, s) \mapsto f(s).$$

Proposition 3.12.3 *Let S be a finitely generated semigroup. Then the evaluation map is continuous with respect to the pointwise topology on* $\mathrm{End}(S)$.

Proof Let $(f_i, s_i)_{i \in I}$ be a net of elements of $\mathrm{End}(S) \times S$ converging to (f, s). Consider the profinite metric d on S. Since S is finitely generated, the topology of S is induced by d, as seen in Proposition 3.11.1. Notice that

$$d(f_i(s_i), f(s)) \leqslant \max\{d(f_i(s_i), f_i(s)), d(f_i(s), f(s))\} \qquad (3.3)$$

for all $i \in I$. On one hand, we have $d(f_i(s_i), f_i(s)) \leqslant d(s_i, s)$ by Proposition 3.11.2, and so $\lim_{i \in I} d(f_i(s_i), f_i(s)) = 0$, since $(s_i)_{i \in I}$ converges to s. On the other hand, we have $\lim_{i \in I} d(f_i(s), f(s)) = 0$, because $(f_i)_{i \in I}$ converges to f with respect to the pointwise topology. Putting all together, it follows from (3.3) that $\lim_{i \in I} f_i(s_i) = f(s)$. ∎

We are now ready to prove Theorem 3.12.1.

Proof of Theorem 3.12.1 Since S is profinite, it is totally disconnected and consequently so is the product space S^S and its subspace $\mathrm{End}(S)$. Moreover, by Lemma 3.12.2, $\mathrm{End}(S)$ is compact. Thus, by Theorem 3.9.3, all we need is to prove that $\mathrm{End}(S)$ is a topological monoid, that is, to prove that the composition of morphisms is continuous. Let (f_i, g_i) be a net converging to (f, g) in $\mathrm{End}(S) \times \mathrm{End}(S)$. By the definition of the pointwise topology on $\mathrm{End}(S)$, the net $(g_i(s))$ converges to $g(s)$. Since, by Proposition 3.12.3, the evaluation map is continuous for the pointwise topology on $\text{End}(S)$, we deduce that $f_i(g_i(s))$ converges to $f(g(s))$ for every $s \in S$. This concludes the proof. ∎

Until the end of this section we continue to see how the hypothesis that S is finitely generated is reflected on certain features of $\text{End}(S)$.

Proposition 3.12.4 *The set of continuous surjective endomorphisms of a finitely generated profinite semigroup S is a closed submonoid of* $\text{End}(S)$.

Proof Let $(\varphi_i)_{i \in I}$ be a net of continuous surjective endomorphisms of S, and suppose that it converges in $\text{End}(S)$ to the continuous endomorphism φ. Let $s \in S$. For each $i \in I$, there is some $t_i \in S$ such that $s = \varphi_i(t_i)$. Let t be a cluster point of the net (t_i). By Proposition 3.12.3, we know that $s = \varphi(t)$. This shows that φ is onto, establishing the proposition. ∎

A topological semigroup S is *Hopfian* if every continuous endomorphism of S which is onto is a continuous automorphism of S.

Proposition 3.12.5 *Every finitely generated profinite semigroup is Hopfian.*

Proof Because $\text{End}(S)$ is a profinite monoid, we may consider the ω-power φ^ω. Since φ^ω is an idempotent endomorphism, its restriction to $\varphi^\omega(S)$ is the identity on $\varphi^\omega(S)$. But, since φ is surjective, we know by Proposition 3.12.4 that φ^ω is also surjective, and so φ^ω is the identity on S. This means that φ is an isomorphism (cf. Proposition 3.7.4). ∎

3.13 Exercises

3.13.1 Section 3.2

3.1 Let U be a subset of a topological space X. Show that $x \in X$ belongs to the closure $\overline{U}$ of U if and only if every neighborhood of x meets U.

3.2 Show that a topological space is Hausdorff if and only if every convergent net has only one limit.

3.3 Show that $x \in \overline{U}$ if and only if $x = \lim u_i$ for some net (u_i) of elements of U. Conclude that a set is closed if and only if it contains all the limits of convergent nets of its elements.

3.4 Let I be a directed set and let $J \subseteq I$ be cofinal. Show that J is directed.

3.5 Let a directed set I. Consider a decomposition of I as a finite union of sets $I = J_1 \cup J_2 \cup \cdots \cup J_n$. Show that at least one of the sets J_k is cofinal.

3.6 Show that if a net $(x_i)_{i \in I}$ of elements of the topological space X converges to x, then every subnet of $(x_i)_{i \in I}$ converges to x.

3.7 Show that if $(x_i)_{i \in I}$ and $(y_j)_{j \in J}$ are nets, then there is a directed set K, and maps $\varphi : K \to I$ and $\psi : K \to J$, such that $(x_{\varphi(k)})_{k \in K}$ is a subnet of $(x_i)_{i \in I}$ and $(y_{\psi(k)})_{k \in K}$ is a subnet of $(y_j)_{j \in J}$. (Hint: consider the direct product $K = I \times J$.)

3.8 Consider a net $(x_i)_{i \in I}$ in a topological space. A *tail* of a net $(x_i)_{i \in I}$ is a set of the form $T_i = \{x_j \mid j \geqslant i\}$. Show that the following conditions are equivalent:

(i) There is some subnet of $(x_i)_{i \in I}$ converging to y.
(ii) y is in the closure of every tail of $(x_i)_{i \in I}$.
(iii) y is a cluster point of $(x_i)_{i \in I}$.

(Hint: for showing (iii) $\Rightarrow$ (i), consider the set of pairs (i, U) formed by an element $i \in I$ and a neighborhood U of y such that $x_i \in U$, and endow such a set with an adequate partial order.)

3.9 Show that a map $f : X \to Y$ between topological spaces is continuous if and only if $f^{-1}(F)$ is a closed subset of X whenever F is a closed subset of Y.

3.10 Show that a map $f : X \to Y$ between topological spaces is continuous if and only if $f(\overline{U}) \subseteq \overline{f(U)}$ for any $U \subseteq X$, where $\overline{U}$ denotes the closure of U.

3.11 Show that if $f : X \to Y$ is a function between topological spaces X and Y, then f is continuous if and only if $f(x_i) \to f(x)$ for every net (x_i) converging to x in X.

3.12 Let X and Y be topological spaces, and let U be a dense subset of X. Show that if f and g are continuous functions $X \to Y$ such that the restriction of f to U equals the restriction of g to U, then $f = g$.

3.13 Prove that any metric space is Hausdorff.

3.14 Let X be a metric space with distance d. Show that the formula $d'(x, y) = \min\{d(x, y), 1\}$ defines a distance on X which induces the same topology as d.

3.15 Let X be a metric space, and let U be a subset of X which contains the limits of all convergent sequences formed by elements of U. Show that U is closed.

3.16 Show that a function $f : X \to Y$ between topological spaces, with X metrizable, is continuous if and only if $\lim f(x_n) = f(x)$ for every sequence (x_n) converging to x in X.

3.17 The equivalences given in Exercise 3.8 have a natural analog for metric spaces and sequences: just replace "topological space" for "metric space", "net" for "sequence" and "subnet" for "subsequence". One way of seeing that is to invoke the following property: for every sequence (x_n) in a metric space, there is a subnet of (x_n) converging to y if and only if there is some subsequence of (x_n) converging to y. Prove this property.

3.18 Give an example of a subnet of a sequence (x_n), in a metric space, which is not a subsequence of (x_n).

3.19 Show that a denumerable product of metrizable spaces is metrizable. (Hint: Let $X = \prod_{n\geq 1} X_n$ where X_n is a metric space with a distance d_n. By Exercise 3.14 we may assume that $d_n(x, y) \leq 1$. For $x = (x_n)$ and $y = (y_n)$ show that

$$d(x, y) = \sup_{n\geq 1} \frac{d_n(x_n, y_n)}{n}$$

is a distance on X and that the topology defined by d is the product topology.)

3.20 Show that in a metrizable space each point x has a countable neighborhood basis, that is a family B_n, with $n \geq 1$, of neighborhoods of x such that every neighborhood of x contains some B_n (such a space is called *first countable*).

3.21 The *cofinite topology* on a set X is the topology for which the nonempty open sets are the complements of finite sets. (a subset of X is *cofinite* when its complement in X is finite).

1. Verify that it is a topology.
2. Show that it is not metrizable if X is not denumerable.

3.22 Show that the set $\mathbb{R}^{\mathbb{R}}$ of maps from $\mathbb{R}$ to itself, with the product topology, is not metrizable. (Hint: Consider the set A of maps f such that $f(x) = 1$ for all real numbers x except a finite number of reals. Show that 0 is in the closure of A but that no sequence of elements of A converges to the constant function 0.)

3.23 Let (X, d_X) be a metric space. Define an equivalence on the set of Cauchy sequences on X by $x \sim y$ if $d_X(x_n, y_n)$ converges to 0 as $n \to \infty$. Let Y be the corresponding set of equivalence classes. Define a distance d_Y on Y by $d_Y([x], [y]) = \lim d_X(x_n, y_n)$. Show that this distance is well defined and that the set of equivalence classes of Cauchy sequences is the completion of X.

3.24 A topological space X is *normal* if for every pair of disjoint closed sets $A, B \subseteq X$ there is a pair of disjoint open sets U, V such that $A \subseteq U$ and $B \subseteq V$. Show that every compact space is normal.

3.25 Show that a closed subset of a compact space is compact.

3.26 Show that the image of a compact space by a continuous map is compact, provided the image is Hausdorff.

3.27 Show that a space X is compact if and only if every net in X has a convergent subnet.

3.28 Show that a compact subspace of a Hausdorff space is closed.

3.29 Show that if $g : X \to Y$ is a continuous map between compact spaces and F is a closed subset of X, then $g(F)$ is a closed subset of Y.

3.30 Suppose that $f : X \to Y$ is an onto continuous map between compact spaces. Let Z be a topological space. Show that $g : Y \to Z$ is continuous if and only if $g \circ f : X \to Z$ is continuous.

3.31 Show that if $h : X \to Y$ is a continuous bijection between compact spaces X and Y, then the inverse map $h^{-1} : Y \to X$ is also continuous. Give examples showing that it is not redundant to assume that X and Y are compact.

3.32 Show that every compact metric space is complete.

3.33 Let X be a compact space. Show that a net of elements of X is convergent if and only if it has a unique cluster point.

3.13.2 Section 3.3

3.34 Show that every monoid M embeds in the monoid M^M.

3.35 Let X, Y be subsets of a topological semigroup S. One denotes by $X \cdot Y$, or simply XY, the set $\{xy \mid x \in X, y \in Y\}$. Show that $\overline{X} \cdot \overline{Y} \subseteq \overline{XY}$, with the equality $\overline{X} \cdot \overline{Y} = \overline{XY}$ holding if S is compact. Give an example where $\overline{X} \cdot \overline{Y} \neq \overline{XY}$.

3.36 Show that if S is a topological semigroup and X is a nonempty subset of S, then $\overline{X^+}$ is a closed subsemigroup of S.

3.37 A semigroup S is *commutative* when $xy = yx$ for every $x, y \in S$. Let X be a generating subset of a compact semigroup S. Show that S is commutative if and only if $xy = yx$ for all $x, y \in X$.

3.13.3 Section 3.5

3.38 Show that if ρ is a closed equivalence relation on a topological space X, then its classes are closed.

3.39 Let $X = [-1, 1]$ with the usual topology and let ρ be the equivalence on X whose classes are the pairs $\{-x, x\}$ for $0 < x < 1$ and the singletons $-1, 0, 1$. Show that the classes of ρ are closed but that ρ is not closed.

3.40 Let X be a topological space. Show that the following conditions are equivalent for an equivalence relation ρ on X.

(i) ρ is open.
(ii) ρ is clopen.
(iii) The classes of ρ are open.
(iv) The classes of ρ are clopen.

3.41 Suppose that ρ is a closed equivalence relation on a compact space X. Show that if C is a ρ-class and U is an open set containing C, then the union of the ρ-classes contained in U is also an open set containing C.

3.42 Show that if X is compact and ρ is an equivalence on X, then the quotient space X/ρ is compact if and only if ρ is closed. (Hint: look at Exercise 3.41.)

3.43 Suppose that $f : X \to Y$ is a continuous map between compact spaces. Explain in detail why the canonical bijection $f' : X/\operatorname{Ker} f \to \operatorname{Im} f$ is a homeomorphism of compact spaces.

3.44 Let S be a compact semigroup, and let T be a semigroup. Suppose that $\varphi : S \to T$ is an onto morphism of semigroups. Endow T with a Hausdorff topology. For that topology, T is a compact semigroup if φ is continuous.

3.45 Let ρ and θ be (closed) congruences of a (compact) semigroup. Show that the (compact) semigroup $S/(\rho \cap \theta)$ is isomorphic to a (closed) subsemigroup of the (compact) semigroup $S/\rho \times S/\theta$.

3.13.4   Section 3.6

3.46 Verify that if e is idempotent and $x \leqslant_{\mathcal{R}} e$, then $x = ex$.

3.47 Here we ask to fill the details in Example 3.6.2:

1. Show that $f \mathrel{\mathcal{L}} g$ if and only if $\operatorname{Ker} f = \operatorname{Ker} g$, and that $f \mathrel{\mathcal{R}} g$ if and only if $\operatorname{Im} f = \operatorname{Im} g$, whenever $f, g : X \to X$ are functions.
2. Deduce that in T_X one has $f \mathrel{\mathcal{R}} g$ if and only if $\operatorname{Ker} f = \operatorname{Ker} g$, and $f \mathrel{\mathcal{L}} g$ if and only if $\operatorname{Im} f = \operatorname{Im} g$, where $\operatorname{Ker} f = \{(x, y) \in X \times X \mid x \cdot f = y \cdot f\}$ and $\operatorname{Im} f = \{x \cdot f \mid x \in X\}$ when $f \in T_X$.
3. The definitions of kernel and image extend to partial transformations by identifying $f \in PT_X$ with $\tilde{f} \in T_{X \uplus \{0\}}$ extending f, with 0 not in X such that $0 \cdot \tilde{f} = 0$ and $x \cdot \tilde{f} = 0$ for all x such that $x \cdot f$ is undefined. We then have $\operatorname{Im} f = \operatorname{Im} \tilde{f} \cap X$, and $\operatorname{Ker} f$ may be defined as $\operatorname{Ker} f = \operatorname{Ker} \tilde{f} \cap D_f \times D_f$, where D_f is the domain of f. Deduce that $f \mathrel{\mathcal{R}} g$ in PT_X if and only if $\operatorname{Ker} f = \operatorname{Ker} g$, and that $f \mathrel{\mathcal{L}} g$ in PT_X if and only if $\operatorname{Im} f = \operatorname{Im} g$.

3.48 Suppose that S is a subsemigroup of the monoid PT_X of right partial transformations on the set X. For each element f of S, the cardinality of the image

$$X \cdot f = \{x \cdot f \mid x \in X, \ x \cdot f \text{ is defined}\}$$

is the *rank* of f in S, and may be denoted by $\operatorname{rank}_S(f)$. Show that if $f \leqslant_{\mathcal{J}} g$ then $\operatorname{rank}_S(f) \leqslant \operatorname{rank}_S(g)$. (Therefore, the elements in a $\mathcal{J}$-class J of S have the same rank. This common rank is called the *rank* of the $\mathcal{J}$-class J.)

3.49 Show that the relations $\mathcal{R}$ and $\mathcal{L}$ commute.

3.50 Show that in a commutative semigroup one has $\mathcal{H} = \mathcal{R} = \mathcal{L} = \mathcal{D} = \mathcal{J}$.

3.51 Let e, f be idempotents of a semigroup S. Show that the following conditions are equivalent:

(i) $e \, \mathcal{D} \, f$.
(ii) $e = xy$ and $f = yx$ for some $x, y \in S$.
(iii) $e = xy$ and $f = yx$ for some $x \in R(e) \cap L(f)$ and $y \in L(e) \cap R(f)$.

(Hint for (i) $\Rightarrow$ (iii): observe that $e = xy$ for some x, y such that $e \, \mathcal{R} \, x \, \mathcal{L} \, y$, $x = ex = xf$ and $y = fy$.)

3.52 Let s, s' be $\mathcal{R}$-equivalent elements of a semigroup S. Set $s = s'u'$ and $s' = su$, with $u, u' \in S^1$. Then the mappings $\rho_u : q \to qu$ and $\rho_{u'} : q' \to q'u'$ are mutually inverse bijections from $L(s)$ onto $L(s')$ and from $L(s')$ onto $L(s)$, respectively (this statement is known as *Green's Lemma*).

3.53 Show that the $\mathcal{H}$-classes contained in the same $\mathcal{D}$-class have equal cardinal.

3.54 Show that any two $\mathcal{L}$-classes in the same $\mathcal{D}$-class contain an equal number of $\mathcal{H}$-classes.

3.55 Let S be a semigroup and let $s, t \in S$. Then $st \in R(s) \cap L(t)$ if and only if $R(t) \cap L(s)$ contains an idempotent (*Clifford-Miller's Lemma*).

3.56 Let D be a $\mathcal{D}$-class of a semigroup S. The following conditions are equivalent.

(i) D contains a regular element.
(ii) All elements of D are regular.
(iii) D contains an idempotent.
(iv) Every $\mathcal{L}$-class (resp. every $\mathcal{R}$-class) of D contains an idempotent.

3.57 Show that in a compact monoid, all Green relations are closed.

3.58 Show that a compact semigroup S has minimal right (resp. left) ideals, and that the minimal right (resp. left) ideals are precisely the $\mathcal{R}$-classes (resp. $\mathcal{L}$-classes) of S contained in the minimum ideal $K(S)$.

3.59 Show that an $\mathcal{H}$-class of a semigroup S is a group if and only if it contains an idempotent.

3.60 Show that in a semigroup S, the $\mathcal{H}$-class of an element s of S is a group if and only if $s^2 \, \mathcal{H} \, s$.

3.61 Show that every $\mathcal{H}$-class contained in the minimum ideal of a compact semigroup is a group.

3.62 To distinguish between different semigroups, denote by $\mathcal{K}_N$ the Green relation $\mathcal{K}$ in a semigroup N. Suppose S is a subsemigroup of T. Show that:

1. One may have $u \mathcal{R}_T v$ but not $u \mathcal{R}_S v$, for some S, T and $u, v \in S$.
2. If $u, v \in S$ are regular in S, then $u \, \mathcal{K}_T \, v$ if and only if $u \, \mathcal{K}_S \, v$, for each Green relation $\mathcal{K}$.

3.63 Let S be the semigroup of matrices of the form

$$\begin{bmatrix} a & 0 \\ b & 1 \end{bmatrix}$$

where a, b are strictly positive rational numbers. Show that the relation $\mathcal{D}$ is the equality, and that S has a single $\mathcal{J}$-class.

3.64 Let $\varphi : M \to N$ be a continuous morphism from a compact semigroup M onto a compact semigroup N. Show that every closed subgroup of N is the image of a closed subgroup of M.

3.65 Show that in a stable monoid, the $\mathcal{J}$-class of the identity is a group.

3.66 Let $\pi : S \to T$ be a morphism of semigroups. Suppose that H, K are maximal subgroups of S contained in the same $\mathcal{D}$-class. Show that $\pi(H)$ is a maximal subgroup of T if and only if $\pi(K)$ is a maximal subgroup of T.

3.67 Show that for every $\mathcal{H}$-class H of a semigroup S, and every $x \in S^1$, we have $Hx = H$ if and only if $Hx \cap H \neq \emptyset$.

3.68 Show that if H is an $\mathcal{H}$-class of a semigroup S, then the set $\Gamma(H)$ of the translations $\rho_x : h \in H \mapsto hx \in H$, with $x \in S^1$ such that $Hx = H$, is a group of Card(H) permutations of elements of H acting freely on the left of H. (The expression *acting freely* stands for: if $x, y \in T(H)$ satisfy $hx = hy$ for some $h \in H$, then $\rho_x = \rho_y$.)

3.69 Show that the groups $\Gamma(H)$ of translations $\rho_x : h \to hx$ (with x belonging to the monoid $T(H) = \{x \in S^1 \mid Hx = H\}$) corresponding to the various $\mathcal{H}$-classes H contained in the same $\mathcal{D}$-class D are isomorphic.

3.70 Show that, in every semigroup, if H is a maximal subgroup, then $\Gamma(H)$ is isomorphic to H.

3.71 Show that if H is an $\mathcal{H}$-class of a compact semigroup S, then we have the following:

1. The Schützenberger group $\Gamma(H)$ is a compact group for the pointwise topology, and the natural map $T(H) \to \Gamma(H)$ is a continuous morphism. (The pointwise topology in the monoid of right transformations on H is defined basically in the same way as the pointwise topology on the set of functions $H \to H$, namely, in the space of right transformations on H we have $\tau_i \to \tau$ if and only if $h \cdot \tau_i \to h \cdot \tau$ for every $h \in H$.)
2. The topological spaces H and $\Gamma(H)$ are homeomorphic.
3. If H is a group, then H and $\Gamma(H)$ are isomorphic compact groups.
4. If the $\mathcal{H}$-classes H, K are contained in the same $\mathcal{D}$-class D, then $\Gamma(H)$ and $\Gamma(K)$ are isomorphic compact groups. (The Schützenberger group $G(D)$ of D, as a compact group, is defined as the isomorphism class of the compact groups of the form $\Gamma(H)$ with H an $\mathcal{H}$-class of D.)

3.13.5 Section 3.7

3.72 Let M be a monoid, and let $u \in M$. A *right inverse* of u is an element v of M such that $uv = 1$. We also clearly have the dual notion of left inverse. Show that if M is a compact monoid, then the following conditions are equivalent:

(i) v is a right inverse of u.
(ii) v is a left inverse of u.
(iii) v is an inverse of u.

3.13.6 Section 3.8

3.73 Explain the following property of the projective limit $S = \varprojlim_{i \in I} S_i$, given by the projective system $(S_i, \psi_{j,i}, I)$: if $s, t \in S$ are such that $\psi_i(s) = \psi_i(t)$ for all $i \in I$, then $s = t$.

3.74 Show that if J is cofinal in the directed set I, then $\varprojlim_{j \in J} S_j = \varprojlim_{i \in I} S_i$.

3.75 Let $S = \varprojlim_{i \in I} S_i$ be a topological projective limit with projections $\psi_i : S \to S_i$.

1. Show that if X is a subset of S such that $\psi_i(X) = S_i$ for every $i \in I$, then X is dense in S.
2. Deduce that if $\zeta : T \to \varprojlim_{i \in I} S_i$ is a continuous morphism of compact semigroups such that $\psi_i \circ \zeta$ is onto for every $i \in I$, then ζ is onto.

3.76 Show that every projective limit $S = \varprojlim_{i \in I} S_i$ of topological semigroups, with connecting morphisms $\psi_{i,j} : S_i \to S_j$ and projections $\psi_i : S \to S_i$, is isomorphic to the projective limit $\varprojlim_{i \in I} \psi_i(S)$ with onto connecting morphisms $\psi_{i,j} : \psi_i(S) \to \psi_j(S)$.

3.13.7 Section 3.9

3.77 Show that in a profinite semigroup S, one has $s_i \to s$ if and only if $\varphi(s_i) \to \varphi(s)$ for every continuous morphism $\varphi : S \to F$ into a finite semigroup F.

3.78 Show that the image of a connected space by a continuous function is connected.

3.79 Show that a union of connected sets with nonempty intersection is connected.

3.80 Show that a product of totally disconnected spaces is totally disconnected.

3.81 Let X be a compact space and let $x \in X$. Show that the intersection of all clopen sets containing x is the connected component of x.

3.82 Show that a zero-dimensional Hausdorff space X is totally disconnected and that the converse holds if X is compact.

3.83 Show that if S is a profinite semigroup, then S^1 is also a profinite semigroup.

3.84 Let $G = \varprojlim G_i$ be a profinite group with onto connecting morphisms $\varphi_{i,j} :$ $G_i \to G_j$ between finite groups. Let $\varphi_i : G \to G_i$ be the projections. Set $H_i = \varphi_i(H)$, where H is a closed normal subgroup of G. Show that $G/H = \varprojlim G_i/H_i$.

3.85 Show that every Schützenberger group of a profinite semigroup is a profinite group. (See Exercise 3.71.)

3.13.8 Section 3.12

3.86 Let S and T be topological semigroups, and let $\alpha : A \to S$ be a generating map of S, as a topological semigroup. Show that if $f : S \to T$ and $g : S \to T$ are continuous morphisms such that $f \circ \alpha = g \circ \alpha$, then $f = g$.

3.87 Show that if S is a finitely generated profinite semigroup, then $\text{End}(S)$ is a metrizable topological space, if we consider the pointwise topology.

3.88 Give an example of a profinite group G such that:

1. The group of continuous automorphisms of G is not compact with respect to the pointwise topology.
2. The monoid of continuous endomorphisms of G has no topology for which it is a compact monoid.

3.14 Solutions

3.14.1 Section 3.2

3.1 Assume first that $x \notin \overline{U}$. Then there is a closed set V containing U and not x. Thus $X \setminus V$ is a neighborhood of x which does not meet U. Conversely if V is a neighborhood of x which does no meet U, then V contains an open neighborhood W of x which does neither meet U. The complement of W is a closed set containing U not containing x. Thus $x \notin \overline{U}$.

3.2 Let $(x_i)_{i \in I}$ be a net in a Hausdorff space X. Assume that $x_i \to x$ and $x_i \to y$ and suppose that $x \neq y$. Since X is Hausdorff, there are disjoint neighborhoods U, V of x and y. Let $i \in I$ be such that $x_j \in U$ for all $j \geqslant i$ and let $k \in I$ be such that $x_\ell \in V$ for all $\ell \geqslant k$. Let m be a common upper bound of i, k. Then $x_m \in U \cap V$, a contradiction.

For the converse, arguing by contradiction, assume that two points $x, y \in X$ cannot be separated by disjoint neighborhoods. Consider the set I consisting of pairs (U, V) of neighborhoods respectively of x and y. Order I by the reverse of component-wise inclusion to ge a directed set. For each $i = (U, V) \in I$, choose $x_i \in U \cap V$. Then, by construction, the net (x_i) converges to both x and y.

3.3 Let $(x_i)_{i \in I}$ be a net of elements of U converging to x. If x is not in $\overline{U}$, then the complement of $\overline{U}$ is an open neighborhood of x. Thus there is an $i \in I$ such that $X \setminus \overline{U}$ contains all x_j for $j \geqslant i$, a contradiction with $x_i \in U$ for all i. Conversely, let x be an element of $\overline{U}$. Consider the set $\mathcal{N}$ of neighborhoods of x ordered by reverse inclusion. Then $\mathcal{N}$ is directed. For each $N \in \mathcal{N}$, let x_N be an element of $N \cap U$, which exists as seen in Exercise 3.1. Then $x_N \to x$ and thus $x \in \overline{U}$.

3.4 Let $j, k \in K$. Since I is directed there is some $i \in I$ such that $j, k \leqslant i$. Since J is cofinal there is some $\ell \in J$ such that $i \leqslant \ell$. Thus ℓ is a common upper bound of j, k.

3.5 By an inductive argument, one is reduced to the case $n = 2$. Suppose that J_1 is not cofinal. Then there is $i_0 \in I$ for which there is no $j \in J_1$ such that $i_0 \leqslant j$. Let $i \in I$. Since I is directed, we may take a common upper bound j of i and i_0. By the assumption on i_0, we have $j \in J_2$, thus J_2 is cofinal.

3.6 Let $(x_i)_{i \in I}$ be a convergent net, and let $x = \lim x_i$. Suppose that $(x_{\varphi(j)})_{j \in J}$ is a subnet. Given a neighborhood U of x, there is $i \in I$ such that $x_k \in U$ whenever $i \leqslant k$. Since $\varphi(J)$ is cofinal in I, there is $j_0 \in J$ such that $k \leqslant \varphi(j_0)$. Since φ is increasing, it follows that $j_0 \leqslant j$ implies $k \leqslant \varphi(j)$ and so $x_{\varphi(j)} \in U$. Hence, the net $(x_{\varphi(j)})_{j \in J}$ converges to x.

3.7 Endow $K = I \times J$ with the component-wise order. Then K is a directed set. The component projections $\varphi : K \to I$ and $\psi : K \to J$ are increasing and onto, thus $(x_{\varphi(k)})_{k \in K}$ and $(y_{\psi(k)})_{k \in K}$ are subnets as we searched for.

Since a set U is closed if and only $U = \overline{U}$, we conclude that U is closed if and only if it contains all the limits of convergent nets of elements of U.

3.8 (i) $\Rightarrow$ (ii) Suppose that the subnet $(x_{\varphi(j)})_{j \in J}$ converges to y. Fix $i \in I$. Let U be a neighborhood of y. As $(x_{\varphi(j)})_{j \in J} \to y$, there is $j_0 \in J$ such that $x_{\varphi(j)} \in U$ whenever $j \geqslant j_0$. Since $\varphi(J)$ is cofinal in I, there is $j' \in J$ such that $\varphi(j') \geqslant i$. Let j'' be a common upper bound of j_0 and j'. Then $x_{\varphi(j'')} \in U$ and $\varphi(j'') \geqslant \varphi(j') \geqslant i$, thus $x_{\varphi(j'')} \in T_i \cap U$, which shows (ii) by the property established in Exercise 3.1.

(ii) $\Rightarrow$ (iii) Let U be a neighborhood of y. In view of Exercise 3.1, it suffices to observe that, for every $i \in I$, the set $T_i \cap U$ is nonempty.

(iii) $\Rightarrow$ (i) Following the hint, we consider the set Λ of pairs (i, U) such that $x_i \in U$, with $i \in I$ and U a neighborhood of y. Endow Λ with the partial order given by $(i, U) \leqslant (j, V)$ if and only if $i \leqslant j$ and $U \supseteq V$. With this partial order, Λ is a directed set: if (i, U) and (j, V) belong to Λ, and if k is a common upper bound of i and j, then, because y is a cluster point of $(x_i)_{i \in I}$, there is some $\ell \geqslant k$ such that $x_\ell \in U \cap V$, whence $(\ell, U \cap V)$ is a common upper bound of (i, U) and (j, V).

Define $\varphi : \Lambda \to I$ by letting $\varphi(i, U) = i$. Clearly, φ is increasing and $\varphi(\Lambda)$ is cofinal in I. Therefore, we may consider the subnet $(x_{\varphi(i,U)})_{(i,U) \in \Lambda}$ of $(x_i)_{i \in I}$.

Let W be a neighborhood of y. Then there is $i_W \in I$ for which we have $x_{i_W} \in W$. The pair (i_W, W) belongs to Λ. As the relation $(i_W, W) \leqslant (i, U)$ implies $x_{\varphi(i,U)} \in U \subseteq W$, we conclude that the net $(x_{\varphi(i,U)})_{(i,U) \in \Lambda}$ converges to y.

3.9 Observe that $X \setminus f^{-1}(F) = f^{-1}(Y \setminus F)$ and that the closed sets are the complements of open sets.

3.10 Assume first that f is continuous and consider $x \in \overline{U}$ for some $U \subseteq X$. Let V be a neighborhood of $f(x)$. Then $f^{-1}(V)$ is a neighborhood of x and thus $U \cap f^{-1}(V) \neq \emptyset$. For $y \in U \cap f^{-1}(V)$, we have $f(y) \in f(U) \cap V$, which shows that $f(U) \cap V \neq \emptyset$. Thus $f(x) \in \overline{f(U)}$. Conversely, let us show that $f^{-1}(V)$ is closed for every closed set $V \subseteq Y$. This will clearly imply that f is continuous. Set $U = f^{-1}(V)$ and consider $x \in \overline{U}$. Then $f(x) \in f(\overline{U}) \subseteq \overline{f(U)}$. But since $f(U) \subseteq V$, we have $\overline{f(U)} \subseteq \overline{V}$. Thus we obtain $f(x) \in \overline{V} = V$ and finally $x \in U$. Thus U is closed.

3.11 Assume first that f is continuous. Let (x_i) be a net converging to x. Let U be an open neighborhood of $f(x)$. Then $f^{-1}(U)$ is an open neighborhood of x and thus there is an $i \in I$ such that U contains all x_j for $j \geqslant i$. Thus U contains all $f(x_j)$ for $j \geqslant i$ which implies that $f(x_i) \to f(x)$.

Conversely, consider $U \subseteq X$ and $x \in \overline{U}$. Let $(x_i)_{i \in I}$ be a net of elements of U with limit x. By hypothesis $f(x_i) \to f(x)$. Thus $f(x) \in \overline{f(U)}$. This implies that f is continuous by Exercise 3.10.

3.12 Let $x \in X$. Because U is dense, there is a net (u_i) of elements of U such that $x = \lim u_i$, in view of Exercise 3.3. Since f and g are continuous, we have $f(x) = \lim f(u_i) = \lim g(u_i) = g(x)$ by Exercise 3.11.

3.13 Let x, y be two distinct points and let $\varepsilon = d(x, y)/2$. Then we have $B(x, \varepsilon) \cap B(y, \varepsilon) = \emptyset$. Indeed, let $z \in B(x, \varepsilon) \cap B(y, \varepsilon)$. Then we obtain $d(x, y) \leqslant d(x, z) + d(z, y) < \varepsilon + \varepsilon = d(x, y)$, a contradiction.

3.14 The two distances define the same balls $B(x, \varepsilon)$ for $\varepsilon < 1$.

3.15 Let x be an element of the closure $\overline{U}$ of U. For each $n \geqslant 1$, let x_n be an element of $\overline{U} \cap B(x, 1/n)$. Then $x = \lim x_n$ and thus $x \in U$.

3.16 We prove that for every subset U of X, one has $f(\overline{U}) \subseteq \overline{f(U)}$. This implies that f is continuous by Exercise 3.10. If $x \in \overline{U}$, there is by Exercise 3.15 a sequence x_n of elements of U converging to x. Thus $f(x) = \lim f(x_n)$, which shows that $f(x) \in \overline{f(U)}$.

3.17 Since a subsequence is a subnet, we only need to show the "only if". Suppose that the subnet $(x_{n_i})_{i \in I}$ of the sequence $(x_n)_{n \geqslant 1}$ converges to y. Then, y is an accumulation point of $(x_n)_{n \geqslant 1}$, as seen in Exercise 3.8. Therefore, we may define recursively a strictly increasing mapping $k \mapsto n_k$ as follows: let $n_1 = 1$, and for each integer $k > 1$, let n_k be an integer greater than n_{k-1} and such that $x_{n_k} \in B(y, 1/k)$. Then $(x_{n_k})_k$ converges to y.

3.18 In the metric space of the real numbers, consider the sequence $x_n = 1/n$, with $n \geqslant 1$. For each positive real ε, let $n_\varepsilon = \min\{k \mid 1/k \leqslant \varepsilon\}$. Then $(x_{n_\varepsilon})_{\varepsilon > 0}$ is a subnet of (x_n) which is not a sequence, since the indexing set of positive real numbers is not denumerable.

3.19 It is clear that d is a distance. To show that it defines the product topology, consider first an open set U for the metric topology and $x \in U$. Then we have $B(x, \varepsilon) \subseteq U$ for some $\varepsilon > 0$. Choose a positive integer n such that $1/n < \varepsilon$ and let $V = \{y \in X \mid d'_i(x_i, y_i) < \varepsilon \text{ for } 1 \leqslant i \leqslant n\}$, which is an open set of the product topology. Then, for every $y_i \in V$, we have $d(x, y) = \max\{d'_1(x_1, y_1)/1, \ldots, d'_n(x_n, y_n)/n, 1/n\} < \varepsilon$ and thus $y \in B(x, \varepsilon) \subseteq U$. This shows that $V \subseteq U$. The proof that for every open set U of the product topology and every $x \in U$ there is an open set V of the metric topology contained in U and containing x is similar.

3.20 Set $B_n = B(x, 1/n)$.

3.21

1. The family of cofinite sets is closed under finite intersection and arbitrary union.
2. Assume by contradiction that $\mathcal{B}_n = \{B_n \mid n \geqslant 1\}$ is a basis of open neighborhoods of $x \in X$. Set $F_n = X \setminus B_n$ and $F = \{x\} \cup (\bigcup_{n \geqslant 1} F_n)$. Since each F_n is finite, F is denumerable. Since X is not denumerable, there is some $y \in X \setminus F$. Consider $U = X \setminus \{y\}$. Then U is an open neighborhood of x. Then there is some $n \geqslant 1$ such that $B_n \subseteq U$. This implies that $y \in F_n$ and thus $y \in F$, a contradiction.

3.22 Let U be an elementary open set of the form $U = \{f \mid f(x_i) \in U_i, 1 \leqslant i \leqslant n\}$ for $x_1, \ldots, x_n \in \mathbb{R}$ and $U_1, \ldots, U_n \subseteq \mathbb{R}$ open sets. Then $A \cap U$ contains the function g defined by $g(x) = 0$ if $x \in \{x_1, \ldots, x_n\}$ and $g(x) = 1$ otherwise. Thus 0 is in the closure of A. Assume by contradiction that f_n is a sequence of elements of A converging to the constant function 0. Let $J_n = \{x \in \mathbb{R} \mid f_n(x) \neq 1\}$. Then $J = \bigcup_{n \geqslant 1} J_n$ is denumerable and thus there is $y \in \mathbb{R}$ such that $y \notin J$. The set $U = \{f \mid 0 < f(y) < 1\}$ is a neighborhood of 0 which does not contain any of the f_n, a contradiction.

3.23 First d_Y is well defined. Indeed, let $(x_n), (y_n)$ be Cauchy sequences in X. Then, using $d_X(x_n, y_n) \leqslant d_X(x_n, x_m) + d_X(x_m, y_m) + d_X(y_m, y_n)$, we obtain $d_X(x_n, y_n) - d_X(x_m, y_m) \leqslant d_X(x_n, x_m) + d_X(y_n, y_m)$, which implies that $(d_X(x_n, y_n))$ is a Cauchy sequence of reals and thus converges. Next, it is straightforward to verify that it is a distance and that Y is complete. Finally, we may embed X into Y as a dense subspace by identification of $x \in X$ with the equivalence class of the constant sequence $(x, x, \ldots)$.

3.24 Let us first show that for every $x \in A$ there is a pair of disjoint open sets $U, V \subseteq X$ such that $x \in U$ and $B \subseteq V$. Indeed, since X is Hausdorff, we can find for every $y \in B$ disjoint open sets $U_y, V_y \subseteq X$ such that $x \in U_y$ and $y \in V_y$. The family $(V_y)_{y \in B}$ is an open cover of B and we can find a finite subcover corresponding to $y_1, \ldots, y_n$. Then $U = \bigcap_{i=1}^{n} U_{y_i}$ and $V = \bigcup_{i=1}^{n} U_{y_i}$ are open sets separating x and B.

By the previous point, there is for every $y \in B$ a pair U_y, V_y of disjoint open sets such that $y \in V_y$ and $A \subseteq U_y$. Then $(V_y)_{y \in B}$ is an open cover of B and thus there is a finite subcover corresponding to $y_1, \ldots, y_n$. The pair $U = \bigcap_{i=1}^{n} U_{y_i}$ and $V = \bigcup_{i=1}^{n} V_{y_i}$ is a solution.

3.25 Let A be a closed subset of a compact space X. Let $\mathcal{C}$ be an open cover of A. Every $U \in \mathcal{C}$ is of the form $V_U \cap A$ for some open set $V_U \subseteq X$. Then the family $(X \setminus A) \cup \{V_U \mid U \in \mathcal{C}\}$ is an open cover of X. A finite subcover gives by intersection with A a finite subcover of $\mathcal{C}$. Since every subspace of a Hausdorff space is also Hausdorff, the conclusion follows.

3.26 Let $f : X \to Y$ be a continuous map from X to Y with X compact. Let (U_i) be an open cover of $f(X)$. Then $(f^{-1}(U_i))$ is an open cover of X and thus has a finite subcover $(f^{-1}(V_n))$. Then (V_n) is a finite subcover of $f(X)$. The assumption that $f(X)$ is Hausdorff is necessary because it is part of our definition of compactness.

3.27 Let $(x_i)_{i \in I}$ be a net of elements of a compact space X. By Exercise 3.8, whose notation we follow, what we want to show is that the intersection $\bigcap_{i \in I} \overline{T_i}$ is nonempty. If $i_1, \ldots, i_n$ is a finite collection of elements of I, and if i_0 is a common upper bound of all elements in the collection, then the intersection $T_{i_1} \cap \ldots \cap T_{i_k}$ clearly contains T_{i_0}, and so

$$\overline{T_{i_1}} \cap \ldots \cap \overline{T_{i_k}} \supseteq \overline{T_{i_1} \cap \ldots \cap T_{i_k}} \supseteq \overline{T_{i_0}} \neq \emptyset.$$

Hence, the family $(\overline{T_i})_{i \in I}$ has the finite intersection property, and so $\bigcap_{i \in I} \overline{T_i} \neq \emptyset$.

Conversely, consider a family $\mathcal{U}$ of closed sets having the finite intersection property. Let $\mathcal{V}$ be the family of finite intersections of elements of $\mathcal{U}$ ordered by reverse inclusion. By hypothesis, every $V \in \mathcal{V}$ is nonempty. Select $x_V \in V$. Then (x_V) is a net. A limit of a subnet of this net is in all $U \in \mathcal{U}$. Thus X is compact.

3.28 Let Y be a compact subspace of a compact space X. Suppose that (y_i) is a net of elements of Y converging to an element y of X. Accordingly to Exercise 3.27, one must have $y \in Y$. This shows that $Y = \overline{Y}$.

3.29 The set $g(F)$ is compact by Exercise 3.26, thus closed by Exercise 3.25.

3.30 We only need to check the "if" part. Suppose that $g \circ f$ is continuous, and let V be a closed subset of Z. In view of Exercise 3.9, we know that $W = f^{-1}(g^{-1}(V))$ is a closed set of X, and all we need to show is that $g^{-1}(V)$ is also closed. Since f is onto, one has $f(W) = g^{-1}(V)$. The conclusion then follows from the property established in Exercise 3.29.

3.31 The continuity of h^{-1} is in an immediate application of Exercise 3.30, since the identity $Id_X = h^{-1} \circ h$ is continuous.

Consider the interval $[0, 1]$ with the topology of the real numbers. Take any bijection $h : Y \to [0, 1]$. If Y is endowed with the discrete topology, then h is continuous, but h^{-1} is not. On the other hand, if Y is endowed with the trivial topology, then h^{-1} is continuous, but h is not.

3.32 Let $(x_n)_n$ be a Cauchy sequence in the compact metric space X. Fix $\varepsilon > 0$. Then, there is an integer p such that $i, j \geqslant p$ implies $d(x_i, x_j) \leqslant \varepsilon/2$. On the other hand, in view of Exercise 3.27 (see also Exercise 3.17), there is a subsequence $(x_{n_m})_m$ converging to some x, and the map $m \mapsto n_m$ can be chosen to be strictly increasing. Take k such that $k \geqslant p$. Since $n_k \geqslant k$, we have

$$d(x_k, x) \leqslant d(x_k, x_{n_k}) + d(x_{n_k}, x) \leqslant \frac{\varepsilon}{2} + \frac{\varepsilon}{2} = \varepsilon,$$

and so $x_n \to x$.

3.33 The "only if part" is contained in Exercise 3.6.

Conversely suppose that the net $(x_i)_{i \in I}$ has a unique cluster point x. By Exercise 3.8, whose notation we follow, this means $\bigcap_{i \in I} \overline{T_i} = \{x\}$. We claim that $x_i \to x$. Suppose the claim is false. Then, there is an open neighborhood U of x such that for every $i \in I$ there is $j \geqslant i$ satisfying $x_j \notin U$. Denoting by U^C the complement of U, we have $T_i \cap U^C \neq \emptyset$ for every $i \in I$. In particular, the set $F_i = \overline{T_i} \cap U^C$ is nonempty, for all i. Since U is open, the set F_i is closed. Moreover, if $i_1, \ldots, i_n$ is a finite collection of elements of I, and if i_0 is a common upper bound of all elements in the collection, we have

$$\bigcap_{k=1}^{n} F_{i_k} = U^C \cap (\overline{T_{i_1}} \cap \ldots \cap \overline{T_{i_k}}) \supseteq U^C \cap (\overline{T_{i_1} \cap \ldots \cap T_{i_k}}) \supseteq U^C \cap \overline{T_{i_0}} = F_{j_0} \neq \emptyset.$$

Therefore, the family F_i has the finite intersection property. Since X is compact, it follows that $\bigcap_{i \in I} F_i \neq \emptyset$, that is $U^C \cap (\bigcap_{i \in I} \overline{T_i}) \neq \emptyset$. This contradicts having both $x \in U$ and $\bigcap_{i \in I} \overline{T_i} = \{x\}$. To avoid the contradiction, the claim that $(x_i)_{i \in I}$ converges to x must hold.

3.14.2 Section 3.3

3.34 Consider the mapping $\varphi : M \to M^M$ such that $\varphi(m)(x) = mx$ for all $x \in M$. Then $\varphi(mm')(x) = mm'x = (\varphi(m) \circ \varphi(m'))(x)$ for all $x \in M$, thus φ is a morphism of monoids. Since $m = \varphi(m)(1)$, this morphism is injective.

3.35 Let $s \in \overline{X}$ and $t \in \overline{Y}$. Take nets $(s_i)_{i \in I}$ and $(t_j)_{j \in J}$, respectively of elements of X and Y, with $s_i \to s$ and $t_j \to t$. According to Exercise 3.7, by taking subnets, we may suppose $I = J$. From the continuity of the semigroup operation we get $s_i t_i \to st$, thus $st \in \overline{XY}$, establishing $\overline{X} \cdot \overline{Y} \subseteq \overline{XY}$.

Assume now S to be compact. Then $S \times S$ is also compact. As $\overline{X} \times \overline{Y}$ is closed in $S \times S$, and the multiplication of S is a continuous map $S \times S \to S$, it follows that $\overline{X} \cdot \overline{Y}$ is closed in S, according to Exercise 3.29.

Take the closed sets $X = \{n + 1/2 \mid n \in \mathbb{Z}_+\}$ and $Y = \{1/n \mid n \in \mathbb{Z}_+\} \cup \{0\}$ of the multiplicative topological semigroup $\mathbb{R}$. Note that $(n + 1/2)/n \in XY$ and $(n + 1/2)/n \to 1 \notin XY$. Therefore XY is not closed and $\overline{X} \cdot \overline{Y} \neq \overline{XY}$.

3.36 The inclusions $\overline{X^+} \cdot \overline{X^+} \subseteq \overline{X^+ \cdot X^+} \subseteq \overline{X^+}$ hold, the first one by Exercise 3.35.

3.37 Necessity is obvious. Conversely, take $s, t \in S$ and nets $(s_i)_{i \in I}$ and $(t_i)_{i \in J}$ of elements of X^+ respectively converging to s and t. According to Exercise 3.7, we may suppose $I = J$. Note that $s_i t_i \to st$ and $t_i s_i \to ts$. By the assumption on X, the equality $t_i s_i = s_i t_i$ holds for all i, thus $st = ts$.

3.14.3   Section 3.5

3.38 Let $x_n \in X$ be a net of elements all in the same class C converging to some $x \in X$. If $y \in C$, then (x_n, y) is a net of elements of ρ converging to (x, y). Since ρ is closed, the limit is in ρ and thus $x \in C$. Therefore C is closed.

3.39 The first statement is clear. Next, the sequence of pairs $(1 + \frac{1}{n}, 1 - \frac{1}{n})$ is formed of elements of ρ but its limit is the pair $(-1, 1)$, which is not in ρ.

3.40 (i) $\Rightarrow$ (iv) We show that if C is a class of ρ, then $X \setminus C$ is closed. Let (x_i) be a net of elements of $X \setminus C$ converging to x. Since $(x_i, x) \to (x, x)$ and ρ is a neighborhood of (x, x), we have $(x_j, x) \in \rho$ for some j, thus $x \notin C$. Therefore every class of ρ is open. But then every class C is also closed since it is the complement of the union of the other classes. Thus every class of ρ is clopen.

(iv) $\Rightarrow$ (iii) is obvious.

(iii) $\Rightarrow$ (ii) If all the classes are open, then ρ, which is the union of the products $C \times C$ over the classes C, is also open; and its complement, which is the union of the products $C \times D$ over the classes C, D such that $C \neq D$, is also open. Thus ρ is clopen.

(ii) $\Rightarrow$ (i) is obvious.

3.41 Let U_ρ be the union of the sets of the form $[z]_\rho$, with $[z]_\rho \subseteq U$. Suppose that (x_i) is a net of elements of $X \setminus U_\rho$ converging to x. What we want to show is that $x \notin U_\rho$. Suppose on the contrary that $x \in U_\rho$. As $U_\rho \subseteq U$ and U is open, we may suppose that $x_i \in U$ for all i. Since $x_i \notin U_\rho$, we may choose $y_i \in [x_i]_\rho \setminus U$, for every i. Because X is compact, we may take a cluster point (x, y) of (x_i, y_i). As ρ is closed, we know that $y \in [x]_\rho$. On the other hand, from U being open and $y_i \notin U$

for all i, we get $y \notin U$. Hence, we have $[x]_\rho \not\subseteq U$, contradicting $x \in U_\rho$. To avoid the contradiction, we must have $x \notin U_\rho$.

3.42 Since X is compact and $q_\rho : X \to X/\rho$ is continuous, we are reduced to showing that ρ is closed if and only if X/ρ is Hausdorff (see Exercise 3.26).

Suppose that ρ is a closed relation. Let C, D be distinct ρ-classes. Since the disjoint sets C and D are closed (Exercise 3.38), there are disjoint open subsets U, V of X such that $C \subseteq U$ and $D \subseteq V$ (Exercise 3.24). According to Exercise 3.41, we may suppose that U and V are both unions of ρ-classes. Hence, we have $U = q_\rho^{-1}(q_\rho(U))$, $V = q_\rho^{-1}(q_\rho(V))$, $C \in q_\rho(U)$ and $D \in q_\rho(V)$. In particular, from $U \cap V = \emptyset$ we get $q_\rho(U) \cap q_\rho(V) = \emptyset$. Moreover, from the definition of the quotient topology, the sets $q_\rho(U)$ and $q_\rho(V)$ are open. Therefore, the open sets $q_\rho(U), q_\rho(V)$ separate the elements C, D of X/ρ.

Conversely, suppose that X/ρ is Hausdorff (and therefore, compact, in view of Exercise 3.26). Let (x_i, y_i) be a net of elements of ρ converging to (x, y). By the continuity of q_ρ, we have $[x_i]_\rho \to [x]_\rho$ and $[y_i]_\rho \to [y]_\rho$. As $[x_i]_\rho = [y_i]_\rho$ for all i, and since in a Hausdorff space a convergent net has a unique limit, we must have $[x]_\rho = [y]_\rho$, thus establishing that ρ is closed.

3.43 First note that indeed $X/\operatorname{Ker} f$ and $\operatorname{Im} f$ are compact: the former in view of Example 3.5.3 and Exercise 3.42, the latter according to Exercise 3.26.

The canonical maps $q : X \to X/\operatorname{Ker} f$ and $f' : X/\operatorname{Ker} f \to \operatorname{Im} f$ are such that $f = f' \circ q$. Since q and f are continuous, it follows from Exercise 3.30 that the bijection f' is continuous. As $X/\operatorname{Ker} f$ and $\operatorname{Im} f$ are both compact, we conclude that f' is a homeomorphism (Exercise 3.31).

3.44 If φ is continuous, then the natural semigroup isomorphism $S/\operatorname{Ker} \varphi \to T$ is a homeomorphism of compact spaces (cf. Exercise 3.26 and Example 3.5.4), and so T is a compact semigroup for the topology that was prescribed to it.

3.45 Let $\varphi : S \to S/\rho \times S/\theta$ be the morphism defined by $\varphi(s) = ([s]_\rho, [s]_\theta)$. Then $\operatorname{Ker} \varphi = \rho \cap \theta$ clearly holds, thus $S/(\rho \cap \theta)$ is isomorphic to a subsemigroup of $S/\rho \times S/\theta$, by Theorem 3.5.2. Moreover, if ρ and θ are closed relations, then $\rho \cap \theta$ is closed, φ is continuous, and $S/\rho \times S/\theta$ is a compact semigroup isomorphic to a closed subsemigroup of $S/\rho \times S/\theta$, as seen in Theorem 3.5.6.

3.14.4 Section 3.6

3.46 $x \leqslant_{\mathcal{R}} e$ means that $x = et$ for some t, thus $ex = e^2 t = et = x$.

3.47
1. If $(a, b) \in \operatorname{Ker} g$, then $f(a) = h(g(a)) = h(g(b)) = f(b)$ whenever $f = h \circ g$, and so clearly $\operatorname{Ker} g \subseteq \operatorname{Ker} f$ if the functions $f, g : X \to X$ are such that $f \leqslant_{\mathcal{L}} g$. It is also immediate that $f \leqslant_{\mathcal{R}} g$ implies $\operatorname{Im} f \subseteq \operatorname{Im} g$. Conversely, if $\operatorname{Ker} g \subseteq \operatorname{Ker} f$, then $h \in X^X$ given by $h(g(x)) = f(x)$ is well defined, thus $f \leqslant_{\mathcal{L}} g$. On the other

hand, if $\operatorname{Im} f \subseteq \operatorname{Im} g$, then for each $x \in X$ there is $h(x) \in X$ with $f = g(h(x))$, showing that $f \leqslant_{\mathcal{L}} h$.

2. T_X is the dual of X^X.

3. Observe that $f = gh$ if and only if $\tilde{f} = \tilde{g}\tilde{h}$, for all $f, g, h \in T_X$. Hence, for all $f, g \in PT_X$, we have $f \mathcal{R} g$ if and only if $D_f = D_g$ and $\operatorname{Ker} \tilde{f} = \operatorname{Ker} \tilde{g}$. Similarly for the relation $\mathcal{L}$.

3.48 Let $h, k \in S^1$ be such that $f = hgk$. Then we have $X \cdot hgk \subseteq X \cdot gk$ and $\operatorname{Card}(X \cdot gk) \leqslant \operatorname{Card}(X \cdot g)$, the latter inequality holding because the mapping $y \in X \cdot g \mapsto y \cdot k \in X \cdot gk$ is a surjective partial function.

3.49 Consider $x \mathcal{R} y \mathcal{L} z$. Let $u, v, r, s \in S^1$ be such that $x = yv, y = xu = rz, z = sy$. Then $sx = syv = zv$ and thus $x \mathcal{L} t \mathcal{R} z$ with $t = sx = zv$.

3.50 If the semigroup S is commutative, then $S^1 u S^1 = u S^1 = S^1 u$ for every $u \in S$, whence $\mathcal{J} \subseteq \mathcal{H}$ in S. Since $\mathcal{H} = \mathcal{R} \cap \mathcal{L} \subseteq \mathcal{R} \vee \mathcal{L} = \mathcal{D} \subseteq \mathcal{J}$ holds in any semigroup, the conclusion follows.

3.51 (i) $\Rightarrow$ (iii) Consider $x \in S$ such that $e \mathcal{R} x \mathcal{L} f$. Note that $x = ex = xf$ (see Exercise 3.46 and its dual). On the other hand, we also have $e = xy$ and $f = zx$ for some x, z. Since $e = xf \cdot y = x \cdot fy$, $f = z \cdot xf$, $xf = xf \cdot f$ and $fy = f \cdot fy$, we may as well replace x by xf and y by fy. That is, we may suppose that $x = xf$ and $y = fy$. It follows that $yx = fyx = zxyx = zex = zx = f$.

(iii) $\Rightarrow$ (ii) Trivial.

(ii) $\Rightarrow$ (i) We have $e = (xy)^2 \leqslant_{\mathcal{R}} xyx \leqslant_{\mathcal{R}} xy = e$, thus $e \mathcal{R} xyx$. Similarly, $f \leqslant_{\mathcal{L}} xyx \leqslant_{\mathcal{R}} f$, establishing $e \mathcal{R} xyx \mathcal{L} f$.

3.52 If $q \in L(s)$, then $S^1 q = S^1 s$ and therefore $S^1 qu = S^1 su = S^1 s'$. This shows that ρ_u maps $L(s)$ into $L(s')$. Next, if $q \in L(s)$ there exist $v, v' \in S^1$ such that $q = vs, s = v'q$. Then since $suu' = s'u' = s$ we have

$$\rho_{u'}\rho_u(q) = quu' = vsuu' = vs = q$$

and thus $\rho_{u'}\rho_u$ is the identity on $L(s)$. Similarly $\rho_u\rho_{u'}$ is the identity on $L(s')$.

3.53 In the setting of Exercise 3.52, the bijections $\rho_u : L(s) \mapsto L(s')$ and $\rho_{u'} : L(s') \mapsto L(s)$ restrict to mutually inverse bijections $H(s) \mapsto H(s')$ and $H(s') \mapsto H(s)$. Indeed, if p and q are $\mathcal{H}$-equivalent elements of $L(s)$ then, since $q = quu'$ and $p = puu'$, we have $qu \mathcal{R} q \mathcal{H} p \mathcal{R} pu$, thus $qu \mathcal{H} pu$. Therefore, $\operatorname{Card}(H(s)) = \operatorname{Card}(H(s'))$ if $s \mathcal{R} s'$. Dually, $\operatorname{Card}(H(s')) = \operatorname{Card}(H(s''))$ if $s' \mathcal{L} s''$. Since $\mathcal{D} = \mathcal{RL} = \mathcal{LR}$, we are done. (more precisely, we have mutually inverse maps $q \in H(s) \mapsto vqu \in H(s'')$ and $q' \in H(s'') \mapsto v'qu' \in H(s)$, with u, u' as above and v, v' such that $s'' = vs'$ and $s' = v's''$.)

3.54 In the setting of Exercise 3.52, the maps $\tilde{\rho}_u : H(q) \mapsto H(qu)$ and $\tilde{\rho}_{u'} : H(q') \mapsto H(q'u')$ are mutually inverse bijections from the set of $\mathcal{H}$-classes of $L(s)$ to the set of $\mathcal{H}$-classes of $L(s')$.

3.55 If $R(t) \cap L(s)$ contains an idempotent e, then $e = tu$, $t = eu'$, $e = vs$ and $s = v'e$. Then $stu = s(tu) = se = (v'e)e = v'e = s$. Thus $st \in R(s)$. Similarly, $st \in L(t)$. Thus $st \in R(s) \cap L(t)$.

Conversely, assume that $st \in R(s) \cap L(t)$. Since, by Green's Lemma, the right multiplication by t is a bijection φ from $L(s)$ onto $L(st)$, there is an $e \in L(s)$ such that $et = t$. Since φ preserves R-classes, we have $e \in R(t)$. Let $u \in S^1$ be such that $e = tu$. Then $tutu = etu = tu$ and thus $e = tu$ is an idempotent in $R(t) \cap L(s)$.

3.56 (i) $\Rightarrow$ (iv) Suppose that $s = sxs$ for some $x \in S$. Note that $s \mathrel{R} sx$. On the other hand, $sx = sxsx$, thus sx is an idempotent in the R-class $R(s)$. Symmetrically, xs is an idempotent in the L-class $L(s)$.

(iv) $\Rightarrow$ (iii) and (ii) $\Rightarrow$ (i) Trivial.

(iii) $\Rightarrow$ (ii) Let $s \in D$ and let e be an idempotent in $R(s)$. Then $s = eu$ and $e = sv$, for some u, v. Thus $s = e^2 u = sveu = svs$ showing that s is regular.

3.57 Let (x_n) be a net of elements of M all R-equivalent to x and converging to some y. Then $x = x_n u_n$ and $x_n = x v_n$ for some $u_n, v_n \in M$. Up to taking a subnet, we may assume that (u_n), (v_n) converge to u, v. Then $x = yu$ and $y = xv$ showing that $y \mathrel{R} x$. Thus R is closed. A similar argument can be used for the other Green relations.

3.58 It suffices to consider the case of right ideals, as the case for left ideals is dual. Let u be an element of $K = K(S)$. We claim that uS^1 is a minimal right ideal. Suppose that I is a right ideal contained in uS^1. If $v \in I$, then we have $vS^1 \subseteq I \subseteq uS^1$. This implies $S^1 v S^1 \subseteq S^1 u S^1 = K$. By the minimality of K, it follows that $S^1 v S^1 = S^1 u S^1$, that is, $v \mathcal{J} u$. Because S is stable, from $v \mathcal{J} u$ and $vS^1 \subseteq uS^1$ we obtain $vS^1 = uS^1$ and so $I = uS^1$. This shows that uS^1 is a minimal right ideal, and that the elements of uS^1 are precisely the elements in the R-class of u. Therefore, the R-classes contained in $K(S)$ are minimal right ideals of S. Finally, suppose that R is a minimal right ideal. Let $r \in R$. For $u \in K$, we have $ruS^1 \subseteq R$, so that $ruS^1 = R$, by the minimality of R. In particular, $R \subseteq S^1 u S^1 = K(S)$, and so R must be an R-class of $K(S)$.

3.59 The condition is clearly necessary. Conversely, let us show that the H-class H of an idempotent is the group of invertible elements of the monoid eSe. If $x \in H$, then $x \in eS \cap Se$ and thus $x \in eSe$. Since $xu = e$ for some $u \in S$, we have $x(eue) = e$. Similarly, since $vx = e$, we have $(eve)x = e$ and thus x belongs to the group of invertible elements of eSe. Conversely, it is easy to see that an invertible element of eSe is in H.

3.60 This is a special case of Clifford-Miller's Lemma: in the setting of Exercise 3.55, take $s = t$.

3.61 Let H be an H-class contained in $K(S)$. Take $s \in H$. The idempotent s^ω belongs to the J-class $K(S)$, because $K(S)$ is a closed subsemigroup of S. Since $s^\omega \in sS \cap Ss$ and S is stable, it follows that $s^\omega \in H$, and so H is a group.

3.62

1. Every Green relation in the group $\mathbb{Z}$ is the universal relation, while in the subsemigroup of positive integers every Green relation is the equality.

2. It suffices to consider the case $\mathcal{K} = \mathcal{R}$, since the other cases are similar or follow from it. Note that $u \, \mathcal{R}_S \, v$ trivially implies $u \, \mathcal{R}_T \, v$, irrespectively of u, v being regular or not. Conversely, suppose that u, v are regular elements of S such that $u\mathcal{R}_T v$. Let $x \in S$ be such that $u = uxu$. Then ux is an idempotent such that $ux\mathcal{R}_S u$, thus $ux\mathcal{R}_T v$. It follows that $v = uxv$ (cf. Exercise 3.46), whence $v \leqslant_{\mathcal{R}_S} u$. By symmetry, we get $v\mathcal{R}_S u$.

3.63 Set

$$x = \begin{bmatrix} a & 0 \\ b & 1 \end{bmatrix}, y = \begin{bmatrix} c & 0 \\ d & 1 \end{bmatrix}, z = \begin{bmatrix} e & 0 \\ f & 1 \end{bmatrix}, t = \begin{bmatrix} g & 0 \\ h & 1 \end{bmatrix}.$$

Then

$$xyz = \begin{bmatrix} ace & 0 \\ bce + de + f & 1 \end{bmatrix}.$$

Thus $xyz = x$ implies $ce = 1$, hence $de + f = 0$ which is impossible. Thus the $\mathcal{R}$-class of x is reduced to x and similarly for its $\mathcal{L}$-class. Next $xyz = t$ has an infinity of solutions given y, t. Indeed, choose first d, e, f such that $de + ff < h$. Then, choose b, c such that $bce = h - de - f$. Finally, choose $a = g/ce$. This shows that S is formed of one $\mathcal{J}$-class.

3.64 Let G be a closed subgroup of N. The subsemigroup $\varphi^{-1}(G)$ is closed and thus compact. Let e be an idempotent in the minimal ideal of $\varphi^{-1}(G)$. Then φ maps the maximal subgroup of M containing e onto G.

3.65 Let e be an idempotent in the $\mathcal{J}$-class J of 1. Then $e \, \mathcal{H} \, 1$ since the monoid is stable, and thus $e = 1$. This implies that J has a single $\mathcal{R}$-class and a single $\mathcal{L}$-class. Thus J is the $\mathcal{H}$-class of the idempotent 1, which is a group.

3.66 Let H' and K' be the maximal subgroups of T such that $\pi(H) \subseteq H'$ and $\pi(K) \subseteq K'$. Let e, f be the idempotents in H, K respectively. We know that $e = xy$ and $f = yx$ for some $x, y \in S$ (Exercise 3.51). As seen in the proof of Proposition 3.6.11, the map $u \in H \mapsto yux \in K$ is a well-defined isomorphism. Similarly, we have the isomorphism $\varphi : u \in H' \mapsto \pi(y)u\pi(x) \in K'$. Observe that $\varphi(\pi(H)) = \pi(yHx) = \pi(K)$. Hence, $\pi(H) = H'$ implies $K' = \varphi(H') = \pi(K)$. Symmetrically, if $\pi(K) = K'$ holds, then $\pi(H) = H'$ also holds.

3.67 This holds by Green's Lemma (Exercise 3.52), as seen in the solution of Exercise 3.53.

3.68 Suppose that $x, y \in T(H)$ satisfy $hx = hy$ for some $h \in H$. Let $u \in H$. Then $u = zh$ for some $z \in S^1$, thus $ux = zhx = zhy = uy$, entailing $\rho_x = \rho_y$.

Let $x, y \in T(H)$. For every $h \in H$, one has $h \cdot \rho_{xy} = hxy = (h \cdot \rho_x) \cdot \rho_y$, thus $\rho_x \rho_y = \rho_{xy}$, and so $\Gamma(H)$ is a monoid of right transformations on H. Moreover, by Green's Lemma (cf. Exercise 3.52), the transformation ρ_x has an inverse. This shows that $\Gamma(H)$ consists of permutations of H acting (freely) on its right.

Fix $h \in H$. For each $u \in H$, choose $\psi(u) \in S^1$ such that $u = h\psi(u)$. Note that $\psi(u) \in T(H)$ (cf. Exercise 3.67). The mapping $u \in H \mapsto \rho_{\psi(u)} \in \Gamma(H)$ is clearly injective. To see that it is surjective, observe that, when $x \in T(H)$, the equality $hx = h\psi(hx)$ yields $\rho_x = \rho_{\psi(hx)}$, since the action of $\Gamma(H)$ on H is free.

3.69 Let first H, K be two $\mathcal{H}$-classes contained in the same $\mathcal{L}$-class L. Then $T(H) = T(K)$. Indeed, fix some $h \in H$. For any $x \in T(H)$, there is $u \in S^1$ such that $hxu = h$. By Green's Lemma, the right translations by x and u are mutually inverse bijections from L onto itself preserving $\mathcal{H}$-classes. Thus $x \in T(K)$. Let $v \in S^1$ be such that $vh \in K$. Then the left multiplication by v is a bijection between H and K such that, for $\rho_x \in \Gamma(H)$ and $h \in H$, $v(\rho_x(h)) = \rho_x(vh)$. In other words, the permutation groups $\Gamma(H)$ to $\Gamma(K)$ only differ in the names of the elements on which they act.

A similar result holds for $\mathcal{H}$-classes included in the same $\mathcal{R}$-class and concerning the left Schützenberger group $\Lambda(H)$ formed of the left translations $\lambda_y : h \to yh$ on H realized by the elements $y \in S^1$ such that $yH = H$.

We claim that the left and right Schützenberger groups relative to the same $\mathcal{H}$-class H are isomorphic. Fix $h \in H$. For all $x^+, y^+, x^-, y^- \in S^1$ such that $hx^+ = x^-h \in H$ and $hy^+ = y^-h \in H$, we have, by the freeness of the actions, $\rho_{x^+} = \rho_{y^+} \Leftrightarrow hx^+ = hy^+ \Leftrightarrow x^-h = y^-h \Leftrightarrow \lambda_{x^-} = \lambda_{y^-}$. This gives well-defined and mutually inverse functions $f_h : \Gamma(H) \to \Lambda(H)$ and $g_h : \Lambda(H) \to \Gamma(H)$, satisfying $f_h(\rho_{x^+}) = \lambda_{x^-}$ and $g_h(\lambda_{x^-}) = \rho_{x^+}$, with x^+, x^- as above. Since $hx^+y^+ = x^-hy^+ = x^-y^-h$, we have $f_h(\rho_{x^+y^+}) = \lambda_{x^-y^-}$, that is to say $f_h(\rho_{x^+}\rho_{y^+}) = f_h(\rho_{x^+})f_h(\rho_{y^+})$. Hence f_h is an isomorphism $\Gamma(H) \to \Lambda(H)$.

Assuming that H, K are $\mathcal{H}$-classes in the same $\mathcal{D}$-class, let I be the $\mathcal{H}$-class given by the intersection of the $\mathcal{R}$-class of H and the $\mathcal{L}$-class of K. Then $\Gamma(H) \simeq \Lambda(H) \simeq \Lambda(I) \simeq \Lambda(K) \simeq \Gamma(K)$ where $\simeq$ denotes isomorphism.

3.70 Since H is a subgroup, we have $H \subseteq T(H)$ and so the map $x \mapsto \rho_x$ is a well-defined morphism $H \to \Gamma(H)$. Let e be the idempotent of H. If $x \in H$ is such that ρ_x is the identity, then $e = ex = x$, and so the morphism is injective. On the other hand, each $x \in T(H)$ has the same action on H by right translations as $ex \in H$.

3.71

1. Clearly, $T(H)$ is closed, thus a compact monoid. Suppose that $(x_i)_{i \in I}$ is a net of elements of $T(H)$ converging to x. By continuity of multiplication, for each $h \in H$ one has $hx_i \to hx$, thus $\rho_{x_i} \to \rho_x$ in the pointwise topology. This shows that the natural morphism $\rho : T(H) \mapsto \Gamma(H)$ is continuous. Therefore, $\Gamma(H)$ is a compact group, according to Exercise 3.44.

2. Fix $h \in H$. For each $u \in H$, choose $\psi(u) \in T(H)$ such that $u = h\psi(u)$. In the solution of Exercise 3.68 it is seen that the mapping $u \in H \mapsto \rho_{\psi(u)} \in \Gamma(H)$

is bijective. Therefore, it remains to see that it is continuous, which we do next. For each $v \in H$, we may take $\varphi(v) \in S^1$ such that $v = \varphi(v)h$. Then we have $v \cdot \rho_{\psi(u_i)} = \varphi(v)h\psi(u_i) = \varphi(v)u_i \to \varphi(v)u = \varphi(v)h\psi(u) = v \cdot \rho_{\psi(u)}$, thus $\rho_{\psi(u_i)} \to \rho_{\psi(u)}$ in the pointwise topology.

3. Continuity of the multiplication of S immediately entails continuity of the isomorphism $x \in H \mapsto \rho_x \in \Gamma(H)$ (see the solution of Exercise 3.70).

4. Suppose H, K are contained in the same $\mathcal{L}$-class L. As seen in the solution of Exercise 3.69, we have an isomorphism $\Gamma(H) \to \Gamma(K)$ given by the correspondence $\rho_x|_H \mapsto \rho_x|_K$, where $\rho_x|_H$ and $\rho_x|_K$ are the restrictions to H and K, respectively, of the right translation $\rho_u : u \in L \to ux \in L$, whenever $x \in T(H) = T(K)$. Since the natural mapping $T(K) \to \Gamma(K)$ is the composition of the aforementioned isomorphism with the natural mapping $T(H) \to \Gamma(H)$, we deduce that such isomorphism is continuous, according to Exercise 3.31. Following the steps of the solution of Exercise 3.69, it suffices to check that the isomorphism $\Gamma(H) \to \Lambda(H)$, established there, is continuous. Recycling the notation in the second last paragraph of the solution of Exercise 3.69, suppose that $\rho_{x_i^+} \to \rho_{x^+}$ in $\Gamma(H)$, and let $f_h(\rho_{x_i^+}) = \lambda_{x_i^-}$ and $f_h(\rho_{x^+}) = \lambda_{x^-}$. Then $x_i^- hz = hx_i^+ z \to hx^+ z = x^- hz$ for every $z \in T(H)$, entailing $\lambda_{x_i^-} \to \lambda_{x^-}$ and showing that f_h is continuous.

3.14.5   Section 3.7

3.72 It clearly suffices to show that a right inverse of u is an inverse of u. From the equality $uv = 1$, we get $u \mathrel{\mathcal{R}} 1$ and $v \mathrel{\mathcal{L}} 1$. But we also have $u \leqslant_{\mathcal{L}} 1$ and $v \leqslant_{\mathcal{L}} 1$. Since M is stable, we conclude that u, v and 1 belong to the maximal subgroup of M containing 1. Therefore, u and v are inverses of each other.

3.14.6   Section 3.8

3.73 This property is immediate once we see that S can be taken as in Proposition 3.8.3. More formally, let S' be the closed subsemigroup of $\prod_{i \in I} S_i$ formed by the $(s_i)_{i \in I}$ such that $\psi_{i,j}(s_i) = s_j$ whenever $i \geqslant j$. Then we have an isomorphism $\theta : S \to S'$ such that $\psi_i = \psi_i' \circ \theta$ for all i, where $\psi' : S' \to S_i$ is the component projection.

3.74 Note that the map $\pi : \varprojlim S_i \to \varprojlim S_j$ given by $\pi((s_i)_{i \in I}) = (s_j)_{j \in J}$ is a well defined continuous morphism of semigroups.

Conversely, we define a map $\rho : \varprojlim S_j \to \varprojlim S_i$ as follows. Take an element $(s_j)_{j \in J}$ of $\varprojlim S_j$. For each $i \in I$, since J is cofinal in I, there is some $j \in J$ such that $i \leqslant j$. For such j, define $s_i = \psi_{j,i}(s_j)$. Note that this definition is consistent: on one hand, if $i \in J$, then we may take $j = i$ and observe that $\psi_{i,i}$ is the identity, on the other hand, the definition of s_i is independent of the choice of j, because if j' is another element of J such that $i \leqslant j'$, and if $k \in J$ is a common upper bound of j and j', then

$$\psi_{j',i}(s_{j'}) = \psi_{j',i}(\psi_{k,j'}(s_k)) = \psi_{k,i}(s_k) = \psi_{j,i}(\psi_{k,j}(s_k)) = \psi_{j,i}(s_j).$$

We may therefore define $\rho((s_j)_{j \in J}) = (s_i)_{i \in I}$. Clearly, ρ is a continuous morphism of semigroups.

Observing that $\pi \circ \rho$ is the identity on $\varprojlim S_j$ and that $\rho \circ \pi$ is the identity on $\varprojlim S_j$, we conclude that $\varprojlim S_i$ and $\varprojlim S_j$ are isomorphic topological semigroups.

3.75

1. We may consider S as in the statement of Proposition 3.8.3. The sets of the form $\psi_i^{-1}(U)$, with U an open set of S_i, form a subbasis of S, by the definition of the product topology. Since $X \cap \psi_i^{-1}(U) \neq \emptyset$ when the open set U is nonempty, this shows that X is dense in S.

2. The set $\zeta(T)$ is dense in S by the established property, and it is closed according to Exercise 3.29.

3.76 Let $S' = \varprojlim_{i \in I} \psi_i(S)$, and let $\psi_i' : S' \to \psi_i(S)$ be the associated projections. By the definition of projective limit, there is a unique continuous morphism $\theta : S \to S'$ such that Diagram (3.4) commutes for every $i, j \in I$.

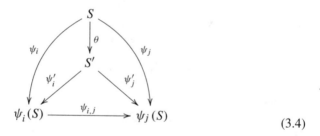

$$(3.4)$$

Since $\psi_i'(\theta(S)) = \psi_i(S)$ for every $i \in I$, the morphism θ is onto, according to Exercise 3.75. On the other hand, if $s, t \in S$ satisfy $\theta(s) = \theta(t)$, then $\psi_i(s) = \psi_i(t)$ for every $i \in I$, thus $s = t$ as seen in Exercise 3.73. Therefore, θ is an isomorphism.

3.14.7 Section 3.9

3.77 The "only if" part follows immediately from the continuity of φ. Conversely, suppose that $\varphi(s_i) \to \varphi(s)$ for every continuous morphism $\varphi : S \to F$ into a finite semigroup F. We know that $S = \varprojlim_{k \in K} S_k$, for some projective system of finite S_k. Consider the projections $\psi_k : S \to S_k$. Then, by our hypothesis, we have $\psi_k(s_i) \to \psi_k(s)$. As we are dealing with the pointwise topology, it follows that $(\psi_k(s_i))_{k \in K} \to (\psi_k(s))_{k \in K}$, that is to say, $s_i \to s$.

3.78 Let $f : X \to Y$ be a continuous map from the connected space X onto Y. Assume that U, V are two disjoint open sets such that $Y = U \cup V$. Then $X = f^{-1}(U) \cup f^{-1}(V)$, a contradiction.

3.79 Let $X = \bigcup_{i \in I} X_i$ with X_i connected and $\bigcap_{i \in I} X_i \neq \emptyset$. Let U, V be disjoint open sets such that $X = U \cup V$. Then, for each $i \in I$, we have either $X_i \subseteq U$ or $X_i \subseteq V$. Since the intersection of the X_i is not empty, the same case occurs for every i, say the first one. Then $X \subseteq U$ and thus $V = \emptyset$.

3.80 Suppose C is a nonempty connected subset of a product $\prod_{i \in I} X_i$ of totally disconnected spaces X_i. Then for every i, the projection $\pi_i(C)$ is connected and hence is reduced to one point. This implies that C is a singleton.

3.81 Let Q be the intersection of all clopen sets containing x and let C be the connected component of x.

Let F be a clopen set containing x. Since $C = (C \cap F) \cup (C \setminus F)$ is a partition of C in closed sets, and since $C \cap F \neq \emptyset$, we have $C \subseteq F$. This shows that $C \subseteq Q$ (we did not use the compactness of X).

Conversely, let us show that Q is connected, which will imply $Q \subseteq C$. For this, consider a pair X_1, X_2 of disjoint closed sets in Q with $Q = X_1 \cup X_2$ and $x \in X_1$. Since Q is closed, X_1 and X_2 are closed in X. Since X is compact, it is normal (see Exercise 3.24) and thus there are open sets U, V such that $X_1 \subseteq U$ and $X_2 \subseteq V$. Now $Q \subseteq U \cup V$. Let S be the family of clopen sets containing Q and let S' be the family of their complements. The family S' is an open cover of $X \setminus (U \cup V)$. Let $X \setminus K_1, \ldots, X \setminus K_n$ with be a finite subcover and let $K = K_1 \cap \ldots \cap K_n$. Then K is clopen and $Q \subseteq K \subseteq U \cup V$. Since

$$\overline{U \cap K} \subseteq \overline{U} \cap K = \overline{U} \cap (U \cup V) \cap K = U \cap K$$

the intersection $U \cap K$ is also clopen. As $x \in U \cap K$, we have $Q \subseteq U \cap K$ and $X_2 \subseteq Q \subseteq U \cap K \subseteq U$ and we obtain $X_2 = \emptyset$. Thus Q is connected.

3.82 Let $x, y \in X$ be distinct. Since X is Hausdorff, there are disjoint neighborhoods of x and y. Since X is zero dimensional, these neighborhoods contain clopen neighborhoods U, V of x, y. Thus $U, X \setminus U$ is a partition of X in open sets separating x, y. We conclude that X is totally disconnected.

Conversely, assume that X is a totally disconnected compact space. Consider a nonempty closed set $U \subseteq X$ and $x \notin U$. By Exercise 3.81, for each point $p \in U$ there is a clopen subset $K_p \subseteq X$ that contains x but not p. The complements of K_p form an open cover of the compact set U and, therefore, finitely many of them suffice for that purpose, say those determined by the points $p_1, \ldots, p_n$. Then $K = \bigcap_{i=1}^{n} K_{p_i}$ is a clopen subset of X containing x and disjoint from U. Hence, X is zero-dimensional.

3.83 Trivially, it suffices to consider the case where S is not a monoid. The compact semigroup S^1 is also zero-dimensional, since 1 is an isolated point of S^1, and the conclusion follows from Theorem 3.9.3. (Alternatively, one may show directly that an equality of the form $S^1 = \varprojlim S_i^1$ may be induced from a projective limit $S = \varprojlim S_i$ of finite semigroups with onto connecting morphisms.)

3.84 By Proposition 3.8.7, the projections φ_i are onto. Let $K = \prod_{i \in I} G_i/H_i$ and consider the natural continuous morphism $\zeta : G \to K$, which is injective, as seen in the proof of Proposition 3.10.2. The morphism $\varphi_{i,j}$ induces a morphism $\bar{\varphi}_{i,j} : G_i/H_i \to G_j/H_j$, given by $\bar{\varphi}_{i,j}(gH_i) = \varphi_{i,j}(g)H_j$. The morphisms $\bar{\varphi}_{i,j}$ define a projective system whose limit $\varprojlim G_i/H_i$ contains $\zeta(G)$. Since the projection $\varphi_i : G \to G_i$ is onto, the projection $\bar{\varphi}_i : \varprojlim G_i/H_i \to G_i/H_i$ satisfies $\bar{\varphi}_i(\zeta(G)) = G_i/H_i$. Therefore, ζ is an isomorphism, in view of Exercise 3.75.

3.85 Let H be an $\mathcal{H}$-class in a profinite semigroup S. It was seen in Exercise 3.71 that H and $\Gamma(H)$ are homeomorphic spaces. Since S is profinite, and H is closed, the topology of H is compact zero-dimensional, and so $\Gamma(H)$ is a profinite group, by Theorem 3.9.3.

3.14.8 Section 3.12

3.86 Let $X = \alpha(A)$, so that the restrictions of f and g to X coincide. As f, g are morphisms of semigroups, the maps f, g also coincide in the subsemigroup X^+. Since $\overline{X^+} = S$ and f, g are continuous, they must in fact coincide in S.

3.87 Let A be a finite generating set of the profinite semigroup S. Consider the function $\rho : \text{End}(S) \to S^A$ mapping each continuous endomorphism of S to its restriction $A \to S$. Suppose that (f_i) is a net of elements of $\text{End}(S)$ converging to f. Then in particular we have $f_i(a) \to f(a)$ for every $a \in A$, that is, $\rho(f_i) \to \rho(f)$. Therefore, the map ρ is continuous. It is also an injective map (cf. Exercise 3.87), and so $\text{End}(S)$ is (homeomorphic) to a topological subspace of S^A. Since A is finite, the space S^A is metrizable (Proposition 3.2.17), and therefore so is $\text{End}(S)$.

3.88 Let $G = \prod_{n \in \mathbb{N}} G_n$ where G_n is (a copy of) the group $\mathbb{Z}/2\mathbb{Z}$, for each $n \geqslant 0$. For each $n \geqslant 1$, consider the continuous automorphism $\sigma_n : G \to G$ defined by

$$\sigma_n(x_0, x_1, x_2, x_3, x_4, \ldots) = (x_1, x_2, \ldots, x_n, x_0, x_{n+1}, \ldots),$$

with $x_i \in \mathbb{Z}/2\mathbb{Z}$.
1. Note that the sequence $(\sigma_n)_n$ converges pointwise to the endomorphism

$$\sigma(x_0, x_1, x_2, x_3, x_4, \ldots) = (x_1, x_2, x_3, x_4, x_5, \ldots),$$

which is not injective. Hence, $\text{Aut}(G)$ is not closed in $\text{End}(G)$ and, therefore, it cannot be compact.

2. By Exercise 3.72, it suffices to exhibit a continuous non-invertible endomorphism with left inverse. Consider the continuous endomorphism $\mu : G \to G$ defined by

$$\mu(x_0, x_1, x_2, x_3, x_4, \ldots) = (0, x_0, x_1, x_2, x_3, \ldots),$$

with $x_i \in \mathbb{Z}/2\mathbb{Z}$. Then σ is a left inverse of μ. However, μ has no inverse, since it is not onto.

3.15 Notes

For all unproved results on topological spaces, see any classical textbook on topology, for example Willard (2004). Our convention that compact spaces must be Hausdorff is not universal in the literature, but it is also quite common.

In the perspective of Universal Algebra, algebraic structures are defined not to be empty (Burris and Sankappanavar 1981). We adopt this perspective, when demanding semigroups not to be empty.

Our presentation of Green's relations is classical (see Lallement (1979) for a more detailed presentation). A classical reference concerning profinite groups is Ribes and Zalesskii (2010).

A proof of the fact that all Cantor spaces are homeomorphic can be found in Willard (2004, Theorem 30.3).

The term zero-dimensional is by reference to a notion of dimension in topological spaces, actually the Lebesgue dimension, see Munkres (1999). Theorem 3.9.3 may be found in Almeida (2005a). Lemma 3.9.4 is from Hunter (1988) and is known as *Hunter's Lemma*. The implication (v) $\Rightarrow$ (i) in Theorem 3.9.3 appears in an early paper of Numakura (1957).

We mention a notable difference between the realms of profinite semigroups and profinite groups. It is very easy to find examples of non-continuous semigroup morphisms between finitely generated profinite semigroups: for example, take the morphism from $\widehat{\mathbb{N}}$ onto the multiplicative monoid $\{0, 1\}$ that assigns the elements of $\mathbb{N}$ to 1 and the elements of $\widehat{\mathbb{N}} \setminus \mathbb{N}$ to 0. In contrast, we have the very deep and difficult result of Nikolov and Segal (2003), depending on the classification of finite simple groups, that states the continuity of every group morphism between two finitely generated profinite groups.

Theorem 3.12.1 is due to Hunter (1983). It was rediscovered in Almeida (2005a) and slightly extended in Steinberg (2011). The compact-open topology is one of the possible topologies introduced as stronger than pointwise convergence and weaker than uniform convergence (see Willard (2004)).

The normal spaces of Exercise 3.24 are a classical notion. Exercise 3.24 is Theorem 17.10 in Willard (2004). The intersection of all clopen sets containing x introduced in Exercise 3.81 is called the *quasi-component* of x. Exercise 3.81 is Theorem 6.1.23 in (Engelking 1989). Exercise 3.82 is Theorem 29.7 in Willard (2004).

The term *Hopfian* is classical in the context of discrete groups. It is well known that finitely generated free groups are Hopfian, but the term is also used for arbitrary topological semigroups (Steinberg 2011). Proposition 3.12.5 is Proposition 2.5.2 in Ribes and Zalesskii (2010).

The eggbox diagrams in Figs. 3.1 and 3.3 were obtained using the software GAP (GAP 2020; Delgado et al. 2006; Delgado and Morais 2006).

Chapter 4
Free Profinite Monoids, Semigroups and Groups

4.1 Introduction

In this chapter we study free profinite monoids and free profinite semigroups, also looking at free profinite groups. Actually, free profinite semigroups and free profinite monoids are almost the same thing, the latter obtained from the former by adjoining a neutral element, topologically isolated.

We construct the free profinite monoid as the projective limit of the finite quotients of the free monoid. We look at some aspects of the elements of the free profinite monoid, viewing them as generalizations of the usual words. For that reason, those elements are called pseudowords.

We study the closure of recognizable sets in a free profinite monoid and the corresponding notion in a free profinite group, closely related to the so-called profinite topology of the free group.

We begin by an account of the notions related to words, languages and free monoids in Sect. 4.2. We state without proof Kleene's theorem (Theorem 4.2.8).

In Sect. 4.3 we give similarly an account of the basic notions concerning free groups. We state without proof the Nielsen-Schreier Theorem (Theorem 4.3.1). We introduce Stallings automata and use them to prove the theorem of Hall asserting that any finitely generated subgroup of a free group is a free factor of a subgroup of finite index (Theorem 4.3.4).

Section 4.4 introduces free profinite monoids and free profinite semigroups. We relate the recognizable languages to the clopen subsets of the free profinite monoid (Theorem 4.4.9), and in the process we investigate some features of pseudowords that help building an intuitive use of them. In Sect. 4.5 we interpret pseudowords as operations over profinite semigroups. Free profinite groups are reviewed in the following section (Sect. 4.6).

In Sect. 4.7, we introduce the notion of presentation of a profinite semigroup, monoid or group.

© Springer Nature Switzerland AG 2020
J. Almeida et al., *Profinite Semigroups and Symbolic Dynamics*, Lecture Notes in Mathematics 2274, https://doi.org/10.1007/978-3-030-55215-2_4

In Sect. 4.8 we study profinite codes, which correspond to injective morphisms from a free profinite monoid to another. The main result is that, somewhat surprisingly, any finite code is a profinite code (Theorem 4.8.3).

In Sect. 4.9 we give a glimpse about the field of pseudovarieties of semigroups and of varieties of languages, and its relation to (free) profinite semigroups, with a look on how that helps understand pseudowords. This relation has been one of the main forces motivating the study of profinite semigroups, but it is not necessary for the following chapters.

4.2   Free Monoids and Semigroups

The *free semigroup* generated by a nonempty set A, denoted by A^+, is the semigroup of finite nonempty sequences of elements of A with the operation between them being concatenation. If we also allow the empty sequence, denoted by ε, then we obtain a monoid having ε as neutral element, the *free monoid* generated by A, denoted by A^*. So, we have $A^* = A^+ \cup \{\varepsilon\}$.

An element of the free monoid A^* is called a *word* over A, the neutral element ε is the *empty word*, the set A is said to be an *alphabet*, and the elements of A are the *letters* of A. The *length* of w is the number of letters (repetitions included) which compose w. We denote by $|w|$ the length of w. The subsets of free monoids are usually called *languages*. The theory of formal languages deals with the study of subsets of free monoids.

Example 4.2.1 Let $A = \{a, b\}$. Then $ab^2a^2 = ab \cdot ba \cdot a$ is a word of length 5 belonging to the language $\{a, ab, ba\}^*$, where $\{a, ab, ba\}^*$ is the submonoid of A^* generated by $\{a, ab, ba\}$.

Remark 4.2.2 Suppose that X is a subset of a monoid M. The notation introduced in Chap. 3 for the submonoid of M generated by X may conflict with the notation X^* for the free monoid generated by X. But context will prevent ambiguity. If X is a subset of the free monoid, then in general we use the notation X^* in the sense used in Chap. 3, as we did in Example 4.2.1.

Note that A generates the semigroup A^+. Moreover, if $\varphi : A \to S$ is a map from A into a semigroup S, then there is a unique morphism of semigroups $\bar{\varphi} : A^+ \to S$ extending A. This is what we mean by the *universal property* of the free semigroup A^+. Similarly, we have a universal property for the free monoid A^*, and in the absence of risk of confusion, we may also denote by $\bar{\varphi}$ the unique morphism of monoids $A^* \to M$ extending a map φ from A into the monoid M. The morphism φ is explicitly given by $\bar{\varphi}(a_1 \cdots a_n) = \varphi(a_1) \cdots \varphi(a_n)$ for every $a_1, \ldots, a_n \in A$. Therefore, $\bar{\varphi}$ is completely determined by φ. That is why to denote $\bar{\varphi}$ simply by φ is an acceptable abuse of notation.

Example 4.2.3 Let $A = \{a, b\}$. The correspondence $\varphi : a \mapsto 0, b \mapsto 1$ defines a unique morphism $\varphi : A^* \to \mathbb{Z}/2\mathbb{Z}$. For this morphism, one has $\varphi(w) = 1$ if and only if the word w has an odd number of b's.

We say that a language $L \subseteq A^*$ is *recognizable* when there is a morphism $\varphi : A^* \to M$ into a finite monoid M recognizing L, that is, such that $L = \varphi^{-1}\varphi(L)$. We also say that L is recognized by a finite monoid M if it is recognized by some morphism from A^* into M. Note that $L = \varphi^{-1}\varphi(L)$ if and only if $L = \varphi^{-1}(P)$ for some subset P of M, a trivial fact that quite often is convenient to have in mind.

We focus on recognizable languages over *finite* alphabets. Let us see three simple examples.

Example 4.2.4 Let $A = \{a, b\}$. The language $L = A^* b A^*$ is the set of words over A with at least one letter b. It is recognized by the morphism $\varphi : a \mapsto 1, b \mapsto 0$ from A^* onto the multiplicative monoid $M = \{0, 1\}$, with the usual multiplication. More precisely, one has $L = \varphi^{-1}(0)$.

We revisit Example 4.2.3.

Example 4.2.5 Let $A = \{a, b\}$. The set $L = (a \cup ba^*b)^*$, the set of words with an even number of b's, is recognized by the morphism $\varphi : a \mapsto 0, b \mapsto 1$ from A^* onto the additive group $\mathbb{Z}/2\mathbb{Z}$, already met in Example 4.2.3.

In the next example we see that the singletons are recognizable. For that purpose, we first introduce an algebraic construction which is interesting by itself. To each ideal I of a semigroup S one associates its *Rees congruence* ϱ_I, defined by $(u, v) \in \varrho_I$ if and only if $u, v \in I$ or $u = v$. In other words, the Rees quotient $S \to S/\varrho_I$ collapses all elements of I into a zero, and only identifies distinct elements that are in I. The quotient S/ϱ_I is usually denoted S/I.

Example 4.2.6 Let w be a word over the finite alphabet A. Take the ideal $I \subseteq A^*$ formed by the words of A^* with length greater than $|w|$. As A is finite, so is $A^* \setminus I$, whence the Rees quotient A^*/I is finite. The quotient morphism $q : A^* \to A^*/I$ is such that $\{w\} = q^{-1}(q(w))$, thus $\{w\}$ is recognized by A^*/I.

The recognizable languages deserve special attention in the theory of formal languages. One important reason is that recognizable languages capture "finite" properties of words, i.e., that can be recognized by finite devices, like finite monoids, or, alternatively, finite automata (introduced some lines ahead).

Another important reason for studying recognizable languages is *Kleene's theorem*, for whose statement we first need some preparation. The *rational* languages on a finite alphabet A are the languages which can be obtained from subsets of A by applying finitely many times the *rational operations*, which are

(i) the union operation $(L, K) \mapsto L \cup K$,
(ii) the product operation $(L, K) \mapsto LK$,
(iii) the *star operation*, which associates to each language L the submonoid L^* of A^* generated by L.

Example 4.2.7 For every finite alphabet A, and every word w over A, the languages $\{w\}$, A^*, wA^*, A^*w and A^*wA^* are rational.

We now enunciate Kleene's theorem, a basic building block of the theory of formal languages.

Theorem 4.2.8 (Kleene) *A language on a finite alphabet is rational if and only if it is recognizable.*

As a consequence, the class of recognizable languages is closed under the rational operations. In particular, the product $LK = \{uv \mid u \in L, v \in K\}$ of two recognizable languages is recognizable (see Exercise 4.4).

In the other direction, if $L \subseteq A^*$ is recognized by $\varphi : A^* \to M$, then $A^* \setminus L$ is also recognized by φ, and so, in view of Kleene's theorem, it is clear that the class of rational languages is closed under the Boolean operations.

Finite Automata

Finite monoids recognizing languages are often associated with finite automata. An *automaton* $\mathcal{A} = (Q, I, T)$ on the alphabet A is a labeled directed multigraph on a set Q of vertices called its *states*. One defines also two distinguished sets of states, the set I of *initial states* and the set T of *terminal states*. More formally, the set of edges of $\mathcal{A}$ is identified with a subset of $Q \times A \times Q$. Under this identification, the edge (q, a, r) goes from q to r and is labeled by the letter a. The automaton (Q, I, T) on the alphabet A is said to be *finite* when Q and A are finite.

Example 4.2.9 In Fig. 4.1 we see a finite automaton over $A = \{a, b\}$ with set of states $Q = \{1, 2, 3, 4, 5\}$. The edges $(2, a, 3)$ and $(2, b, 3)$ are graphically overlapped. The initial states $1, 2$ are marked with a small incoming arrow, and the final states $2, 4, 5$ are marked with a small outgoing arrow (overlapped with the incoming arrow in the case of state 2).

The edges of the automaton $\mathcal{A} = (Q, I, T)$ on A are labeled by letters of A and consequently the paths are labeled by words in A^*. At each vertex we admit an empty loop labeled by ε. The language *recognized* by the automaton $\mathcal{A}$ is the set of labels of paths (including empty loops) from an initial state to a terminal state. The *transition monoid* of the automaton is the image of the morphism φ from A^* into the monoid of binary relations on Q defined by $(p, q) \in \varphi(w)$ if there is a path

Fig. 4.1 An example of a finite automaton with two initial states and three final states

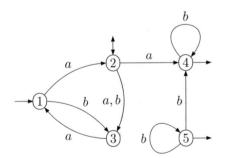

Fig. 4.2 An automaton
recognizing A^*bA^*

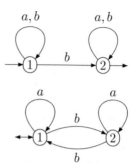

Fig. 4.3 An automaton
recognizing $(a \cup ba^*b)^*$

from p to q labeled w. One says that φ is the *transition morphism* of $\mathcal{A}$. Note that φ recognizes the language L recognized by the automaton: indeed, $w \in L$ if and only if $\varphi(w)$ meets $I \times T$.

Example 4.2.10 The set $L = A^*bA^*$ of Example 4.2.4 is recognized by the automaton of Fig. 4.2, whose transition monoid is the same monoid $\{0, 1\}$ of Example 4.2.4. The unique initial state is 1 (indicated by an incoming arrow) and the unique terminal state is 2 (indicated by an outgoing arrow).

Example 4.2.11 The set $L = (a \cup ba^*b)^*$ of Example 4.2.5 is recognized by the automaton of Fig. 4.3. State 1 is the unique initial and terminal state.

A *deterministic automaton* is an automaton $\mathcal{A} = (Q, i, T)$ with a unique initial state i such that the transition monoid of $\mathcal{A}$ consists of right transformations on Q. The image of $q \in Q$ by $\varphi(w)$ for $w \in A^*$ is denoted $q \cdot w$.

It is easy to see that a language is recognized by a finite monoid if and only if it is recognized by a finite automaton, if and only if it is recognized by a finite deterministic automaton (Exercise 4.5).

The automaton $\mathcal{A} = (Q, I, T)$ is said to be *trim* if for every state $q \in Q$ there is a path (possibly an empty loop) from i to q, and from q to a state of T, provided $T \neq \emptyset$. In case $T = \emptyset$, the automaton is considered trim if it has only one state.

The automaton $\mathcal{A} = (Q, I, T)$ is said to be *reduced* when the mapping associating to each $q \in Q$ the set $L_q = \{u \in A^* \mid q \cdot u \in T\}$ is injective.

Example 4.2.12 The automata in Example (4.2.10) and (4.2.11) are trim and reduced. Only the former is deterministic. The automaton in Example (4.2.9) is also not deterministic, and it is neither trim nor reduced: the state 5 is not accessible from an initial state, and $L_4 = L_5 = b^*$.

The *minimal automaton* $\mathcal{A}(L) = (Q, i, T)$ of a nonempty language $L \subseteq A^*$ is a deterministic automaton on A which is isomorphic to the following model. The states of $\mathcal{A}(L)$ are the nonempty languages of the form

$$u^{-1}L = \{v \in A^* \mid uv \in L\}$$

with $u \in A^*$; we have an edge labeled a from $u^{-1}L$ to $a^{-1}(u^{-1}L) = (ua)^{-1}L$ whenever $u^{-1}L$ and $(ua)^{-1}L$ are nonempty; the initial state is the language $\varepsilon^{-1}L = L$; and the terminal states are the languages $u^{-1}L$ containing the empty word ε. Finally, on the alphabet A, the minimal automaton of the empty language is the automaton $(i, i, \emptyset)$, with a single state i, where each letter of A labels a loop at i.

It turns out that the minimal deterministic automaton $\mathcal{A}(L)$ is the unique deterministic automaton recognizing L that is trim and reduced (Exercise 4.6). The terminology *minimal automaton* is justified by the fact that, in a natural sense, the automaton $\mathcal{A}(L)$ is a morphic image of every deterministic trim automaton recognizing L (cf. Exercise 4.6). From this minimality property, one deduces that L is recognizable if and only if $\mathcal{A}(L)$ is finite.

The *syntactic monoid* of a language $L \subseteq A^*$ is the transition monoid, denoted $M(L)$, of the minimal automaton $\mathcal{A}(L)$. The corresponding transition morphism $\eta_L : A^* \to M(L)$ is called the *syntactic morphism* of L. In particular, L is recognized by η_L.

By the above discussion, we see that the following holds.

Proposition 4.2.13 *Consider a language $L \subseteq A^*$. The following conditions are equivalent.*

1. *L is recognizable.*
2. *$\mathcal{A}(L)$ is finite.*
3. *$M(L)$ is finite.*

The monoid $M(L)$ and the morphism η_L have the following minimality property, which may be seen as a companion of the minimality of $\mathcal{A}(L)$.

Proposition 4.2.14 *Consider a language $L \subseteq A^*$ and its syntactic morphism $\eta_L : A^* \to M(L)$. If L is recognized by an onto morphism $\varphi : A^* \to M$, then there is a unique morphism $\psi : M \to M(L)$ such that the diagram*

commutes.

The proof of Proposition 4.2.14 is left as an exercise (Exercise 4.8).

Example 4.2.15 The syntactic monoid of A^*bA^* is the multiplicative monoid $\{0, 1\}$, while that of $(a \cup ba^*b)^*$ is the cyclic group $\mathbb{Z}/2\mathbb{Z}$ (cf. Examples 4.2.10 and 4.2.10).

Languages in Free Semigroups

Sometimes it is preferable to consider languages formed by nonempty words only, over an alphabet A, that is, subsets of A^+. Such a language is then said to be recognizable, as a subset of A^+, if and only if there exists a morphism $\varphi : A^+ \to S$ into a finite semigroup S such that $L = \varphi^{-1}\varphi(L)$. This is an example of the obvious analogies between the free monoid setting and the free semigroup setting. To distinguish these two contexts, sometimes one says that the languages contained in free monoids are $*$-languages, and that the languages contained in free semigroups are $+$-languages. In the second setting, we no longer admit empty loops at the states of an automaton, and the transition semigroup already met in Chap. 3, plays the role that the transition monoid plays in the setting of $*$-languages.

Quite often, there is not much difference between dealing with $*$-languages and $+$-languages, but that is not always the case.

Example 4.2.16 Consider finite and cofinite languages, where by a *cofinite* $+$-language (resp. $*$-language), we mean a language having a finite complement in the environment free semigroup (resp. free monoid). The class of finite and cofinite $+$-languages is stable under taking the inverse image by morphisms between free semigroups over finite alphabets. In contrast, if φ is the endomorphism of $\{a, b\}^*$ such that $\varphi(a) = a$ and $\varphi(b) = \varepsilon$, then the inverse image by φ of the finite $*$-language $\{a\}$ is the $*$-language b^*ab^*, which is neither finite nor cofinite.

4.3 Free Groups

We give a brief introduction to free groups, assuming the elementary notions of general group theory to be known.

Let A be an alphabet and let $\bar{A}$ be a disjoint copy of A consisting of the symbols $\bar{a}$ for $a \in A$. We extend the map $a \mapsto \bar{a}$ to an involution on $A \cup \bar{A}$, that is to say, $\bar{\bar{a}} = a$ for every $a \in A \cup \bar{A}$.

We consider two words on the alphabet $A \cup \bar{A}$ to be *equivalent* if one can be obtained from the other by a sequence of insertions or deletions of a factor $a\bar{a}$ for $a \in A \cup \bar{A}$. A word w over the alphabet $A \cup \bar{A}$ is *reduced* if it has no factor $a\bar{a}$ for $a \in A \cup \bar{A}$. It is easy to verify that every word on the alphabet $A \cup \bar{A}$ is equivalent to a unique reduced word (Exercise 4.9).

The *free group* on a set A is the group having as elements the reduced words on $A \cup \bar{A}$, with the product $x \cdot y = z$ such that z is the reduced word equivalent to the word xy, and with inverse given by $(a_1 a_2 \cdots a_n)^{-1} = \bar{a}_n \cdots \bar{a}_2\, \bar{a}_1$. We denote by $FG(A)$ the free group on A.

The free group has the following universal property: every map from A into a group G has a unique extension to a group morphism from $FG(A)$ into G.

The following result, known as the *Nielsen-Schreier Theorem* is well-known (see the Notes Section for a reference).

Theorem 4.3.1 *Every subgroup of a free group is free.*

Thus, for every subgroup H of $FG(A)$, there is a set B and an isomorphism $\varphi : FG(B) \to H$. The set $\varphi(B)$ is called a *basis* of H.

We will now give a brief introduction to a method allowing one to associate to every finitely generated subgroup of the free group a labeled graph. This will give us in particular a practical method to compute a basis of a finitely generated subgroup of a free group.

By a *finite inverse A-labeled automaton* (or *Stallings automaton*) we mean a finite connected directed multigraph $\mathcal{A}$ whose edges are labeled with elements of the finite alphabet A such that no two edges with the same label leave from the same vertex or arrive at the same vertex. The label of an undirected path is obtained by taking the product in the group $FG(A)$ of the labels of the successive edges of the path, which are taken as inverses if the edge appears in the wrong direction at that point of the path.

Additionally, a vertex v_0 is distinguished to be both initial and final and, for technical reasons, this is the only vertex that is allowed to have total degree 1 (that is only one edge either entering or leaving v_0). Note that the labels of undirected loops at the vertex v_0 form a subgroup $H(\mathcal{A})$ of $FG(A)$. A set of free generators of $H(\mathcal{A})$ can be obtained by choosing a spanning tree and for each edge not in the tree, reading a loop at v_0 by following the tree until the beginning of the edge, following the edge, and then returning to v_0 along the tree. In particular, the group $H(\mathcal{A})$ is finitely generated.

Conversely, to any finitely generated subgroup H of the free group, we may associate a Stallings automaton $\mathcal{A}$ such that $H = H(\mathcal{A})$. Choose a basis X of H. Build the flower automaton of H, with loops around a vertex v_0 labeled by the elements of X. We successively fold edges with the same label leaving or arriving at the same vertex (such an operation is called a *Stallings folding*). Since a Stallings folding does not modify the group of paths around v_0, the result is a Stallings automaton $\mathcal{A}$ such that $H = H(\mathcal{A})$. It can be shown that the result does not depend on the edges chosen to fold at each step. Thus, the correspondence $\mathcal{A} \mapsto H(\mathcal{A})$ is a one-to-one correspondence between finite inverse A-labeled automata and finitely generated subgroups of $FG(A)$.

Example 4.3.2 Let $A = \{a, b\}$ and let H be the subgroup of $FG(A)$ generated by the set $X = \{aa, aba\}$. The flower automaton corresponding to X is represented in Fig. 4.4 on the left with $v_0 = 0$. We first identify the two edges leaving 0 with label a to obtain the graph in the middle and next the two edges arriving at 0 labeled a to obtain the graph on the right which is the Stallings automaton of H.

The case of subgroups of finite index of $FG(A)$ deserves special attention. It is connected with the following class of Stallings automata. A *permutation automaton*,

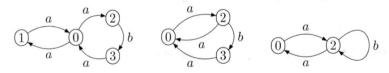

Fig. 4.4 A Stallings automaton

also called a *group automaton*, on the alphabet A is a finite, trim, deterministic automaton $\mathcal{A} = (Q, v_0, v_0)$ on A, where v_0 is a state of $\mathcal{A}$, such that all letters of A act as permutations on the vertices of $\mathcal{A}$, that is to say, the transition monoid of $\mathcal{A}$ is a group, which we may say to be the *transition group* of $\mathcal{A}$.

Proposition 4.3.3 *Let A be a finite alphabet. If $\mathcal{A} = (Q, v_0, v_0)$ is a permutation automaton on A, then $H(\mathcal{A})$ is a subgroup of $FG(A)$ of finite index equal to* Card(Q).

Conversely, if H is a subgroup of finite index d of $FG(A)$, then H is finitely generated and its Stallings automaton $\mathcal{A}$ is a permutation automaton with d states, whose transition group is the representation of H on the right cosets of H.

The proof of Proposition 4.3.3 is relegated to Exercise 4.11.

A *free factor* of the free group $FG(A)$ is a subgroup H such that for some basis X of H there is a subset Y of $FG(A) \setminus H$ such that $X \cup Y$ is a basis of $FG(A)$.

We use Stallings automata to prove the following result.

Theorem 4.3.4 (Hall) *For every finitely generated subgroup H of $FG(A)$ and every $x \in FG(A) \setminus H$, there is a subgroup of finite index K such that H is a free factor of K and $x \notin K$.*

Proof Consider the automaton of the subgroup H and fold into it a path starting at v_0 labeled x. Complete the automaton into a permutation automaton $\mathcal{A}$ by adding edges. The subgroup $K = H(\mathcal{A})$ determined by $\mathcal{A}$ has the required properties. ∎

Example 4.3.5 Consider again the subgroup H of Example 4.3.2 and the word $x = b$. We start with the automaton of Fig. 4.5 on the left. In this example, without adding vertices, the only completion into a permutation automaton is represented on the right. Thus, K is the subgroup of index 3 with basis $\{aa, aba, bb, bab\}$.

Recall the notion of residually finite topological group. The notion applies as well to an arbitrary group by considering the discrete topology. The following is an immediate corollary of Theorem 4.3.4.

Corollary 4.3.6 *Every free group is residually finite.*

Proof Consider a free group $FG(A)$. It suffices to show that, for every $x \in FG(A) \setminus \{1\}$, there is a morphism into a finite group that separates x from 1. One can start by letting B be the set of letters that appear in x, viewed as a reduced word, and consider the morphism $FG(A) \to FG(B)$ that is the identity on B and maps $A \setminus B$ to 1. Thus, we may assume that A is finite. We may then take H to be the trivial subgroup in Theorem 4.3.4 to obtain a subgroup of finite index K such that $x \notin K$.

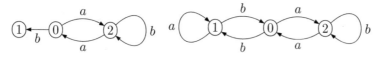

Fig. 4.5 The Stallings automata of H and K

By Lemma 4.6.2, K is recognized by a morphism into a finite group, and such a morphism has the required property.   ∎

4.4   Free Profinite Monoids and Semigroups

Free monoids/semigroups have "free profinite" counterparts, introduced throughout this section. These counterparts are built as projective limits of finite quotients of the corresponding free monoids/semigroups. In this context, the following definition appears naturally: a congruence ρ on a semigroup S is said to be of *finite index* if S/ρ is finite. More generally, the *index* of ρ is the cardinal of S/ρ.

Construction and Characterization of the Free Profinite Monoids

Let $\mathcal{C}_A$ be the set of congruences of A^* of finite index, endowed with the partial order such that $\theta \leqslant \rho$ if and only if $\rho \subseteq \theta$, for $\theta, \rho \in \mathcal{C}_A$. For this order, $\mathcal{C}_A$ is a directed set (cf. Exercise 3.45). Consider the projective system $(A^*/\theta, q_{\rho,\theta}, \mathcal{C}_A)$ formed by quotients A^*/θ such that θ has finite index, and where the connecting morphisms are the canonical maps $q_{\rho,\theta} : A^*/\rho \to A^*/\theta$. Recall that we view finite semigroups as compact semigroups with the discrete topology. Then, the projective limit of this projective system is a profinite monoid. We denote it by $\widehat{A^*}$, so that we have the following formula:

$$\widehat{A^*} = \varprojlim_{\theta \in \mathcal{C}_A} A^*/\theta.$$

We say that $\widehat{A^*}$ is the *free profinite monoid* on the alphabet A, and that A is the *basis* of $\widehat{A^*}$. The reason for this terminology will be gradually given. Let ι be the natural mapping $A \to \widehat{A^*}$, defined by $\iota(a) = ([a]_\theta)_{\theta \in \mathcal{C}_A}$.

Proposition 4.4.1   *The mapping* $\iota : A \to \widehat{A^*}$ *is a generating mapping of* $\widehat{A^*}$.

Proof   The submonoid of $\widehat{A^*}$ generated by $\iota(A)$ is the image of the unique monoid morphism $\bar{\iota} : A^* \to \widehat{A^*}$ extending ι. Let $q_\theta : \widehat{A^*} \to A^*/\theta$ be the natural projection. Since the set of elements of the form $q_\theta \circ \iota(a) = [a]_\theta$, with $a \in A$, generates A^*/θ, one sees that $q_\theta \circ \bar{\iota}$ is onto, for every $\theta \in \mathcal{C}_A$. Therefore, the image of $\bar{\iota}$ is dense in $\widehat{A^*}$, according to Exercise 3.75.   ∎

The next proposition describes the *universal property* of $\widehat{A^*}$, illustrated in Fig. 4.6.

Fig. 4.6 The universal property of $\widehat{A^*}$

Proposition 4.4.2 *The map $\iota : A \to \widehat{A^*}$ is such that for any map $\varphi : A \to M$ into a profinite monoid there exists a unique continuous morphism $\hat{\varphi} : \widehat{A^*} \to M$ such that $\hat{\varphi} \circ \iota = \varphi$.*

Proof If the morphism $\hat{\varphi}$ exists, then it must be unique because ι is a generating mapping of $\widehat{A^*}$, as showed in Proposition 4.4.1 (cf. Exercise 3.86).

To show the existence of $\hat{\varphi}$, let us suppose first that M is finite. Consider the unique monoid morphism $\bar{\varphi} : A^* \to M$ extending φ. Let $\rho = \operatorname{Ker}\bar{\varphi}$, and denote by p_ρ the component projection from $\widehat{A^*} = \lim_{\overleftarrow{\theta \in C_A}} A^*/\theta$ onto A^*/ρ. Then $p_\rho \circ \bar{\iota}$ is the quotient morphism $q_\rho : A^* \to A^*/\rho$. On the other hand, the morphism $\bar{\varphi} : A^* \to M$, extending φ, factorizes as $\bar{\varphi} = \bar{\varphi}' \circ q_\rho$, for an isomorphism $\bar{\varphi}' : A^*/\rho \to M$, in view of the isomorphism theorem for monoids (cf. Theorem 3.5.2 and its proof). That is, both inner triangles in Diagram 4.1 commute.

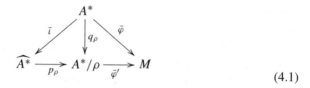

$$(4.1)$$

We may therefore take $\hat{\varphi} = \bar{\varphi}' \circ p_\rho$, establishing the proposition for the case where M is finite.

More generally, suppose that M is profinite. Consider a projective system of finite monoids $(M_i, \psi_{j,i}, I)$ for which $M = \lim_{\overleftarrow{i \in I}} M_i$ holds. For each $j \in J$, as M_j is finite, we know by the already proved case that the composition $\psi_j \circ \varphi$ entails a unique continuous morphism $\hat{\varphi}_j : \widehat{A^*} \to M_j$ such that $\hat{\varphi}_j \circ \iota = \psi_j \circ \varphi$ (see Diagram 4.2). For every connecting morphism $\psi_{j,i}$, one has

$$\psi_{j,i} \circ \hat{\varphi}_j \circ \iota = \psi_{j,i} \circ \psi_j \circ \varphi = \psi_i \circ \varphi = \hat{\varphi}_i \circ \iota.$$

Then, thanks to the uniqueness of $\hat{\varphi}_i$, we get $\psi_{j,i} \circ \hat{\varphi}_j = \hat{\varphi}_i$ (so that at this point we know that the non-dashed part in Diagram 4.2 is commutative).

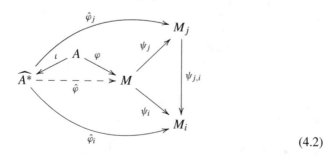

$$(4.2)$$

Hence, by the definition of projective limit, there is a unique continuous morphism $\hat{\varphi} : \widehat{A^*} \to M$ such that $\hat{\varphi}_i = \psi_i \circ \hat{\varphi}$ for all $i \in I$. Note that

$$\psi_i \circ (\hat{\varphi} \circ \iota) = \hat{\varphi}_i \circ \iota = \psi_i \circ \varphi$$

for all $i \in I$. Therefore, we have $\hat{\varphi} \circ \iota = \varphi$ (cf. Exercise 3.73), and so the whole Diagram 4.2 is commutative. The morphism $\hat{\varphi}$ is the one we searched for, thus concluding the proof. ∎

With "abstract nonsense" arguments as those used to show the uniqueness of the projective limit of a projective system, one easily sees that the universal property completely determines $\widehat{A^*}$, up to isomorphism of topological monoids.

Note that, according to Proposition 3.11.1, if A is finite, then $\widehat{A^*}$ is metrizable. From hereon, with some rare signalized exceptions, we shall always assume that the alphabet A is finite.

Words and Pseudowords

The elements of the free profinite monoid $\widehat{A^*}$ are said to be the *pseudowords* over A. One reason for this terminology is that the words over A may be seen as elements of $\widehat{A^*}$, according to the following proposition.

Proposition 4.4.3 *The unique morphism $\bar{\iota} : A^* \to \widehat{A^*}$ extending the natural mapping $\iota : A \to \widehat{A^*}$ is injective.*

Proof Take $u, v \in A^*$ such that $\bar{\iota}(u) = \bar{\iota}(v)$. As seen in Example 4.2.6, there is a morphism φ from A^* onto a finite monoid M such that $\{u\} = \varphi^{-1}\varphi(u)$. In accordance to Proposition 4.4.2, there is a continuous morphism $\widehat{A^*} \to M$ such that $\hat{\varphi} \circ \bar{\iota} = \varphi$. Therefore, $\varphi(u) = \hat{\varphi} \circ \bar{\iota}(u) = \hat{\varphi} \circ \bar{\iota}(v) = \varphi(v)$, and since $\{u\} = \varphi^{-1}(\varphi(u))$, we deduce that $u = v$. ∎

From hereon, we identify each element u of A^* with $\bar{\iota}(u)$, according to Proposition 4.4.3. By this identification, the free monoid A^* becomes a dense submonoid of the free profinite monoid $\widehat{A^*}$, as seen in Proposition 4.4.1. Under this perspective, in which the inclusion $A^* \subseteq \widehat{A^*}$ makes sense, the universal property established in Proposition 4.4.3 says that every mapping $\varphi : A \to M$ into a finite monoid M (or, by other words, every morphism $\varphi : A^* \to M$ into a finite monoid M) has a unique extension to a continuous morphism $\hat{\varphi} : \widehat{A^*} \to M$.

A frequent method of extracting information about pseudowords consists in looking at some suitable continuous morphism from $\widehat{A^*}$ into a profinite monoid, quite often finite.

Example 4.4.4 Let $A = \{a, b\}$. Consider the unique continuous morphism $\varphi : A \to \mathbb{Z}/3\mathbb{Z}$ mapping each letter of A to 1 mod 3. As $\widehat{A^*}$ is a profinite monoid, the sequence of words $(ab)^{n!}$ converges in $\widehat{A^*}$ to a pseudoword, the ω-power $(ab)^\omega$. We use φ to show that the pseudowords $(ab)^\omega$, $(ab)^{\omega+1}$ and $(ab)^\omega b(ab)^\omega$ are distinct. As $(ab)^\omega$ is idempotent, one has $\varphi((ab)^\omega) = 0$ mod 3, thus $\varphi((ab)^{\omega+1}) = \varphi((ab)^\omega) + \varphi(a) + \varphi(b) = 2$ mod 3 and $\varphi((ab)^\omega \cdot b \cdot (ab)^\omega) = \varphi((ab)^\omega) + \varphi(b) + \varphi((ab)^\omega) = 1$ mod 3, and we are done.

In the case where the alphabet has just one letter a, the free profinite monoid is isomorphic with the additive monoid $\widehat{\mathbb{N}}$ of profinite natural integers of Chap. 2. Indeed any finite monoid with one generator is isomorphic to some monoid $N_{i,p}$ and, as we have seen, $\widehat{\mathbb{N}} = \varprojlim N_{i,p}$. The natural isomorphism from $\widehat{\mathbb{N}}$ to $\widehat{a^*}$ is the map $n \mapsto a^n$. Note that this is consistent with the notation a^ω for the unique idempotent of free profinite monoid $\widehat{a^*}$ distinct of the identity. More generally, in an arbitrary profinite monoid M we have for each $s \in M$ a continuous morphism $v \mapsto s^v$ from $\widehat{\mathbb{N}}$ onto M: it is the unique continuous morphism from the free profinite monoid $\widehat{\mathbb{N}}$ onto M that maps 1 to s.

The *length* $|x|$ of a pseudoword $x \in \widehat{A^*}$ is a profinite natural integer, defined as follows: the map $x \in \widehat{A^*} \mapsto |x| \in \widehat{\mathbb{N}}$ is the unique continuous morphism from $\widehat{A^*}$ into $\widehat{\mathbb{N}}$ such that $|a| = 1$ for every $a \in A$. This definition of length extends in a consistent manner the definition of length of a word.

Example 4.4.5 The length of the pseudoword $b^\omega a b^2 a^\omega$ is the profinite natural integer $\omega + 3 + \omega = \omega + 3$. (Recall that $\omega = \lim n!$ is an idempotent of $\widehat{\mathbb{N}}$.)

We say that the elements of $\widehat{\mathbb{N}} \setminus \mathbb{N}$ are *infinite* profinite natural integers, while those of $\mathbb{N}$ are *finite*. This makes sense because the elements of $\mathbb{N}$ are isolated points of $\widehat{\mathbb{N}}$: indeed, if $k \in \mathbb{N}$, then $\{k\}$ is clopen because it is the inverse image of the canonical projection on $N_{i,p}$, whenever $i > k$. In the next result we see how this property extends to $\widehat{A^*}$. The proof gives an example of how to gain insight on $\widehat{A^*}$ by looking at an appropriate profinite morphic image of $\widehat{A^*}$ (here, $\widehat{\mathbb{N}}$ is the image under consideration).

Proposition 4.4.6 *The elements of A^* are isolated points of $\widehat{A^*}$, and they are the unique elements of $\widehat{A^*}$ with finite length.*

Proof Suppose that $|u| \in \mathbb{N}$. Consider any sequence (u_n) of elements of A^* such that $u_n \to u$, which we know that exists because A^* is dense in $\widehat{A^*}$. Then we have $|u_n| \to |u|$. Since the elements of $\mathbb{N}$ are isolated points in $\widehat{\mathbb{N}}$, by taking subsequences we may as well suppose that $|u_n| = |u|$ for all sufficiently large n. Hence, we have $u_n \to u$ in the subspace X of $\widehat{A^*}$ consisting of the words in A^* with length $|u|$. The space X is finite, thus compact and discrete. It follows that $u_n = u$ for all sufficiently large n. This shows that $u \in A^*$. Since A^* is dense in $\widehat{A^*}$, this also shows that u is an isolated point of $\widehat{A^*}$. ∎

Since the elements of $\widehat{A^*} \setminus A^*$ are precisely those with infinite length, we may say that they are the *infinite pseudowords*, while the elements of A^* are the finite pseudowords. Recall that $\widehat{\mathbb{N}} \setminus \mathbb{N}$ is an ideal of $\widehat{\mathbb{N}}$: it consists of those elements of $\widehat{\mathbb{N}}$ that, for every i, p, project in an element of the minimum ideal of $N_{i,p}$. Therefore, for every A, the set $\widehat{A^*} \setminus A^*$ is an ideal. We first saw in Chap. 2 that $\widehat{\mathbb{N}} \setminus \mathbb{N}$ is a group, isomorphic to $\widehat{\mathbb{Z}}$. In contrast, if $\mathrm{Card}(A) > 1$ then the ideal $\widehat{A^*} \setminus A^*$ is not even a $\mathcal{J}$-class (Exercise 4.18).

Similarly to what we have done concerning the length of a pseudoword, we may define in the same spirit other properties of pseudowords that extend to them some of the intuition we have for words, as seen in the next example.

Example 4.4.7 For each positive integer n, we may consider the unique continuous morphism $\ell_n \colon \widehat{A^*} \to \mathbb{Z}/n\mathbb{Z}$ such that $\ell_n(a) = 1 \bmod n$ for each $a \in A$. We may then say that a pseudoword u of $\widehat{A^*}$ has even length when $\ell_2(w) = 0$, and that it has odd length when $\ell_2(w) = 1$.

To develop some more intuition on what an infinite pseudoword w is, and on what it is not, let us go back to the pseudowords considered in Example 4.4.4.

Example 4.4.8 Since $(ab)^{n!} \to (ab)^\omega$, one may say that $(ab)^\omega$ is an infinite pseudoword that "begins" with the right infinite word $ababab\cdots$, as every finite initial segment of $ababab\cdots$ is a prefix of $(ab)^{n!}$ for every sufficiently large n. Likewise, $(ab)^\omega$ is seen to "end" with the left infinite word $\cdots ababab$.

This reasoning remains valid if we work with $(ab)^{(n+1)!}$ instead of $(ab)^{n!}$, so that we may say that the infinite pseudoword $(ab)^{\omega+1} = \lim(ab)^{n!+1}$ also "begins" with the right infinite word $ababab\cdots$ and "ends" with the left infinite word $\cdots ababab$. And the same happens with the pseudoword $(ab)^\omega b(ba)^\omega$.

Therefore, the three distinct pseudowords $(ab)^\omega$, $(ab)^{\omega+1}$ and $(ab)^\omega b(ba)^\omega$ "begin" and "end" with the same infinite (resp. right and left) infinite word.

When asked what an infinite pseudoword looks like, a tentative informal first answer may be: *an infinite pseudoword starts with a right infinite word, ends with a left infinite word, and has something in the "middle"*... The tricky part of this short explanation is that it is left unanswered what the "middle" is. But the reference to the "middle" emphasizes that a right infinite word and a left infinite word are not enough to characterize an infinite pseudoword.

The Topological Closure of a Recognizable Language

The following statement expresses a fundamental property of free profinite monoids.

Theorem 4.4.9 *The following conditions are equivalent for a language X over the alphabet A.*

(i) *X is recognizable.*
(ii) *the closure $\overline{X}$ of X in $\widehat{A^*}$ is open.*
(iii) *$X = K \cap A^*$ for some clopen set $K \subseteq \widehat{A^*}$.*

Note that the clopen set K in condition (iii) can be chosen as $K = \overline{X}$ (see Remark 4.4.11 below). As part of the full proof of the theorem, we first show the next lemma.

Lemma 4.4.10 *Let $\varphi : A^* \to M$ be a morphism into a finite monoid, and let $\hat{\varphi}$ be the unique continuous morphism $\widehat{A^*} \to M$ extending φ. For every $P \subseteq M$, the clopen set $\hat{\varphi}^{-1}(P)$ is the topological closure of the language $\varphi^{-1}(P) \subseteq A^*$.*

Proof Since φ is the restriction of $\hat{\varphi}$ to A^*, we have $\varphi^{-1}(P) = \hat{\varphi}^{-1}(P) \cap A^*$. Because $\hat{\varphi}$ is continuous and M has the discrete topology, $\hat{\varphi}^{-1}(P)$ is indeed clopen. Therefore, $\overline{\varphi^{-1}(P)} \subseteq \hat{\varphi}^{-1}(P)$ holds. On the other hand, as A^* is dense and $\hat{\varphi}^{-1}(P)$ is open, $\hat{\varphi}^{-1}(P)$ is contained in the topological closure of $\hat{\varphi}^{-1}(P) \cap A^* = \varphi^{-1}(P)$. ∎

Proof of Theorem 4.4.9 Lemma 4.4.10 establishes in particular that the topological closure of every recognizable language is clopen, whence (i) $\Rightarrow$ (ii) holds.

Because the elements of A^* are isolated in $\widehat{A^*}$ (Proposition 4.4.6), we have $X = \overline{X} \cap A^*$, for every language $X \subseteq A^*$, and so (ii) $\Rightarrow$ (iii) holds.

Finally, assume that (iii) holds. By the monoid version of Proposition 3.9.8, there exists a continuous morphism $\psi : \widehat{A^*} \to M$ into a finite monoid M which recognizes K. Let φ be the restriction of ψ to A^*. Then we have the chain of equalities $X = A^* \cap K = A^* \cap \psi^{-1}\psi(K) = \varphi^{-1}\psi(K)$ and so X is recognizable.

Remark 4.4.11 One has $\overline{K \cap A^*} = K$ for every clopen set of $\widehat{A^*}$, since A^* is dense in $\widehat{A^*}$. Thus, in condition (iii), one may choose $K = \overline{X}$. Moreover, by Theorem 4.4.9, a subset of $\widehat{A^*}$ is clopen in $\widehat{A^*}$ if and only if it is the topological closure of a recognizable language.

Corollary 4.4.12 *The closures of recognizable languages form a basis of clopen sets for the topology of $\widehat{A^*}$.*

Proof Since $\widehat{A^*}$ is profinite, it is zero-dimensional by Theorem 3.9.3 and thus the clopen sets form a basis of its topology. The statement follows from Remark 4.4.11.   ∎

Corollary 4.4.13 *The mapping $K \mapsto K \cap A^*$ is an isomorphism of the Boolean algebra of clopen subsets of $\widehat{A^*}$ with the Boolean algebra of recognizable subsets of A^*.*

Proof The mapping in question is order preserving and it is a bijection by Theorem 4.4.9. Hence, it is an isomorphism of Boolean algebras.   ∎

The Boolean algebra of clopen subsets of a compact zero-dimensional space is known as its *Stone dual*. Conversely, from a Boolean algebra B one may construct a compact zero-dimensional space of which it is the dual, up to isomorphism, namely as the space of ultrafilters of B. The two correspondences are mutually inverse up to natural isomorphism. Thus, Corollary 4.4.13 may be read as saying that the underlying topological space of $\widehat{A^*}$ is the Stone dual of the Boolean algebra of recognizable languages over the alphabet A.

Example 4.4.14 Recall that the singletons of A^* are recognizable. This agrees with the fact they are clopen sets of $\widehat{A^*}$. The latter fact may be justified using only Lemma 4.4.10, thus giving a new proof that the induced topology of A^* is discrete, independent from the proof made in Proposition 4.4.6.

The next proposition is based on the fact that the sets of the form $\widehat{A^*}w\widehat{A^*}$, with $w \in A^*$, are clopen. For each $v \in \widehat{A^*}$ and positive integer N, we denote by $c_N(v)$ the set of factors of length N of v.

Proposition 4.4.15 *Let N be a positive integer. If the sequence of pseudowords $(v_k)_k$ converges to v, then $c_N(v_k) = c_N(v)$ for all sufficiently large k.*

Proof Let $w \in A^*$. One has $\widehat{A^*}w\widehat{A^*} = \overline{A^*} \cdot \{w\} \cdot \overline{A^*} = \overline{A^*wA^*}$. Therefore, since A^*wA^* is recognizable, its closure $\widehat{A^*wA^*}$ is a clopen set, by Theorem 4.4.9.

Suppose also that w is a word of length N such that $v \in \widehat{A^*}w\widehat{A^*}$. Since $\widehat{A^*}w\widehat{A^*}$ is open, from $(v_k) \to v$ we get that $v_k \in \widehat{A^*}w\widehat{A^*}$ for all sufficiently large k. Because there is only a finite number of words of length N, we conclude that $c_N(v) \subseteq c_N(v_k)$ for all sufficiently large k.

Suppose that we have a strict inclusion $c_N(v) \subsetneq c_N(v_k)$ for infinitely many k. Again because there are only finitely many words of length N, there is some word w with length N such that $w \in c_N(v_k) \setminus c_N(v)$ for infinitely many k. Hence, there is a subsequence $(v_{k_\ell})_\ell$ such that $v_{k_\ell} \in \widehat{A^*}w\widehat{A^*}$ for all ℓ. As $\widehat{A^*}w\widehat{A^*}$ is closed and $v_{k_\ell} \to v$, we get $v \in \widehat{A^*}w\widehat{A^*}$, contradicting the choice of w. To avoid the contradiction, we must have $c_N(v) = c_N(v_k)$ for all sufficiently large k.   ∎

In other words, in Proposition 4.4.15 it is being stated that the correspondence $w \in \widehat{A^*} \mapsto c_N(w) \in \mathcal{P}(A^*)$ is a continuous function, for the discrete topology of the set $\mathcal{P}(A^*)$ of subsets of A^*, where $\mathcal{P}(X)$ denotes the set of subsets of a set X.

We now focus on the case $N = 1$. The notation $c(u)$ is used for $c_1(u)$, the set of letters appearing as factors of u, called the *content* of u. Clearly, one has $c(uv) = c(u) \cup c(v)$ for all $u, v \in A^*$. Since, according to Proposition 4.4.15, the mapping $c : \widehat{A^*} \to \mathcal{P}(A)$ is continuous with respect to the discrete topology of $\mathcal{P}(A)$, and A^* is dense in $\widehat{A^*}$, we conclude that the equality $c(uv) = c(u) \cup c(v)$ holds for all $u, v \in \widehat{A^*}$. That is to say, the content of pseudowords over A is a continuous homomorphism from $\widehat{A^*}$ onto $\mathcal{P}(A)$, where $\mathcal{P}(A)$ is endowed with the finite monoid structure in which the union of sets is the operation.

With arguments similar to those appearing in the proof of Proposition 4.4.15, namely using the fact that wA^* is rational for each $w \in A^*$, one may show that every infinite pseudoword u over A has a unique prefix of length N, continuously determined by u (see Exercise 4.19 for a formal statement). A dual property concerning suffixes also holds.

A Topological Property of the Multiplication

The property expressed in the following proposition is a consequence of the fact that the class of recognizable languages is closed under the operation of taking the product of two languages. A map $\varphi : X \to Y$ from a topological space X to a topological space Y is *open* if $\varphi(U)$ is open for every open set $U \subseteq X$.

Proposition 4.4.16 *The multiplication $(u, v) \mapsto uv$ is an open mapping from $\widehat{A^*} \times \widehat{A^*}$ to $\widehat{A^*}$.*

Proof Since, by Corollary 4.4.12, the closures of recognizable sets form a basis of the topology of $\widehat{A^*}$, it is enough to prove that for every pair K, L of recognizable sets, the set $\overline{K} \cdot \overline{L}$ is open. But since $\overline{K} \cdot \overline{L} = \overline{KL}$ for any sets $K, L \subseteq A^*$ and since KL is recognizable by Kleene's Theorem, the result follows.   ∎

From Proposition 4.4.16, we extract the following convenient tool for dealing with products of pseudowords.

Proposition 4.4.17 *For every* $u, v \in \widehat{A^*}$ *and every sequence* (w_n) *converging to* uv, *there are sequences* $(u_n), (v_n)$ *such that* $\lim u_n = u$, $\lim v_n = v$ *and* $w_n = u_n v_n$ *for every* n.

Proof We may suppose that the sequence (w_n) is indexed by the set of natural numbers. We recursively define a strictly increasing sequence $(n_k)_{k \in \mathbb{N}}$ of natural numbers as follows. Set $n_0 = 0$. By Proposition 4.4.16, for each $k \geq 1$ the product $B(u, 1/k) B(v, 1/k)$ of the open balls $B(u, 1/k)$ and $B(v, 1/k)$ is an open neighborhood of uv. Thus there is an integer n_k greater than n_{k-1} for which $n \geq n_k$ implies $w_n \in B(u, 1/k) B(v, 1/k)$. Therefore, for each $k \geq 1$, if the integer n is such that $n_k \leq n < n_{k+1}$, we may choose $u_n \in B(u, 1/k)$, $v_n \in B(v, 1/k)$ such that $w_n = u_n v_n$. If $0 \leq n < n_1$, set $u_n = 1$ and $v_n = w_n$. We have built sequences $(u_n), (v_n)$ satisfying $w_n = u_n v_n$ for all $n \geq 0$, and such that if k is any positive integer, we have $u_n \in B(u, 1/k)$ and $v_n \in B(u, 1/k)$ whenever $n \geq n_k$, thus implying $(u_n) \to u$ and $(v_n) \to v$. ∎

A semigroup S is *equidivisible* if, for every $u, v, x, y \in S$, the equality $uv = xy$ implies the existence of $t \in S^1$ such that $ut = x$ and $v = ty$, or such that $xt = u$ and $y = tv$; in other words, any two factorizations of the same element must have a common refinement.

All free monoids are equidivisible. Indeed, assume that $uv = xy$. If $|u| \leq |x|$, then $x = ut$ and $v = ty$. In the other case, $xt = u$ and $y = tv$. Since free monoids embed in free profinite monoids as complements of ideals, this is a particular case of the following statement.

Proposition 4.4.18 *All free profinite monoids are equidivisible.*

Proof Assume that the pseudowords u, v, x, y of $\widehat{A^*}$ are such that the equality $uv = xy$ holds. By Proposition 4.4.17, there are sequences u_n, v_n, x_n, y_n of elements of A^* converging respectively to u, v, x, y and such that $u_n v_n = x_n y_n$. Taking subsequences, by symmetry we may assume that $|u_n| \leq |x_n|$ for all n, the other case being similar. Then we have $x_n = u_n t_n$ and $v_n = t_n y_n$ for some word t_n. Taking again a subsequence, we can assume that $\lim t_n = t$ and obtain $x = ut$ and $v = ty$. ∎

This property enables us to extend to pseudowords some of the intuition we have about words.

Example 4.4.19 If w is a factor of uv, with u, v, w pseudowords over A, then it follows from $\widehat{A^*}$ being equidivisible that either w is a factor of u, or of v, or $u = xp$ and $v = qy$ with $w = pq$, for some pseudowords x, y, p, q.

Free Profinite Monoids as Completions

Recall from Chap. 3 that the *profinite metric* on a profinite monoid M is defined by

$$
d(u, v) = \begin{cases} 1/r(u, v) & \text{if } u \neq v \\ 0 & \text{otherwise} \end{cases}
$$

where $r(u, v)$ is the minimum cardinality of a monoid N for which there is a continuous morphism $\varphi : M \to N$ such that $\varphi(u) \neq \varphi(v)$.

We have shown that for a finitely generated profinite monoid M, the topology is induced by the profinite metric (Proposition 3.11.1). This leads to an alternative definition of the free profinite monoid.

If M is a topological monoid with topology defined by a distance d and such that the multiplication is uniformly continuous, the completion of M as a metric space has a well defined monoid structure called the *completion* of M.

Theorem 4.4.20 *For a finite alphabet A, the completion of A^* for the profinite metric is the free profinite monoid $\widehat{A^*}$.*

This is indeed a simple consequence of the fact that A^* is dense in $\widehat{A^*}$ which is compact and therefore complete.

Free Profinite Semigroups

The *free profinite semigroup* generated by a set A, denoted by $\widehat{A^+}$, is defined in the semigroup realm in an entirely similar manner as the free profinite monoid $\widehat{A^*}$ is in the monoid realm. Moreover, $\widehat{A^+}$ is, as a compact semigroup, isomorphic with the closed subsemigroup $\widehat{A^*} \setminus \{\varepsilon\}$ of $\widehat{A^*}$, where ε is, recall, the empty word, whence an isolated point. The universal property of $\widehat{A^+}$ consists in every map $\varphi : A \to S$, with S a profinite semigroup, admitting a unique extension to a continuous morphism $\hat{\varphi} : \widehat{A^+} \to S$. Not surprisingly, $\widehat{A^*}$ and $\widehat{A^+}$ have many other similarities. For example, $\widehat{A^+}$ inherits from $\widehat{A^*}$ the property of being equidivisible, and the crucial Theorem 4.4.9 has an analog concerning the closure of $+$-languages in $\widehat{A^+}$. The natural metric and Theorem 4.4.20 also have straightforward analogs in the setting of profinite semigroups.

4.5 Pseudowords as Operations

One often sees each word over the alphabet $A = \{x, y\}$ as a binary operation on a semigroup S, via substitution of letters by elements of S: for example, the word xy defines the semigroup operation $(s, t) \in S^2 \mapsto st \in S$, the word $y^2 x^3 y$ defines the operation $(s, t) \in S^2 \mapsto t^2 s^3 t \in S$, and the word y, viewed as an element of A^*, defines the projection $(s, t) \in S^2 \mapsto t \in S$. In this section we import this simple idea to the framework of pseudowords, seeing them as operations on profinite semigroups. Because we wish to this for all profinite semigroups, not just for monoids, we work with the free profinite semigroup, instead of the free profinite monoid.

For each positive integer m, take the alphabet $X_m = \{x_1, \ldots, x_m\}$, with m letters. Consider a pseudoword ρ over X_m. Let S be a profinite semigroup, and let α be an element $(s_1, \ldots, s_m)$ of S^m. By the universal property of $\widehat{X_m^+}$, there is a unique continuous morphism $\hat{\alpha} : \widehat{X_m^+} \to S$ such that $\hat{\alpha}(x_i) = s_i$ for all $i \in \{1, \ldots, m\}$. We define the m-ary operation $\rho_S : S^m \to S$ associating to each $(s_1, \ldots, s_m)$ in S^m the element $\hat{\alpha}(\rho)$ of S. Accordingly, we have the formula $\rho_S(s_1, \ldots, s_m) = \hat{\alpha}(\rho)$. For the sake of simplicity, we may denote $\rho_S(s_1, \ldots, s_m)$ simply by $\rho(s_1, \ldots, s_m)$.

Example 4.5.1 For the pseudoword $\rho = x_1^\omega$ over the alphabet $\{x_1\}$, the unary operation $\rho_S : S \to S$ thus defined is the ω-power operation $s \in S \mapsto s^\omega \in S$.

We may work with any m-letter alphabet, not only X_m, but note that the definition of ρ_S depends on the chosen ordering of the letters of the alphabet.

Example 4.5.2 Consider the pseudoword $\varrho = x^{\omega-1}y^{\omega-1}xy$. If G is a profinite group, then ϱ_G is the commutator operation $(g, h) \mapsto [g, h] = g^{-1}h^{-1}gh$, where we are choosing the alphabetic ordering for $\{x, y\}$.

The next example is useful to have in mind when reading the literature on pseudowords.

Example 4.5.3 If ρ is a pseudoword over $X_m = \{x_1, \ldots, x_m\}$, then we have $\rho(x_1, \ldots, x_m) = \rho$. Indeed, if $\alpha = (x_1, \ldots, x_m)$, then $\hat{\alpha}$ is the identity on $\widehat{X_m^+}$.

The following property, which will be used in the final section of Chap. 6, is one of our motivations for considering here pseudowords as operations.

Proposition 4.5.4 *Let S be a profinite semigroup. Consider a finite nonempty subset $Y = \{s_1, \ldots, s_m\}$ of S. The closed subsemigroup of S generated by Y is the set*

$$\{\rho_S(s_1, \ldots, s_m) \mid \rho \in \widehat{X_m^+}\}. \tag{4.3}$$

Proof Let $\alpha = (s_1, \ldots, s_m)$. By the definition of its elements, the set (4.3) is $\hat{\alpha}(\widehat{X_m^+})$. Finally, since $\hat{\alpha} : \widehat{X_m^+} \to S$ is a continuous morphism of compact semigroups, we have $\hat{\alpha}(\widehat{X_m^+}) = \hat{\alpha}(\overline{X_m^+}) = \overline{\hat{\alpha}(X_m)^+} = \overline{Y^+}$. ∎

In order to make an effective use of this perspective of pseudowords as operations, one needs to collect some more properties. First, we mention that, for each $\rho \in \widehat{X_m^+}$, the assignment $S \mapsto \rho_S$ commutes with continuous morphisms, that is, if $\psi : S \to T$ is a continuous morphism of profinite semigroups, then we have

$$\psi(\rho_S(s_1, \ldots, s_m)) = \rho_T(\psi(s_1), \ldots, \psi(s_m))$$

for every $(s_1, \ldots, s_m) \in S^m$ (Exercise 4.23). Also, for every profinite semigroup, the operation $\rho_S : S^m \to S$ is continuous: this is an immediate consequence of a stronger result, encapsulated in the next proposition.

Proposition 4.5.5 *For each positive integer m and profinite semigroup S, the evaluation map $\epsilon : \widehat{X_m^+} \times S^m \to S$, given by*

$$\epsilon(\rho, (s_1, \ldots, s_m)) = \rho(s_1, \ldots, s_m)$$

whenever $\rho \in \widehat{X_m^+}$ and $s_1, \ldots s_m \in S$, is continuous.

Proof Suppose that we have, for some directed set K, a net $(\rho_k)_{k \in K}$ converging to ρ in $\widehat{X_m^+}$, and a net $(s_{k,1}, \ldots, s_{k,m})_{k \in K}$ converging to $(s_1, \ldots, s_m)$ in S^m. We want to show that $\rho_k(s_{k,1}, \ldots, s_{k,m}) \rightarrow \rho(s_1, \ldots, s_m)$. It suffices to show that, given an arbitrary continuous morphism ψ from S into a finite semigroup T, one has $\psi(\rho_k(s_{k,1}, \ldots, s_{k,m})) \rightarrow \psi(\rho(s_1, \ldots, s_m))$ (cf. Exercise 3.77). That is, since the operations ρ_k and ρ commute with continuous morphisms, and the topology of T is the discrete one, we want to show that, for such ψ, the equality

$$\rho_k(\psi(s_{k,1}), \ldots, \psi(s_{k,m})) = \rho(\psi(s_1), \ldots, \psi(s_m))$$

holds for all $k \geqslant p$, for some $p \in K$. For each $\ell \in \{1, \ldots, m\}$, since $s_{k,\ell} \rightarrow s_\ell$, there is $k_\ell \geqslant 1$ such that $\psi(s_{k,\ell}) = \psi(s_\ell)$ for all $k \geqslant k_\ell$. Let p be a common upper bound of $k_1, \ldots, k_m$. Because when $k \geqslant p$ we have the equality $\rho_k(\psi(s_{k,1}), \ldots, \psi(s_{k,m})) = \rho_k(\psi(s_1), \ldots, \psi(s_m))$, we are now reduced to show that the equality

$$\rho_k(\psi(s_1), \ldots, \psi(s_m)) = \rho(\psi(s_1), \ldots \psi(s_m)) \tag{4.4}$$

holds whenever $k \geqslant q$, for some $q \geqslant p$. Let $\alpha = (\psi(s_1), \ldots, \psi(s_m))$. Then the equality (4.4) may be rewritten as $\hat{\alpha}(\rho_k) = \hat{\alpha}(\rho)$, and this equality indeed holds for all $k \geqslant q$, for some $q \geqslant p$, since the morphism $\hat{\alpha} : \widehat{X_m^+} \rightarrow T$ is continuous. ∎

Occasionally, it may be convenient to also interpret the empty word ε as an m-ary operation on profinite monoids, by letting $\varepsilon(t_1, \ldots, t_m)$ be the neutral element of T, for each profinite monoid and $(t_1, \ldots, t_m) \in T^m$.

4.6 Free Profinite Groups

We denote by $FG(A)$ the free group generated by an alphabet A. In a similar way as the free profinite monoid $\widehat{A^*}$ is defined, the *free profinite group* generated by A, denoted $\widehat{FG(A)}$, is the projective limit of the projective system formed by the isomorphism classes of A-generated finite groups. Note in particular that, under our general assumption that A is a finite set, the profinite group $\widehat{FG(A)}$ is metrizable.

It is a familiar fact that the free monoid A^* on A is a submonoid of the free group $FG(A)$ on A: an embedding is obtained by mapping each monoid free generator to the corresponding group free generator. In the profinite world, a different phenomenon occurs: the canonical morphism from $\widehat{A^*}$ to $\widehat{FG(A)}$ is not an embedding but instead it is surjective. The free monoid on A is 'smaller' than the free group on A since one has to add the inverses that do not exist in A^*. On the contrary, the free profinite monoid is 'larger' than the free profinite group since there are more morphisms from a free monoid into finite monoids than into finite groups.

The topology on the free group $FG(A)$ induced by the topology of $\widehat{FG(A)}$ is not discrete. Indeed, for any x in $FG(A)$, the sequence $x^{n!}$ tends to 1. On the other hand, since $\widehat{FG(A)}$ is an A-generated profinite group, A^* is dense in $\widehat{FG(A)}$ and there is an onto morphism from $\widehat{A^*}$ onto $\widehat{FG(A)}$.

The topology induced on $FG(A)$ by the topology of $\widehat{FG(A)}$ is also called the *profinite topology* or the *Hall topology*.

Proposition 4.6.1 *The profinite topology of $FG(A)$ is the smallest topology of $FG(A)$ for which every morphism from $FG(A)$ into a finite group is continuous.*

Proof Let $\varphi : FG(A) \to G$ be a morphism into a finite group G. By the universal property of $\widehat{FG(A)}$, the morphism φ admits a unique continuous extension $\hat{\varphi}$ to $\widehat{FG(A)}$. Since $\varphi^{-1}(P) = \hat{\varphi}^{-1}(P) \cap FG(A)$ whenever $P \subseteq G$, every open set in a topology for which φ is continuous is open in the profinite topology of $FG(A)$.

Conversely, as $\widehat{FG(A)}$ is the projective limit of the finite A-generated groups, the topology of $\widehat{FG(A)}$ is generated by the open sets of the form $\hat{\varphi}^{-1}(P)$, with φ and P as in the preceding paragraph. The equality $\varphi^{-1}(P) = \hat{\varphi}^{-1}(P) \cap FG(A)$ then shows that we have a basis of the profinite topology which is contained in the smallest topology of $FG(A)$ for which every morphism from $FG(A)$ into a finite group is continuous.   ∎

Note that, since A^* is embedded in $FG(A)$, we actually have two topologies on A^* respectively induced by the topologies of $\widehat{A^*}$ and $\widehat{FG(A)}$. To distinguish them, the first one is called the *pro-M topology* and the second one the *pro-G topology*. The first one is strictly stronger (meaning larger) than the second one. In fact, the pro-M topology of A^* is discrete, as we saw in Proposition 4.4.6. The terminology *pro-G topology* is also used for the topology of $FG(A)$ induced by the topology of $\widehat{FG(A)}$.

The image of the free profinite group on one generator a by the map $a^n \mapsto n$ into the group $\widehat{\mathbb{Z}}$ of profinite integers (see Chap. 2) is the whole group $\widehat{\mathbb{Z}}$. The *length* $|x|$ of an element x of $\widehat{FG(A)}$ is a profinite integer, where the map $x \in \widehat{FG(A)} \mapsto |x| \in \widehat{\mathbb{Z}}$ is defined as the unique continuous morphism from $\widehat{FG(A)}$ into $\widehat{\mathbb{Z}}$ such that $|a| = 1$ for every $a \in A$. In particular $|a^{-1}| = -1$.

The notion of recognizable $*$-language has the following generalization: a subset X of a monoid M is *recognizable* if there is some monoid morphism $\varphi : M \to N$ into a finite monoid N which recognizes X, that is, such that $X = \varphi^{-1}(\varphi(X))$. Note that when M is a group, then we may assume that N is also a group since the image by a monoid morphism of a group is a group.

The following result is a topological formulation of elementary group theory.

Lemma 4.6.2 *The following conditions are equivalent for a subgroup H of a free group $FG(A)$:*

(i) H is open in the profinite topology of $FG(A)$;
(ii) H is clopen in the profinite topology of $FG(A)$;

(iii) H has finite index in $FG(A)$;
(iv) H is a recognizable subset of $FG(A)$.

Proof (i) $\Rightarrow$ (iv) Suppose that H is open. Then, H contains a basic neighborhood of 1 which, by Proposition 4.6.1, in the profinite topology of $FG(A)$ is the kernel K of a morphism φ from $FG(A)$ onto a finite group. Since H is a union of cosets of K, it follows that H is recognized by φ.

(iv) $\Rightarrow$ (iii) Suppose that H is recognized by a morphism $\varphi : FG(A) \rightarrow G$ to a finite group G. Then, the subgroup $\varphi^{-1}(1)$ is contained in H and has finite index and, therefore, so does H.

(iii) $\Rightarrow$ (ii) Assume that H has finite index in $FG(A)$ and let K be the intersection of all its conjugates, of which there are only finitely many. Hence, K also has finite index and thus it is a clopen normal subgroup, of which H is a union of cosets. We conclude that H is clopen.

(ii) $\Rightarrow$ (i) Trivial.   ∎

We deduce from Hall's Theorem (Theorem 4.3.4) the following statement.

Proposition 4.6.3 *Every finitely generated subgroup H of $FG(A)$ is closed for the profinite topology of $FG(A)$.*

Proof Let x be an arbitrary element of the complement of H. By Theorem 4.3.4, there is a subgroup K of finite index in $FG(A)$ containing H such that $x \notin K$. By Lemma 4.6.2, K is open and, therefore, so is its complement as it is a union of cosets of K. This shows that the complement of H is open, whence H is closed. ∎

The following is basically a practical reformulation of Proposition 4.6.3.

Corollary 4.6.4 *Let H be a finitely generated subgroup of $FG(A)$. If $\overline{H}$ is the topological closure of H in $\widehat{FG(A)}$, then $H = \overline{H} \cap FG(A)$.*

Proof By Proposition 4.6.3, there is a closed subset F of $\widehat{FG(A)}$ such that $H = F \cap FG(A)$. From $H \subseteq F$, we get $\overline{H} \subseteq F$. Therefore, we have $\overline{H} \cap FG(A) \subseteq F \cap FG(A) = H$. ∎

Recall the profinite metric introduced in Sect. 3.11. For a free group G, we denote r_G and d_G respectively the restrictions to $H \times H$ of the functions $r_{\widehat{FG(A)}}$ and $d_{\widehat{FG(A)}}$ considered at the beginning of Sect. 3.11. Here, we view r_G as a partial function, which is not defined on pairs with equal components. Note that, by the universal property of $\widehat{G}$, for distinct $u, v \in G$, $r_G(u, v)$ is the cardinality of the smallest group F such that there is morphism $\varphi : G \rightarrow F$ with $\varphi(u) \neq \varphi(v)$.

The following result is a consequence of Theorem 4.3.4.

Proposition 4.6.5 *Let H be a finitely generated subgroup of $G = FG(A)$. Then the metric d_H and the restriction of the metric d_G to H have the same Cauchy sequences. In particular the two metrics induce the profinite topology on H.*

Proof By Theorem 4.3.4, H is a free factor of a subgroup of finite index L of G. We prove the result in two steps:

(1) the equality $d_H(u, v) = d_L(u, v)$ holds for all $u, v \in H$;
(2) d_L and the restriction of d_G to L have the same Cauchy sequences.

(1) It suffices to observe that $r_H(u, v) = r_L(u, v)$ for distinct $u, v \in H$. Indeed, the restriction to H of a morphism $L \to F$ distinguishing u and v provides a morphism $H \to F$ with the same property, so that $r_H(u, v) \leqslant r_L(u, v)$. The reverse inequality follows from noticing that, since H is a free factor of L, a morphism $H \to F$ distinguishing u and v may be extended to a morphism $L \to F$.

(2) We first claim that, for distinct $u, v \in L$, the following inequalities hold where k is the index of L in G:

$$r_L(u, v) \leqslant r_G(u, v) \leqslant (k \cdot r_L(u, v))!. \qquad (4.5)$$

The first inequality is obtained as in the preceding paragraph. For the second one, consider a morphism $\varphi : L \to F$ into a finite group F of minimum cardinality such that $\varphi(u) \neq \varphi(v)$. Let $K = \varphi^{-1}(1)$ and let K_G be the intersection of all conjugates of K in G, that is, the largest normal subgroup of G contained in K. Since K_G is precisely the preimage of the identity under the representation of G by permutations of the right cosets of K in G by right multiplication, we get the well known fact that the index ℓ of K_G in G divides $\mathrm{Card}(G/K)!$. As K contains K_G and $uK \neq vK$, we get $uK_G \neq vK_G$. Hence, we have

$$r_G(u, v) \leqslant \ell \leqslant \mathrm{Card}(G/K)! = (k \cdot \mathrm{Card}(L/K))! = (k \cdot r_L(u, v))!$$

which establishes the claim.

From the first inequality in (4.5), we deduce that every Cauchy sequence in L with respect to d_L is also a Cauchy sequence with respect to d_G. For the converse, consider the function $f(n) = (k \cdot n)!$. Then, f is a strictly increasing function and, for ε positive, we have

$$d_G(u, v) < 1/f(\lceil 1/\varepsilon \rceil) \Rightarrow d_L(u, v) < \varepsilon,$$

so that Cauchy sequences in L with respect to d_G are also Cauchy sequences with respect to d_L. ∎

The last conclusion of Proposition 4.6.5 fails for non finitely generated subgroups. The classical example is the commutator subgroup F' of a non-cyclic finitely generated free group F, which is known to be a free group of countable infinite rank. In the topology induced on F' by the profinite topology on F, there are only countably many open subgroups while the number of open subgroups in the profinite topology of F' is uncountable.

As for the free profinite monoid (Theorem 4.4.20), we may use the profinite metric to give an alternative definition of the free profinite group. The proof is left as an exercise (Exercise 4.30).

Theorem 4.6.6 *For a finite alphabet A, the completion of $FG(A)$ for the profinite metric is the free profinite group $\widehat{FG}(A)$.*

Combining Proposition 4.6.5 and Theorem 4.6.6, we obtain the following result.

Theorem 4.6.7 *Any injective morphism $\varphi : FG(B) \to FG(A)$ between finitely generated free groups extends to an injective continuous morphism $\hat{\varphi} : \widehat{FG}(B) \to \widehat{FG}(A)$.*

Proof Let $H = \varphi(FG(B))$. Then φ is an isomorphism between $FG(B)$ and H which extends to an isomorphism between $\widehat{FG}(B)$ and $\widehat{H}$, the completion of H with respect to the profinite topology.

By Proposition 4.6.5 and Theorem 4.6.6, the profinite groups $\widehat{H}$ and the closure $\overline{H}$ of H in $\widehat{FG}(A)$ are isomorphic, which shows that $\hat{\varphi}$ is an injective continuous morphism from $\widehat{FG}(B)$ into $\widehat{FG}(A)$.   ∎

We now turn our attention to the endomorphisms of finitely generated free (profinite) groups. Concerning those that are automorphisms, we have the following proposition.

Proposition 4.6.8 *Consider a finite alphabet A. Let φ be a continuous endomorphism of $FG(A)$, and let $\hat{\varphi}$ be its unique extension to a continuous endomorphism of $\widehat{FG}(A)$. The following conditions are equivalent:*

1. *φ is bijective;*
2. *φ is surjective;*
3. *the subgroup of $FG(A)$ generated by $\varphi(A)$ is $FG(A)$;*
4. *the closed subgroup of $\widehat{FG}(A)$ generated by $\varphi(A)$ is $\widehat{FG}(A)$;*
5. *$\hat{\varphi}$ is surjective;*
6. *$\hat{\varphi}$ is bijective.*

Proof The implications $(1) \Rightarrow (2) \Rightarrow (3) \Rightarrow (4) \Rightarrow (5) \Leftarrow (6)$ are trivial; for example, $(3) \Rightarrow (4)$ is a direct consequence of $FG(A)$ being dense in $\widehat{FG}(A)$. The implication $(5) \Rightarrow (6)$ is valid by Proposition 3.12.5. Note that $(2) \Rightarrow (1)$ is the Hopf property of finitely generated free groups (see Exercise 4.13). Therefore, to complete the proof it suffices to show $(5) \Rightarrow (2)$. Let H be the subgroup of $FG(A)$ generated by $\varphi(A)$. Then φ is surjective if and only if $H = FG(A)$. Then the closed subgroup of $\widehat{FG}(A)$ generated by $\varphi(A)$ is the topological closure $\overline{H}$ of H in $\widehat{FG}(A)$. In particular, $\hat{\varphi}$ is surjective if and only if $\overline{H} = \widehat{FG}(A)$. By Corollary 4.6.4, we have $H = \overline{H} \cap FG(A)$. Hence, φ is surjective if and only if $\hat{\varphi}$ is surjective.   ∎

Example 4.6.9 Consider the alphabet $A = \{a, b\}$ and the corresponding Fibonacci morphism $\varphi : A^* \to A^*$. Since $\varphi(A) = \{ab, a\}$ generates $FG(A)$, both the endomorphism of $FG(A)$ induced by φ and its unique extension to a continuous endomorphism of $\widehat{FG(A)}$ are bijective.

For a finite alphabet A, let ψ be a continuous endomorphism of $\widehat{A^*}$. Denote by ψ_G the unique continuous endomorphism of $\widehat{FG(A)}$ such that $\psi_G(a) = p_G(\psi(a))$ for every $a \in A$, that is, such that the following diagram commutes:

$$
\begin{array}{ccc}
\widehat{A^*} & \xrightarrow{\psi} & \widehat{A^*} \\
\downarrow{\scriptstyle p_G} & & \downarrow{\scriptstyle p_G} \\
\widehat{FG(A)} & \xrightarrow{\psi_G} & \widehat{FG(A)}
\end{array}
$$

Note that this implies that $\psi_G^n \circ p_G = p_G \circ \psi^n$ for every nonnegative integer n, and so, taking limits, we get

$$
\psi_G^\nu \circ p_G = p_G \circ \psi^\nu
$$

for every profinite natural number ν. In particular, if ψ_G is bijective, then, as ψ_G^ω is the identity, we have

$$
p_G = p_G \circ \psi^\omega.
$$

Example 4.6.10 Consider again the Fibonacci morphism $\varphi : A^* \to A^*$, for $A = \{a, b\}$. Denote also by φ its unique continuous extension to an endomorphism of $\widehat{A^*}$. Then φ_G is bijective (cf. Example 4.6.9) and $p_G = p_G \circ \varphi^\omega$. This implies in particular, for example, that the projection of the infinite pseudoword $\varphi^\omega(a)$ in $\widehat{FG(A)}$ is a letter (the letter a).

4.7 Presentations of Profinite Semigroups

The classes of an open equivalence relation in a topological space form an open cover of the space (cf. Exercise 3.40), and so if the space is compact there are only finitely many such classes. Therefore, in a compact semigroup the open congruences have finite index. A congruence of a profinite semigroup is *admissible* if it is the intersection of open congruences. Since an open congruence is also closed (Exercise 3.40), an admissible congruence is closed.

Proposition 4.7.1 *A closed congruence ρ of a profinite semigroup S is admissible if and only if the quotient S/ρ is profinite.*

Proof Assume first that ρ is admissible. Let $(\rho_i)_{i \in I}$ be a family of open congruences such that $\rho = \cap_{i \in I} \rho_i$. Since ρ_i is open, it has finite index and thus each quotient S/ρ_i is finite. Therefore, the product $\prod_{i \in I} S_i/\rho_i$ is profinite. Since S/ρ

is (isomorphic to) a closed subsemigroup of $\prod_{i \in I} S_i / \rho_i$, it follows that S/ρ is profinite.

Conversely, assume that the congruence ρ is closed and that S/ρ is profinite. Because S/ρ is profinite, there is a family $(S_i)_{i \in I}$ of finite semigroups such that S/ρ is a closed subsemigroup of $\prod_{i \in I} S_i$. For each $k \in I$, consider the composition $f_k = p_k \circ q_\rho$ formed by the canonical projections $q_\rho : S \to S/\rho$ and $p_k : \prod_{i \in I} S_i \to S_k$. The kernel ρ_k of f_k is a closed congruence of S of finite index and thus it is open. Since the intersection of the congruences ρ_i is ρ, the latter is an admissible congruence.   ∎

Example 4.7.2 Consider the canonical map p_G from the free profinite monoid $\widehat{A^*}$ onto the free profinite group $\widehat{FG(A)}$. Since $\widehat{FG(A)}$ is profinite as a semigroup, the kernel of p_G is an admissible congruence. It is the intersection of the congruences of all continuous morphisms from $\widehat{A^*}$ onto finite groups.

Example 4.7.3 The kernel of the continuous morphism considered in Example 3.9.7 is an example of a closed congruence on a profinite semigroup which is not admissible, in agreement with Proposition 4.7.1.

It follows from the definition of admissible congruence that any intersection of admissible congruences is admissible. Thus, for every relation R on a profinite semigroup S there is a smallest admissible congruence of S containing R, called the *admissible congruence generated* by R.

Let A be a finite set and let R be a binary relation on the free profinite semigroup $\widehat{A^+}$. The quotient of $\widehat{A^+}$ by the admissible congruence generated by R is said to have the *profinite semigroup presentation* or S-*presentation* $\langle A \mid R \rangle_S$. The presentation is said to be *finite* if R is finite, and then the corresponding profinite semigroup is said to be *finitely presented*. The profinite monoid presentation $\langle A \mid R \rangle_M$ of a profinite monoid is the obvious transposition of what precedes, replacing $\widehat{A^+}$ by $\widehat{A^*}$. In the presentation of a profinite semigroup, it is customary to write $u = v$ instead of (u, v) for an element of R.

Example 4.7.4 The free profinite group has the profinite monoid presentation $\langle A \mid \{a^\omega = 1 \mid a \in A\} \rangle_M$.

The more familiar notion of presentation of ordinary (not profinite) semigroups is related as follows to profinite presentations. For a relation R on A^+, the quotient of A^+ by the congruence generated by R is said to have the *semigroup presentation* $\langle A \mid R \rangle$. The *profinite completion* of a semigroup S is the projective limit of all finite semigroups which are a quotient of S. The profinite semigroup $\langle A \mid R \rangle_S$ is the profinite completion of the semigroup $\langle A \mid R \rangle$ (see Exercise 4.32).

Example 4.7.5 Any finite semigroup is finitely presentable (Exercise 4.31). Since the profinite completion of a finite semigroup S is the same semigroup S viewed as a compact semigroup, we have the equality $\langle A \mid R \rangle = \langle A \mid R \rangle_S$ whenever $R \subseteq A^+$ generates a congruence of finite index in A^+.

In the case of a profinite group, the situation is simpler since every closed congruence is admissible. Indeed, by Proposition 3.10.2, the quotient of a profinite group by a closed normal subgroup is profinite and thus the corresponding closed congruence is admissible by Proposition 4.7.1. The quotient of $\widehat{FG}(A)$ by the closed congruence generated by R is said to have the *profinite group presentation* or G-*presentation* $\langle A \mid R \rangle_{\mathsf{G}}$. When R is finite, the profinite group $\langle A \mid R \rangle_{\mathsf{G}}$ is said to be *finitely presented*.

Let T be a profinite semigroup and let $\pi : \widehat{A^+} \to T$ be a continuous morphism onto T. By the universal property of $\widehat{A^+}$, for every continuous endomorphism φ of T there is a continuous endomorphism Φ of $\widehat{A^+}$, called a *lifting of φ via π* such that the diagram below is commutative.

$$
\begin{array}{ccc}
\widehat{A^+} & \xrightarrow{\;\Phi\;} & \widehat{A^+} \\
\downarrow{\scriptstyle\pi} & & \downarrow{\scriptstyle\pi} \\
T & \xrightarrow{\;\varphi\;} & T
\end{array}
\tag{4.6}
$$

The following statement gives a sufficient condition for a finitely generated profinite semigroup to have a finite presentation. We are considering $\text{End}(T)$ as a profinite monoid, as seen in Chap. 3.

Proposition 4.7.6 *Let T be a profinite semigroup and let $\pi : \widehat{A^+} \to T$ be a continuous morphism onto T. Let φ be an automorphism of T and let Φ be a lifting of φ via π. If $\text{Ker}\,\pi \subseteq \text{Ker}\,\Phi^{\omega}$, then T admits the profinite semigroup presentation $\langle A \mid R \rangle_{\mathsf{S}}$ with*

$$
R = \{\Phi^{\omega}(a) = a \mid a \in A\}.
\tag{4.7}
$$

Proof Since Diagram (4.6) commutes, we have $\pi \circ \Phi^n = \varphi^n \circ \pi$ for each integer n. It follows from the continuity of composition in the profinite monoids $\text{End}(\widehat{A^+})$ and $\text{End}(T)$ that $\pi \circ \Phi^{\omega} = \varphi^{\omega} \circ \pi$. Since φ is an automorphism of T, the idempotent φ^{ω} is the identity and thus

$$
\pi \circ \Phi^{\omega} = \pi.
\tag{4.8}
$$

Let ρ be the admissible congruence generated by R. We claim that $\rho = \text{Ker}\,\Phi^{\omega}$. Indeed, since ρ is a closed congruence containing R and $\widehat{A^+}$ is generated by A as a profinite semigroup, we have $\Phi^{\omega}(u)\rho u$ for every $u \in \widehat{A^+}$. This implies that $\text{Ker}\,\Phi^{\omega} \subseteq \rho$. Conversely, since Φ^{ω} is idempotent, we have $R \subseteq \text{Ker}\,\Phi^{\omega}$ and thus $\rho \subseteq \text{Ker}\,\Phi^{\omega}$ because $\text{Ker}\,\Phi^{\omega}$ is admissible. This proves the claim.

Next, since $\operatorname{Ker}\pi \subseteq \operatorname{Ker}\Phi^\omega$, it follows from Eq. (4.8) that $\operatorname{Ker}\pi = \operatorname{Ker}\Phi^\omega$. Therefore, the equalities

$$T = \widehat{A^+}/\operatorname{Ker}\pi = \widehat{A^+}/\operatorname{Ker}\Phi^\omega$$
$$= \widehat{A^+}/\rho = \langle A \mid R\rangle_S.$$

are valid, concluding the proof.   ∎

A profinite semigroup S is *projective* if whenever $f : S \to T$ and $g : U \to T$ are continuous morphisms of profinite semigroups with g onto, there is some continuous morphism $h : S \to U$ making Diagram (4.9) commutative. (See Exercise 4.34 for an equivalent definition.)

$$\begin{array}{ccc} & & S \\ & {}^{h}\swarrow & \downarrow {}^{f} \\ U & \xrightarrow{\ g\ } & T \end{array} \tag{4.9}$$

Example 4.7.7 The free profinite semigroup $\widehat{A^+}$ is a projective profinite semigroup.

An onto continuous morphism $\pi : S \to T$ is a *retraction* if T is a closed subsemigroup of S and the restriction of π to T is the identity. The semigroup T is a *retract* of S. Note that if Φ is a continuous endomorphism of $\widehat{A^+}$, then Φ^ω is a retraction.

Proposition 4.7.8 *The following conditions are equivalent for an A-generated profinite semigroup S.*

(i) $S = \langle A \mid R\rangle_S$ *with R as in Eq. (4.7) for some continuous endomorphism Φ of $\widehat{A^+}$.*

(ii) *S is projective.*

(iii) *S is a retract of $\widehat{A^+}$.*

Proof (i) $\Rightarrow$ (iii) Let Φ be a continuous endomorphism of $\widehat{A^+}$ and let T the image of Φ^ω. Then T is a retract of $\widehat{A^+}$ and it is enough to show, for this particular Φ, that $T = \langle A \mid R\rangle_S$. This results from the application of Proposition 4.7.6 to the lifting of the identity via Φ^ω given by Φ^ω (see Diagram (4.10), where Id denotes the identity map).

$$\begin{array}{ccc} \widehat{A^+} & \xrightarrow{\ \Phi^\omega\ } & \widehat{A^+} \\ {}^{\Phi^\omega}\downarrow & & \downarrow {}^{\Phi^\omega} \\ T & \xrightarrow{\ Id\ } & T \end{array} \tag{4.10}$$

(ii) $\Rightarrow$ (i) Let $\pi \, : \, \widehat{A^+} \, \to \, S$ be an onto continuous morphism. Since S is projective, there is a continuous morphism $\gamma \, : \, S \, \to \, \widehat{A^+}$ such that $\pi \circ \gamma$ is the identity on S. Consider Diagram (4.11), which is commutative.

$$
\begin{array}{ccc}
\widehat{A^+} & \xrightarrow{\ \gamma \circ \pi \ } & \widehat{A^+} \\
\downarrow{\scriptstyle \pi} & & \downarrow{\scriptstyle \pi} \\
S & \xrightarrow{\ Id \ } & S
\end{array}
\tag{4.11}
$$

Since $\gamma \circ \pi$ is idempotent and $\operatorname{Ker}\pi \subseteq \operatorname{Ker}(\gamma \circ \pi)$, Proposition 4.7.6 implies that $S = \langle A \mid R \rangle_S$ with $R = \{(\gamma \circ \pi)(a) = a \mid a \in A\}$.

(iii) $\Rightarrow$ (ii) Suppose that $r \, : \, \widehat{A^+} \, \to \, S$ is a retraction. Let $f \, : \, S \, \to \, T$ and $g \, : \, U \, \to \, T$ be continuous morphisms of profinite semigroups with g onto. As $\widehat{A^+}$ is projective, there is some continuous morphism $h \, : \, \widehat{A^+} \, \to \, U$ such that Diagram (4.12) is commutative.

$$
\begin{array}{ccc}
\widehat{A^+} & \xrightarrow{\ r \ } & S \\
\downarrow{\scriptstyle h} & & \downarrow{\scriptstyle f} \\
U & \xrightarrow{\ g \ } & T
\end{array}
\tag{4.12}
$$

Consider a continuous injective morphism $i \, : \, S \to \widehat{A^+}$ such that $r \circ i$ is the identity mapping of S, whose existence is guaranteed by the hypothesis that S is a retract of $\widehat{A^+}$. Then the morphism $f' = h \circ i$ is such that $g \circ f' = f$. Thus, S is projective. ∎

All results in this section are valid for the categories of profinite monoids and of profinite groups, with the obvious adaptations in the definitions and in the statements of the results.

One should note that a profinite group is projective as a profinite semigroup if and only if it is projective as a profinite group (see Exercise 4.35). In this book, we have a special interest in looking at maximal subgroups of the free profinite semigroup. In this context, one should bear in mind the following important result, although we shall not need it for the sequel.

Theorem 4.7.9 *Every closed subgroup of $\widehat{A^+}$ is a projective group.*

See the Notes Section for a reference.

4.8   Profinite Codes

We will explore the notion of a code in a free profinite monoid, with the purpose of easily getting free profinite monoids inside profinite monoids.

Let A be an alphabet. A subset X of the free semigroup A^+ is called a *code* if any bijection $\beta : B \to X$ from an alphabet B onto X extends to an injective morphism from B^* into A^*. In that case, β is said to be a *coding morphism for X*. The partial function $\beta^{-1} : A^* \to B^*$ is called the *decoding function*.

A submonoid of A^* is *free* if it is isomorphic to a free monoid B^* on some alphabet. The submonoid M generated by a code X is clearly free.

An arbitrary submonoid of A^* has a unique *minimal generating set* , namely the set X of nonempty elements of M which are not the product of two other nonempty elements of M. It is minimal in the sense that for every set $Y \subseteq A^+$ generating M, one has $X \subseteq Y$ as one may easily verify (see Exercise 4.36). Moreover, M is free if and only if its minimal generating set X is a code. The code X is also called the *basis* of M.

For example, if there are not two elements x, y of X such that x is a prefix of y, then X is a code, and it is said to be a *prefix code*. Dually, one has the so called *suffix codes*.

Let $X \subseteq A^*$ be a finite code and let $\beta : B^* \to A^*$ be a coding morphism with finite alphabets A and B, and let $X = \beta(B)$. The *prefix transducer* associated to β is the following labeled graph $\mathcal{T}$. The set of vertices of $\mathcal{T}$ is the set P of proper prefixes of words in X. The edges of $\mathcal{T}$ are labeled by pairs $(u, v) \in A^* \times B^*$ of words. Such a pair is traditionally denoted $u|v$. There is an edge $p \xrightarrow{a|-} pa$, where the dash $(-)$ represents the empty word, for each proper prefix p and letter a such that pa is a proper prefix, and an edge $p \xrightarrow{a|b} 1$ for each p and letter a with $pa = \beta(b) \in X$.

Note that for each edge $p \xrightarrow{a|v} q$ of the prefix transducer, one has

$$pa = \beta(v)q. \tag{4.13}$$

The paths in $\mathcal{T}$ are labeled by the concatenation of the labels of the edges forming the path. In this way, the transducer $\mathcal{T}$ defines a relation between A^* and B^* formed of the labels of the paths from ε to ε called the relation *realized* by $\mathcal{T}$.

Proposition 4.8.1 *For any coding morphism $\beta : B^* \to A^*$, the prefix transducer $\mathcal{T}$ associated realizes the decoding function.*

Proof Each simple path $\varepsilon \to \varepsilon$ (that is, not passing by ε in between) is labeled by construction with $(\beta(b), b)$ for some letter $b \in B$. Thus $\mathcal{T}$ realizes the associated decoding function. ∎

Example 4.8.2 Consider the code $X = \{00, 10, 100\}$. The decoder given by the construction is represented in Fig. 4.7.

We now give a profinite version of the notion of code. Let X be any finite subset of $\widehat{A^*}$. By the universal property of free profinite monoids, any bijection $\beta : B \to X$ extends uniquely to a continuous morphism of monoids $\hat{\beta} : \widehat{B^*} \to \widehat{A^*}$. A finite set $X \subseteq \widehat{A^*}$ is called a *profinite code* if $\hat{\beta}$ is injective for every bijection $\beta : B \to X$.

Theorem 4.8.3 *Every finite code $X \subseteq A^+$ is a profinite code.*

Fig. 4.7 The prefix
transducer

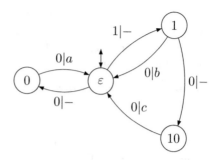

Proof Let $\beta : B^* \to A^*$ be a coding morphism for X. We have to show that for every pair $u, v \in \widehat{B^*}$ of distinct elements, we have $\hat{\beta}(u) \neq \hat{\beta}(v)$, that is, there is a continuous morphism $\hat{\alpha} : \widehat{A^*} \to M$ into a finite monoid M such that $\hat{\alpha}\hat{\beta}(u) \neq \hat{\alpha}\hat{\beta}(v)$. For this, let $\psi : \widehat{B^*} \to N$ be a continuous morphism into a finite monoid N such that $\psi(u) \neq \psi(v)$. Let P be the set of proper prefixes of X and let $\mathcal{T}$ be the prefix transducer associated to β. Let α be the morphism from A^* into the monoid M of $P \times P$-matrices with elements in $N \cup 0$ defined as follows. For $x \in A^*$ and $p, q \in P$, we have

$$\alpha(x)_{p,q} = \begin{cases} \psi(y) & \text{if there is a path } p \xrightarrow{x|y} q \text{ in } \mathcal{T} \\ 0 & \text{otherwise.} \end{cases}$$

We are viewing $N \cup \{0\}$ as a semiring with multiplicative structure given by N with a zero adjoined and additive structure given by N with any constant addition with 0 as an added identity element. Note that indeed α is a morphism and M is a monoid under usual matrix multiplication (square matrices with elements in a semiring form obviously a semiring). Moreover, M is finite and α is a morphism that extends to a continuous morphism $\hat{\alpha} : \widehat{A^*} \to M$. Since, by Proposition 4.8.1, the transducer $\mathcal{T}$ realizes the decoding function of X, we have $\alpha\beta(y)_{1,1} = \psi(y)$ for any $y \in B^*$. By continuity, we have $\hat{\alpha}\hat{\beta}(y)_{1,1} = \psi(y)$ for any $y \in \widehat{B^*}$. Then $\hat{\alpha}$ is such that $\hat{\alpha}\hat{\beta}(u) \neq \hat{\alpha}\hat{\beta}(v)$. Indeed $\hat{\alpha}\hat{\beta}(u)_{1,1} = \psi(u) \neq \psi(v) = \hat{\alpha}\hat{\beta}(v)_{1,1}$. ∎

Example 4.8.4 Let $A = \{a, b\}$ and $X = \{a, ab, bb\}$. The set X is a suffix code. Note that the right infinite word $abbb \cdots$ has two distinct factorizations in terms of elements of X:

$$abbb \cdots = a(bb)(bb) \cdots = (ab)(bb)(bb) \cdots ,$$

for which reason, accordingly to the terminology used in the theory of codes, X is said to be a code with infinite deciphering delay. Nonetheless, X is a profinite code in agreement with Theorem 4.8.3. Note that, in $\widehat{A^*}$, the pseudowords $a(bb)^\omega$ and $ab(bb)^\omega$ are distinct (the first one is a limit of words of odd length and the second one is a limit of words of even length).

Theorem 4.8.3 shows that the closure of the submonoid generated by a finite code is a free profinite monoid. This can be extended to rational codes: briefly suspending at this point our choice to only consider finite alphabets, we have the following result, which we quote without proof (see the Notes).

Theorem 4.8.5 *For any rational code X, the profinite submonoid generated by X is free with basis the closure $\overline{X}$ of X.*

Example 4.8.6 Let $A = \{a, b\}$ and $X = a^*b$. Then the profinite monoid $\widehat{X^*}$ is free with basis the uncountable set $\overline{X} = \{a^\nu b \mid \nu \in \widehat{\mathbb{N}}\}$, endowed with the topology induced from that of $\widehat{A^*}$.

4.9   Relatively Free Profinite Monoids and Semigroups

In this section we give an indication of a more general setting, where free profinite monoids (or semigroups) are a special type of the so called relatively free profinite monoids (or semigroups). This generalization relies on the notion of pseudovariety of monoids (or semigroups), strongly motivated by a connection with the notion of varieties of languages, briefly presented in this section.

The following material is not revisited again in the next chapters. However, it may be helpful to have some insight on one aspect that we look at in this section: pseudowords may be better understood through their morphic images on suitable relatively free profinite monoids or semigroups.

A *pseudovariety* of semigroups (resp. monoids) is a class of finite semigroups (resp. finite monoids) closed under taking subsemigroups, morphic images and finite direct products. We begin with some very simple examples.

Example 4.9.1 The class S of finite semigroups is a pseudovariety of semigroups. The class M of finite monoids is a pseudovariety of monoids, but not of semigroups, since a subsemigroup of a finite monoid may not be a submonoid, even if it is a monoid. The class G of finite groups is both a pseudovariety of semigroups and a pseudovariety of monoids.

Example 4.9.2 A semigroup is *aperiodic* if all its subgroups are trivial. The class of aperiodic finite semigroups is a pseudovariety of semigroups (Exercise 4.37). Similarly, the aperiodic finite monoids form a pseudovariety of monoids.

A semigroup S is *nilpotent* if $S^n = \{0\}$ for some positive integer n and zero 0.

Example 4.9.3 The finite Rees quotients of free semigroups are nilpotent. The class of nilpotent finite semigroups is a pseudovariety of semigroups (Exercise 4.38), here denoted N. The only monoid in N is the trivial semigroup.

A *variety* of +-languages is a class $\mathcal{V}$ of rational +-languages—more precisely, a correspondence $\mathcal{V} : A \mapsto \mathcal{V}A^+$ associating to each finite alphabet A a class of languages over A—which satisfies the following three conditions:

1. $\mathcal{V}A^+$ is closed under Boolean operations;
2. $\mathcal{V}A^+$ is closed under taking *quotients*, that is, if $L \in \mathcal{V}A^+$ then the languages $u^{-1}L = \{w \in A^+ \mid uv \in L\}$ and $Lv^{-1} = \{w \in A^+ \mid uv \in L\}$ also belong to $\mathcal{V}A^+$;
3. $\mathcal{V}$ is a class closed under taking inverse images of morphisms between free semigroups, that is, if $L \in \mathcal{V}A^+$ then $\varphi^{-1}(L) \in \mathcal{V}B^+$ whenever $\varphi : B^+ \to A^+$ is a morphism.

A variety of ∗-languages is defined likewise, *mutatis mutandis*.

Eilenberg's correspondence theorem (see the Notes for more historical context) states that the correspondence V $\mapsto$ $\mathcal{V}$, assigning to each pseudovariety V of semigroups the class $\mathcal{V}$ of rational +-languages recognized by semigroups in V, is a bijection between the set of pseudovarieties of semigroups and the set of varieties of +-languages. A "monoid" version of Eilenberg's theorem also exists, describing a natural correspondence between the pseudovarieties of monoids and the *varieties of ∗-languages*, again *mutatis mutandis*.

Example 4.9.4 The +-variety corresponding to the pseudovariety N of all nilpotent finite semigroups is the class of +-languages that are finite or cofinite (Exercise 4.39).

Example 4.9.5 The ∗-variety corresponding to the pseudovariety M of all finite monoids is the class of rational ∗-languages, by Kleene's theorem.

Let V be a pseudovariety of semigroups. A *pro-V semigroup* is a projective limit of semigroups in V, viewed as topological semigroups. Note that a pro-V semigroup is a profinite semigroup. Of course, the profinite semigroups (resp. monoids, groups) are the pro-S semigroups (resp. pro-M monoids, pro-G groups).

For a pseudovariety of monoids V, and for each finite alphabet A, consider the projective limit of all quotients of A^* that are in V, and denote it by $\widehat{A^*}/V$. This projective limit is a pro-V semigroup; it is in fact the *free pro-V semigroup*, in the sense that, if $\iota_V : A \to \widehat{A^*}/V$ is the generating mapping for $\widehat{A^*}/V$, then, for every continuous mapping $\varphi : A \to S$ into a pro-V semigroup, there is a unique continuous morphism of semigroups $\hat{\varphi} : \widehat{A^*}/V \to S$ such that $\hat{\varphi} \circ \iota_V = \varphi$. The proof of this universal property is similar to, and generalizes, the proof of the universal property of $\widehat{A^*} = \widehat{A^*}/M$.

Monoids of the form $\widehat{A^*}/V$ are called *relatively free profinite monoids*. The free profinite monoid $\widehat{A^*}$ is sometimes said to be the *absolutely* free profinite monoid, to distinguish it from other relatively free profinite monoids.

Example 4.9.6 We already met an example of a relatively free profinite monoid which is not absolutely free: indeed, $\widehat{A^*}/G = \widehat{FG(A)}$, trivially.

When V is a pseudovariety of monoids such that the morphism $A^* \to \widehat{A^*}/V$ extending ι_V is injective—for example, when $V = G$ or $V = M$—we may see A^* as a subset of $\widehat{A^*}/V$. In that case, the *pro-V topology of* A^* is defined as the topology of A^* induced by that of $\widehat{A^*}/V$.

As usual, *mutatis mutandis*, we get the obvious definition of relatively free profinite semigroup over a pseudovariety V of semigroups, of pro-V topology of A^+, etc.

The filling of the details justifying the assertions made in the next example is left as an exercise (Exercise 4.40).

Example 4.9.7 If V is a pseudovariety of semigroups containing N, then A^+ is a dense subsemigroup of $\widehat{A^+}/V$ and the elements of A^+ are isolated in $\widehat{A^+}$. The pro-V topology of A^+ is the discrete topology of A^+.

The free pro-N semigroup $\widehat{A^+}/N$ consists of the semigroup A^+ together with an extra element 0 that becomes an absorbing element of $A^+ \cup \{0\}$. A sequence (u_n) of elements of A^+ is such that $u_n \to 0$ in $A^+ \cup \{0\}$ if and only if $|u_n| \to \infty$.

Note that if $A = \{a\}$ is a singleton, then $a \mapsto 1$ defines a continuous isomorphism between $\widehat{A^+}/N$ and the compact semigroup $(\mathbb{Z}_+)_\infty$ met in Example 3.3.10.

Next is another example, in the spirit of Example 4.9.7, which we present without giving all the details (see the Notes).

Example 4.9.8 Let K be the pseudovariety of finite semigroups S such that $x^\omega y = x^\omega$ for all $x, y \in S$. Then $\widehat{A^+}/K$ is isomorphic to the compact semigroup with underlying set $A^+ \cup A^\mathbb{N}$ and the following characteristics:

1. The semigroup operation is given by:

    (a) for $x, y \in A^+$, $x \cdot y$ is the usual product in A^+;
    (b) $x \cdot y = x$ if $x \in A^\mathbb{N}$ and $y \in A^+ \cup A^\mathbb{N}$;
    (c) $(y_1 \cdots y_n) \cdot (x_0 x_1 x_2 \cdots) = y_1 \cdots y_n x_0 x_1 x_2 \cdots$ when $y_1 \cdots y_n \in A^+$ and $x_0 x_1 x_2 \cdots \in A^\mathbb{N}$, with $y_i, x_j \in A$ for $i \in \{1, \ldots, n\}$ and $j \in \mathbb{N}$.

2. The topology is such that:

    (a) the induced topology in $A^\mathbb{N}$ is the product topology of $A^\mathbb{N}$;
    (b) the induced topology in A^+ is the discrete topology of A^+;
    (c) a sequence $(w_n)_n$ of words of A^+ converges to $x \in A^\mathbb{N}$ if and only, for some letter $\diamond$ not in A, the infinite word $w_n \diamond \diamond \diamond \cdots$ converges to x in the product space $(A \cup \{\diamond\})^\mathbb{N}$.

The +-languages recognized by semigroups of K consist of the Boolean combinations of languages of the form uA^*, with $u \in A^+$.

If V is a pseudovariety of semigroups, denote by p_V the unique continuous morphism of semigroups $p_V : \widehat{A^+} \to \widehat{A^+}/V$ such that $p_V(a) = a$ for every $a \in A$. We can describe intuitive features of the elements of $\widehat{A^+}$ with this kind of morphisms, if knowledge on $\widehat{A^+}/V$ is available.

Example 4.9.9 A pseudoword $w \in \widehat{A^+}$ is infinite if and only if $p_N(w) = \infty$ (cf. Example 4.9.7).

Example 4.9.10 Going back to Example 4.9.8, if $w \in \widehat{A^+}$, then $p_\mathsf{K}(w)$ gives the set of finite prefixes of w, condensed in one element of $\widehat{A^+}/N$. Similarly, for the pseudovariety D dual to K, the projection $p_\mathsf{D}(w)$ gives the set of finite suffixes of w. If w is an infinite pseudoword, then one may say that w "starts" with the right infinite word $p_\mathsf{K}(w)$ and "ends" with the left infinite word $p_\mathsf{D}(w)$, with other information in the "middle".

4.10 Exercises

4.10.1 Section 4.2

4.1 Extend the arguments exhibited in Example 4.2.6 to show that $\{w\}$ is a recognizable language of A^* even if A is infinite, when $w \in A^*$.

4.2 Let M, N be two finite monoids. Let $M \diamond N$ be the set of matrices

$$(m, P, n) = \begin{bmatrix} m & P \\ 0 & n \end{bmatrix}$$

with $m \in M$, $P \subseteq M \times N$ and $n \in N$. Show that the product

$$(m, P, n)(m', P', n') = \begin{bmatrix} mm' & mP' \cup Pn' \\ 0 & nn' \end{bmatrix}$$

turns $M \diamond N$ into a monoid.

4.3 Let $\varphi : A^* \to M$ and $\psi : A^* \to N$ be morphisms from A^* into finite monoids M, N recognizing L, K. Let $\varphi \diamond \psi : A^* \to M \diamond N$ be defined by

$$\varphi \diamond \psi(a) = (\varphi(a), \{(1, \psi(a)), (\varphi(a), 1)\}, \psi(a)).$$

Show that for every $w \in A^*$, one has $\varphi \diamond \psi(w) = (\varphi(w), P, \psi(w))$ with $P = \{(\varphi(u), \psi(v)) \mid w = uv\}$.

4.4 Give a direct proof, without using Kleene's theorem, that the product of two recognizable languages is recognizable. (Hint: use Exercise 4.3.)

4.5 Show that the next conditions are equivalent, for any language $L \subseteq A^*$.

(i) L is recognized by a finite automaton.
(ii) L is recognized by a finite monoid.
(iii) L is recognized by a finite deterministic automaton.

4.6 For each automaton $\mathcal{A} = (Q, i, T)$ on the alphabet A, let $0 = 0_Q$ be an element not in Q. For each $a \in A$, let $q \cdot a = 0$ if $q \cdot a$ is not defined, and let $0 \cdot a = 0$.

Consider automata $\mathcal{A} = (Q, i, T)$ and $\mathcal{A}' = (Q', i', T')$ on A. A *morphism of automata* $\mathcal{A} \to \mathcal{A}'$ is a map $\varphi : Q \uplus \{0_Q\} \to Q' \uplus \{0_{Q'}\}$ such that

- $\varphi(i) = i'$, $\varphi(Q) = Q'$, $\varphi(0_Q) = 0_{Q'}$,
- $\varphi^{-1}(T') = T$,
- $\varphi(q \cdot a) = \varphi(q) \cdot a$, for every $a \in A$.

If φ is bijective, then φ^{-1} is a morphism $\mathcal{A}' \to \mathcal{A}$, and we say that φ is an *isomorphism* of automata.

1. Show that if $\varphi : \mathcal{A} \to \mathcal{A}'$ is a morphism of automata, then $\mathcal{A}$ and $\mathcal{A}'$ recognize the same language.
2. Show that for every trim deterministic automaton $\mathcal{A}$ recognizing a language $L \subseteq A^*$, there is a morphism $\mathcal{A} \to \mathcal{A}(L)$.
3. Conclude that, up to isomorphism of automata, the minimal automaton $\mathcal{A}(L)$ is the unique deterministic, trim, reduced automaton recognizing the language $L \subseteq A^*$.

4.7 Let $L \subseteq A^*$. The *syntactic congruence* of L is the kernel of the syntactic morphism $A^* \to M(L)$. The *syntactic context* of $u \in A^*$ (with respect to L) is the set $C_L(u) = \{(x, y) \in A^* \mid xuy \in L\}$. Show that (u, v) belongs to the syntactic congruence if and only if u and v have the same syntactic context.

4.8 Prove Proposition 4.2.14.

4.10.2   Section 4.3

4.9 Show that every word on $A \cup \bar{A}$ is equivalent to a unique reduced word.

4.10 Show that if H is a subgroup of finite index of $FG(A)$, then every right coset of H is of the form Hu for some $u \in A^*$.

4.11 Prove Proposition 4.3.3.

4.12 Show that if X is a basis of a subgroup H of $FG(A)$ of finite index d, then $\mathrm{Card}(X) - 1 = d(\mathrm{Card}(A) - 1)$. (This equality is known as *Schreier's Index Formula*.)

4.13 A group G is said to be *Hopfian* if a morphism from G to itself which is surjective is an isomorphism. Show that every finitely generated free group is Hopfian.

4.14 Suppose that B is a basis of a finitely generated subgroup H of $FG(A)$. Let X be a generating subset of H. Show that $\mathrm{Card}(X) \geqslant \mathrm{Card}(B)$, and show that if $\mathrm{Card}(X) = \mathrm{Card}(B)$ then X is a basis of H. (Therefore, the cardinality of a basis of H depends only on H; it is called the *rank* of H.)

4.10.3 Section 4.4

4.15 Suppose that $A \subseteq B$. Explain why we may see $\widehat{A^*}$ as a submonoid of $\widehat{B^*}$.

4.16 In this exercise, we see how $\widehat{A^*}$ may be seen as the "projective limit of all finite A-generated monoids". Let M_0 be a set of representatives of the isomorphism classes of finite monoids. Consider the set Φ_A of generating maps $\varphi : A \to M_\varphi$, with M_φ running over M_0. Endow Φ_A with the partial order defined by $\varphi \leqslant \psi$ if and only if M_φ is a morphic image of M_ψ. Explain why $\widehat{A^*} = \varprojlim_{\varphi \in \Phi_A} M_\varphi$.

4.17 Show that the pseudowords $(ab)^\omega a(ab)^\omega$ and $(ab)^\omega b(ab)^\omega$ are distinct.

4.18 Show that if $\mathrm{Card}(A) > 1$ then the ideal $\widehat{A^*} \setminus A^*$ is not a $\mathcal{J}$-class.

4.19 Let N be a positive integer.

1. Show that if u is a (possibly infinite) pseudoword over A with length at least N, then there is a unique word $w = i_N(u)$ of length N such that $u \in w\widehat{A^*}$.
2. Letting $i_N(u) = u$ if u is a word of length less than N, show that we have a continuous morphism of monoids $u \in \widehat{A^*} \mapsto i_N(u) \in A^{\leqslant N}$, where $A^{\leqslant N}$ is the discrete set of words over A with length at most N, endowed with the semigroup operation $\star$ given by $u \star v = i_N(uv)$.

4.20 Let $x, y, u, v \in \widehat{A^*}$, with x, y elements of A^* of the same length. Show that if $xu = yv$ or $ux = vy$, then $x = y$ and $u = v$.

4.21 For each subset X of $\widehat{A^*}$ and pseudoword v over A, consider the sets $Xv^{-1} = \{u \in \widehat{A^*} \mid uv \in X\}$ and $v^{-1}X = \{u \in \widehat{A^*} \mid vu \in X\}$. Show that if $v \in A^*$, then the equalities $\overline{Xv^{-1}} = \overline{X}v^{-1}$ and $\overline{v^{-1}X} = v^{-1}\overline{X}$ hold.

4.22 Show that every infinite pseudoword has an infinite idempotent pseudoword as a factor. (Hint: prove that if S is a finite semigroup with n elements, for any $s_1, s_2, \ldots, s_n \in S$ there exist $t_1, e, t_2 \in S$ such that $e = e^2$ and $s_1 \cdots s_n = t_1 e t_2$.)

4.10.4 Section 4.5

4.23 Let ρ be a pseudoword over $X_m = \{x_1, \ldots, x_m\}$. Give a proof that, for any continuous morphism $\psi : S \to T$ of profinite semigroups, the equality $\psi(\rho_S(s_1, \ldots, s_m)) = \rho_T(\psi(s_1), \ldots, \psi(s_m))$ holds for every $(s_1, \ldots, s_m) \in S^m$.

4.24 Let m be a positive integer. An m-ary *implicit operation* is a family $(\rho_S)_{S \in \mathsf{S}}$ of operations $\rho_S : S^m \to S$, with S running over the whole class S of finite semigroups, such that whenever $\psi : S \to T$ is a morphism of finite semigroups, then one has $\psi(\rho_S(s_1, \ldots, s_m)) = \rho_T(\psi(s_1), \ldots, \psi(s_m))$. Show that the correspondence $\rho \mapsto (\rho_S)_{S \in \mathsf{S}}$ is a bijection between the free profinite semigroup $\widehat{X_m^+}$ and the set of m-ary implicit operations.

4.25 Let S be a compact semigroup, and consider the ω-power map $\pi : S \to S$, given by $\pi(s) = s^{\omega}$.

(i) Show that π is continuous if S is profinite.
(ii) Give an example of a compact semigroup for which π is not continuous.

4.10.5   Section 4.6

4.26 Let X be a subset of $FG(A)$.

1. Show that if X is a recognizable subset of $FG(A)$, then $\overline{X}$ is clopen and $X = \overline{X} \cap FG(A)$, where $\overline{X}$ is the topological closure of X in $\widehat{FG(A)}$.
2. Show that X is a recognizable subset of $FG(A)$ if and only if there is some clopen set $K \subseteq \widehat{FG(A)}$ such that $X = K \cap FG(A)$.
3. In contrast with Theorem 4.4.9, show that, for a subset X of $FG(A)$, the fact that $\overline{X}$ is clopen does not imply that X is recognizable.

4.27 Let H be a subgroup of $FG(A)$. Show that if H has finite index, then the topological closure of $H \cap A^*$ in $\widehat{FG(A)}$ is $\overline{H}$. Give an example where H is finitely generated and $\overline{H \cap A^*} \neq \overline{H}$.

4.28 Show that a subgroup of $FG(A)$ is closed in the profinite topology of $FG(A)$ if and only if it is an intersection of open subgroups.

4.29 Let $A = \{a, b\}$ and let H be the subgroup of $FG(A)$ generated by the elements of the forms $a^n b a^n$ and $b^n a b^n$ with $n \geq 1$. Show that H is not closed in the profinite topology of $FG(A)$.

4.30 Prove Theorem 4.6.6.

4.10.6   Section 4.7

4.31 Show that every finite semigroup has a finite presentation.

4.32 Let S be a semigroup with a presentation $\langle A \mid R \rangle$ for R a relation on A^+. Show that the profinite semigroup with presentation $\langle A \mid R \rangle_S$ is the profinite completion of S.

4.33 Show that the profinite monoid with presentation $\langle a, b \mid ab = 1 \rangle_M$ is the group $\widehat{\mathbb{Z}}$ of profinite integers.

4.34 Show that a profinite semigroup S is projective if and only if it has the *lifting property* for every extension, that is, for every surjective morphism $g : U \to S$ there is a morphism $h : S \to U$ such that $g \circ h$ is the identity.

4.35 A profinite group S is projective in the category of profinite groups if whenever $f : S \to T$ and $g : U \to T$ are continuous morphisms of profinite groups with g onto, there is some continuous morphism $h : S \to U$ such that $g \circ h = f$ (thus making Diagram (4.9) commutative). Show that a profinite group is projective in the category of profinite groups if and only if it is projective in the category of profinite semigroups.

4.10.7 Section 4.8

4.36 Show that for every subsemigroup S of A^+, the set $X = S \setminus S^2$ is the unique minimal generating set of S. Show that S is recognizable if and only if X is recognizable.

4.10.8 Section 4.9

4.37 Show that the family of aperiodic semigroups is a pseudovariety.

4.38 Show that the family of finite nilpotent semigroups is a pseudovariety.

4.39 Recall that a subset of a set X is said to be cofinite when its complement in X is finite. Show that a language $L \subseteq A^+$ is finite or cofinite (in A^+) if and only if L is recognized by a morphism $\varphi : A^+ \to S$ onto a finite nilpotent semigroup S. Conclude that the $+$-variety of $+$-languages recognized by finite nilpotent semigroups is the class of $+$-languages that are finite or cofinite.

4.40 Prove the assertions made in Example 4.9.7.

4.11 Solutions

4.11.1 Section 4.2

4.1 Let B be the set of letters that do not appear in w, and let J be the ideal of A^* consisting of the words of length greater than that of w. Then $I = J \cup A^*BA^*$ is an ideal of A^* and $w \notin I$. Moreover, $A \setminus I$ is finite and the quotient morphism $q : A^* \to A^*/I$ is such that $\{w\} = q^{-1}(q(w))$.

4.2 Since $mm'P'' \cup (mP' \cup Pn')n'' = m(m'P'' \cup P'n'') \cup Pn'n''$ the product is associative. The matrix $\begin{bmatrix} 1 & \emptyset \\ 0 & 1 \end{bmatrix}$ is the identity element.

4.3 A simple induction on the length of w proves the statement.

4.4 Let φ, ψ be morphisms on finite monoids M, N recognizing K, L respectively. Then $\varphi \diamond \psi : A^* \to M \diamond N$ recognises KL since $w \in K$ if and only if $\varphi \diamond \psi(w) = (m, P, n)$ with $P \cap (\varphi(K) \times \psi(L)) \neq \emptyset$.

4.5 (i) $\Rightarrow$ (ii) (This was already explained in the main text, but for the reader's convenience, the justifying argument appears here too.) Let $\mathcal{A} = (Q, I, T)$ be an automaton recognizing L. Consider the morphism φ from A^* into the monoid of binary relations on Q defined by $(p, q) \in \varphi(w)$ if there is a path from p to q labeled w. Then $w \in L$ if and only if $\varphi(w)$ meets $I \times T$. Thus φ recognizes L.

(ii) $\Rightarrow$ (iii) Let $\varphi : A^* \to M$ be a morphism to a finite monoid M recognizing L. Let $\mathcal{A}$ be the automaton with M as set of states and edges the triples (p, a, q) such that $p\varphi(a) = q$. Let $I = \{1\}$ and $T = \varphi(L)$. Then w is in L if and only if $\varphi(w) \in \varphi(L)$, which is equivalent to the existence of a path from 1 to T labeled w in $\mathcal{A}$. Hence, L is recognized by the deterministic automaton $\mathcal{A} = (M, 1, T)$.

(iii) $\Rightarrow$ (i) Trivial.

4.6 By induction on the length of w, one sees that $\varphi(q \cdot w) = \varphi(q) \cdot w$ for every $q \in Q \uplus \{0\}$ and $w \in A^*$.

1. Suppose that $\mathcal{A}$ recognizes L. One has $w \in L$ if and only if $i \cdot w \in T$, if and only if $\varphi(i \cdot w) \in T'$ (as $T = \varphi^{-1}(T')$ and $\varphi(Q) = Q'$), if and only if $\varphi(i) \cdot w \in T'$. Since $i' = \varphi(i)$, this shows that L is recognized by $\mathcal{A}'$.

2. Suppose that $L \neq \emptyset$. Recall that the minimal automaton $\mathcal{A}(L)$ is the automaton (R, j, F) with set of states $R = \{u^{-1}L \mid u \in A^*, u^{-1}L \neq \emptyset\}$, initial state $j = L$, set of terminal states $F = \{u^{-1}L \mid u \in L\}$, and where $(u^{-1}L) \cdot a = (ua)^{-1}L$ if $a \in A$. The extra state 0_R is taken to be the set $\emptyset$.

   For each $q \in Q \uplus \{0\}$, let $L_q = \{v \in A^* \mid q \cdot v \in T\}$. Since $\mathcal{A}$ is trim, one has $L_q = \emptyset$ if and only if $q = 0$, and for every $q \in Q$ there is $u \in A^*$ such that $i \cdot u = q$. For every such u, one has $L_q = u^{-1}L$. Therefore the map $\varphi : Q \uplus \{0\} \to R \uplus \{\emptyset\}$ defined by $\varphi(q) = L_q$, where $q \in Q \uplus \{0\}$, satisfies $\varphi(0) = \emptyset$ and $\varphi(Q) \subseteq R$. Moreover, if $u \in A^*$ satisfies $u^{-1}L \neq \emptyset$, and $w \in u^{-1}L$ then $i \cdot uw \in T$, thus $i \cdot u \in Q$ and $u^{-1}L = L_{i \cdot u} = \varphi(i \cdot u)$. This establishes $\varphi(Q) = R$.

   If $i \cdot u = q$, then $i \cdot ua = q \cdot a$, and so we have $L_{q \cdot a} = (ua)^{-1}L = \varphi(q) \cdot a$, for every $q \in Q \uplus \{0\}$ and $a \in A$.

   Finally, for every $q \in Q$, one has $\varphi(q) \in F$ if and only if $\varepsilon \in L_q$, if and only if $q \in T$, thus $\varphi^{-1}(F) = T$. We conclude that φ is a morphism of automata $\mathcal{A} \to \mathcal{A}(L)$.

   It remains to check the case $L = \emptyset$. Set $\mathcal{A}(\emptyset) = (j, j, \emptyset)$. Since the trim deterministic automaton $\mathcal{A} = (Q, i, T)$ recognizes the language $\emptyset$, one must have $T = \emptyset$, and therefore the map $\varphi : Q \uplus \{0\} \to \{j, 0\}$ such that $\varphi(Q) = \{j\}$ and $\varphi(0) = 0$ is a morphism $\mathcal{A} \to \mathcal{A}(L)$.

3. Suppose that the deterministic automaton $\mathcal{A} = (Q, i, T)$ is trim and reduced. The map φ established in the previous item is injective by the definition of reduced automaton, and so it is an isomorphism.

4.7 Set $\mathcal{A}(L) = (Q, i, T)$. Suppose that $\eta_L(u) = \eta_L(v)$. Let $(x, y) \in C_L(u)$. Then $i \cdot xuy \in T$. Since $\eta_L(xuy) = \eta_L(xvy)$, one has $i \cdot xvy = i \cdot xuy \in T$, thus $xvy \in L$ and $C_L(u) \subseteq C_L(v)$. By symmetry, we conclude that $C_L(u) = C_L(v)$.

Conversely, suppose that $\eta_L(u) \neq \eta_L(v)$. Then we may take $q \in Q$ such that $q \cdot u \neq q \cdot v$. Since $\mathcal{A}(L)$ is reduced, we may suppose without loss of generality that there is $y \in A^*$ such that $q \cdot uy \in T$ and $q \cdot vy \notin T$. And since $\mathcal{A}(L)$ is trim, there is $x \in A^*$ such that $i \cdot x = q$. Then $i \cdot xuy \in T$ and $i \cdot xvy \notin T$, that is $(x, y) \in C_L(v) \setminus C_L(u)$, establishing $C_L(u) \neq C_L(v)$.

4.8 Suppose that $\varphi(u) = \varphi(v)$. It suffices to prove that $\eta_L(u) = \eta_L(v)$. If $x, y \in A^*$ satisfy $xuy \in L$, then $\varphi(xvy) = \varphi(xuy) \in \varphi(L)$. Since L is recognized by φ, we obtain $xvy \in L$. By symmetry, this establishes $\eta_L(u) = \eta_L(v)$ (cf. Exercise 4.7).

4.11.2 Section 4.3

4.9 Write $x \rightarrow y$ if y is obtained from x by deleting a factor $a\bar{a}$ for some $a \in A \cup \bar{A}$. It is enough to show that if $x \rightarrow y$ and $x \rightarrow z$, then $y \rightarrow t$ and $z \rightarrow t$ for some t. We have two cases according to whether the factors $a\bar{a}$ overlap or not. In the first case, $x = ua\bar{a}\bar{a}v$ and both reductions give the same result uav. In the second case, we have $x = ua\bar{a}vb\bar{b}w$, $y = uvb\bar{b}w$ and $z = ua\bar{a}vw$. A second reduction gives $t = uvw$ from y or z.

4.10 Let the onto morphism $\psi : FG(A) \rightarrow G$ be the representation of H on the right cosets of H. Since G is a finite group, we have $\psi(FG(A)) = \psi(A^*)$, because if k is the order of G, then $\psi(a^{-1}) = \psi(a^{k-1})$ for every $a \in A$. Given $u \in FG(A)$, we may take $v \in A^*$ such that $\psi(u) = \psi(v)$. We then have $Hu = H \cdot \psi(u) = H \cdot \psi(v) = Hv$.

4.11 For each permutation automaton $\mathcal{A} = (Q, v_0, v_0)$, consider its transition group $M(\mathcal{A})$ and denote by $\psi_\mathcal{A}$ the unique morphism from $FG(A)$ onto $M(\mathcal{A})$ extending the syntactic morphism $A^* \rightarrow M(\mathcal{A})$. Let $q \cdot w$ denote $q \cdot \psi_\mathcal{A}(w)$.

Set $H = H(\mathcal{A})$. For each $q \in Q$, let $w_q \in A^*$ be the label of a path from v_0 to q. Given $u \in FG(A)$, let $q = v_0 \cdot u$. Then we have $v_0 \cdot uw_q^{-1} = v_0$, whence $Hu = Hw_q$, showing that H has finite index less than or equal to $\mathrm{Card}(Q)$. On the other hand, if $p, q \in Q$ satisfy $Hw_q = Hw_p$, then we have $w_q w_p^{-1} \in H$, which implies $p = v_0 \cdot w_p = v_0 \cdot w_q w_p^{-1} w_p = v_0 \cdot w_q = q$. Therefore, the index of H is $\mathrm{Card}(Q)$.

Conversely, suppose that H is a subgroup of $FG(A)$ of finite index d. Let $w_1, \ldots, w_d \in A^*$ be such that $Q = \{Hw_1, \ldots, Hw_d\}$ is the set of right cosets of H. Consider the deterministic automaton $\mathcal{A} = (Q, v_0, v_0)$ such that $v_0 = H$ and with $\psi_\mathcal{A}(a)$ defined by $Hu \cdot \psi_\mathcal{A}(a) = Hua$, for every $u \in FG(A)$ and $a \in A$. Then we have $H(\mathcal{A}) = \{u \in FG(A) \mid H \cdot u = H\} = H$. Therefore, H is finitely generated and $\mathcal{A}$ is the Stallings automaton of H.

4.12 Let $k = \mathrm{Card}(A)$. By Proposition 4.3.3, the group H is finitely generated, its Stallings automaton $\mathcal{A}$ has d vertices, and the transition monoid of $\mathcal{A}$ is a group. Therefore, the multigraph $\mathcal{A}$ has dk edges. On the other hand, by Euler's formula for planar graphs, a spanning tree of $\mathcal{A}$ has $d - 1$ edges. Therefore, every basis of H has $dk - (d - 1) = d(k - 1) + 1$ elements.

4.13 Let A be a finite set and suppose that φ is a surjective endomorphism of $FG(A)$. Suppose there is an element $w \in FG(A) \setminus \{1\}$ such that $\varphi(w) = 1$. By Corollary 4.3.6, there is a morphism ψ from $FG(A)$ onto a finite group G such that $\psi(w) \neq 1$. Since A is finite, there is a finite number n of morphisms from $FG(A)$ to G. Since φ is surjective, the mapping $\alpha \mapsto \alpha \circ \varphi$ is injective on those morphisms, and whence permutes them. However, for every morphism $\alpha : FG(A) \to G$, we have $\alpha(\varphi(w)) = \alpha(1) = 1$, while $\psi(w) \neq 1$. We thus reach a contradiction with the assumption that there are only n morphisms from $FG(A)$ to G. (Note that the only properties required for this proof are the residual finiteness of the free group and that it be finitely generated.)

4.14 Since X is a generating subset of H, there is a surjective morphism $\alpha : FG(X) \to H$. Suppose that $\mathrm{Card}(B) \geqslant \mathrm{Card}(X)$. As H is free with basis B, there is a surjective morphism $\beta : H \to FG(X)$ such that $\beta(B) = X$. Therefore, $\beta \circ \alpha$ is an onto endomorphism of $FG(X)$. Since $FG(X)$ is Hopfian (Exercise 4.13) the composition $\beta \circ \alpha$ is an isomorphism. Because α is onto, it follows that α, β are isomorphisms. Therefore, X is a basis of H and the equality $\beta(B) = X$ yields $\mathrm{Card}(B) = \mathrm{Card}(X)$.

4.11.3 Section 4.4

4.15 Let $\alpha : A \to \widehat{B^*}$ be the inclusion map (we already know that $B \subseteq \widehat{B^*}$). Consider its unique extension to a continuous morphism $\hat{\alpha} : \widehat{A^*} \to \widehat{B^*}$. Take any map $\beta : B \to A$ such that $\beta(a) = a$ for all $a \in A$, and let $\hat{\beta}$ be the unique continuous morphism $\widehat{B^*} \to \widehat{A^*}$ extending β. Then $\hat{\beta} \circ \hat{\alpha}$ is the identity of $\widehat{A^*}$, since its restriction to A is the identity on A.

4.16 Recall that $\widehat{A^*} = \varprojlim_{\theta \in \mathcal{C}_A} A^*/\theta$, where $\mathcal{C}_A$ is the set of congruences of A^* with finite index. Identify each $\theta \in \mathcal{C}_A$ with the generating map $q_\theta : A \to A/\theta$. Embed $\mathcal{C}_A$ in Φ_A via this identification, noting that $\theta \leqslant \rho$ in $\mathcal{C}_A$ if and only if $q_\theta \leqslant q_\rho$. By the isomorphism theorem (cf. Theorem 3.5.2), Φ_A is a directed set of which $\mathcal{C}_A$ is a cofinal set, thus $\widehat{A^*} = \varprojlim_{\varphi \in \Phi_A} M_\varphi$ (cf. Exercise 3.74).

4.17 If φ is the continuous morphism $\widehat{\{a, b\}^*} \to \mathbb{Z}/2\mathbb{Z}$ with $\varphi(a) = 1$ and $\varphi(b) = 0$, then $\varphi((ab)^\omega) = \varphi(ab)^\omega = 0$ and $\varphi((ab)^\omega a(ab)^\omega) = 1 \neq 0 = \varphi((ab)^\omega b(ab)^\omega)$.

4.18 Let a and b distinct letters of A. Let φ be the continuous morphism $\widehat{A^*} \to \{0, 1\}$, with $\{0, 1\}$ under usual multiplication, such that $\varphi(a) = 0$ and $\varphi(1) = 1$.

Then $\varphi(a^\omega) = 0$ is not $\mathcal{J}$-equivalent to $\varphi(b^\omega) = 1$, whence a^ω and b^ω are not $\mathcal{J}$-equivalent.

4.19

1. The elements of $\widehat{A^*}$ which are not words of length less than N belong to the topological closure of the union of the sets of the form wA^*, with $w \in A^N$. This closure is contained in the union of the closed sets of the form $w\widehat{A^*}$, with $w \in A^N$, because, as A^N is finite, such union is itself closed. This settles the existence of the prefix $i_N(u)$. Moreover, if $w, w' \in A^N$ satisfy $w\widehat{A^*} \cap w'\widehat{A^*} \neq \emptyset$, then we have $w\widehat{A^*} \cap w'\widehat{A^*} \cap A^* \neq \emptyset$, since A^* is dense and, by Theorem 4.4.9, the sets $w\widehat{A^*}$ and $w'\widehat{A^*}$ are open, as they are the closures of the recognizable sets wA^* and $w'A^*$. Because the elements of A^* are isolated, we then have $wA^* \cap w'A^* \neq \emptyset$, thus $w = w'$.

2. Let the sequence (u_k) converge in $\widehat{A^*}$ to u. We claim that $i_N(u_k) \to i_N(u)$. If $u \in A^*$, then $u_k = u$ for all sufficiently large k, so we may assume that $u \in \widehat{A^*} \setminus A^*$ and $u_k \notin A^{\leqslant N}$. By Theorem 4.4.9, the set $i_N(u) \cdot \widehat{A^*}$ is a neighborhood of u, whence $u_k \in i_N(u) \cdot \widehat{A^*}$ for all large enough k. By the definition of i_N, we get $i_N(u_k) = i_N(u)$ for all large enough k, showing that i_N is continuous. The verification that $(A^{\leqslant N}, \star)$ is a monoid is simple routine. Finally, since A^* is dense in $\widehat{A^*}$ and we already established the continuity of i_N, to conclude that i_N is a morphism it suffices to see that $i_N(uv) = i_N(u) \star i_N(v)$ for all $u, v \in A^*$.

4.20 By symmetry, it suffices to consider the equality $xu = yv$. Since $\widehat{A^*}$ is equidivisible, there is $t \in \widehat{A^*}$ with $xt = y$ and $u = tv$, or with $x = ty$ and $tu = v$. Since $\widehat{A^*} \setminus A^*$ is an ideal, if $xt = y$ or $x = ty$ then $t \in A^*$, and in fact $t = 1$, since $|x| = |y|$. (Note that this is also an alternative proof of the uniqueness of the prefix of length N of an infinite pseudoword, for every positive integer N.)

4.21 By symmetry, it suffices to prove the equality $\overline{Xv^{-1}} = \overline{X}v^{-1}$.

Let (u_n) be a convergent sequence of elements of $\overline{X}v^{-1}$, and let $u = \lim u_n$. As $u_n v \in \overline{X}$ for all n, and since the multiplication is continuous, we have $uv = \lim u_n v \in \overline{X}$, thus $u \in \overline{X}v^{-1}$. Therefore, $\overline{X}v^{-1}$ is closed. Since one clearly has $Xv^{-1} \subseteq \overline{X}v^{-1}$, we deduce that $\overline{Xv^{-1}} \subseteq \overline{X}v^{-1}$. (Note that the hypothesis $v \in A^*$ remains to be used.)

Conversely, let $u \in \overline{X}v^{-1}$. Then there is a sequence (x_n) of elements of X such that $uv = \lim x_n$. By Proposition 4.4.17, there are factorizations $x_n = u_n v_n$ with $u_n \to u$ and $v_n \to v$. As $v \in A^*$, the point v is isolated and $v_n = v$ for all enough large n. Therefore, we have $u_n \in Xv^{-1}$ for all large n, thus $u \in \overline{Xv^{-1}}$.

4.22 Let S be a finite semigroup with n elements and let $s_1, \ldots, s_n \in S$. If the n products $p_j = s_1 \cdots s_j$ for $1 \leqslant j \leqslant n$ are all distinct, one of them, say p_k is idempotent and $s_1 \cdots s_n = p_k \cdot p_k \cdot s_{k+1} \cdots s_n$ has the desired form. If $p_i = p_j$ with $i < j$, then $p_i = p_j \cdot (s_{i+1} \cdots s_j)^k$ for all $k \geqslant 1$, and so if $e = (s_{i+1} \cdots s_j)^\omega$ then $s_1 \cdots s_n = s_1 \cdots s_i \cdot e \cdot s_{j+1} \cdots s_n$, as required.

Set $\widehat{A^*} = \lim_{\leftarrow i \in I} S_i$ where (S_i, I) is a denumerable projective system of finite semigroups with morphisms $\varphi_i : \widehat{A^*} \to S_i$ (since we are assuming that A is finite, I may be chosen to be denumerable). Let $x \in \widehat{A^*}$ be an infinite pseudoword. Set $x = \lim x_n$, with $x_n \in A^*$. For every $i \in I$, we have $x_n = u_{n,i} v_{n,i} w_{n,i}$ with $\varphi_i(v_{n,i})$ idempotent for an infinite number of n. Up to a subsequence of $(u_{n,i}, v_{n,i}, w_{n,i})_n$, we may assume by compactness that the sequences $(u_{n,i})_n$, $(v_{n,i})_n$, $(w_{n,i})_n$ converge to u_i, v_i, w_i, respectively, thus $x = u_i v_i w_i$ and $\varphi_i(v_i)^2 = \varphi_i(v_i)$, for all $i \in I$. The sequence $(u_i, v_i, w_i)_i$ has some subsequence $(u_{i_j}, v_{i_j}, w_{i_j})_{j \in J}$ converging to a triple u, v, w, so that $x = uvw$. It remains to check that v is idempotent. Fixed $i \in I$, we know that $\varphi_i(u_{i_j})$ is a morphic image of $\varphi_{i_j}(u_{i_j})$ if $i_j \geqslant i$, and so $\varphi_i(v_{i_j})$ is idempotent if $i_j \geqslant i$. As $\varphi_i(v) = \lim_{j \in J} \varphi_i(v_{i_j})$, we deduce that $\varphi_i(v) = \varphi_i(v^2)$. Since $i \in I$ is arbitrary, we conclude that v is indeed idempotent.

4.11.4   Section 4.5

4.23 Let $\alpha = (s_1, \ldots, s_m)$ and $\beta = (\psi(s_1), \ldots, \psi(s_m))$. For each $x_i \in X_m$, one has $\hat{\beta}(x_i) = \psi(s_i) = \psi(\hat{\alpha}(x_i))$. Because $\psi \circ \hat{\alpha}$ and $\hat{\beta}$ are both continuous morphisms defined on $\widehat{X_m^+}$, the equality $\psi \circ \hat{\alpha} = \hat{\beta}$ holds. Therefore, we have $\psi(\rho_S(s_1, \ldots, s_m)) = \psi(\hat{\alpha}(\rho)) = \hat{\beta}(\rho) = \rho_T(\psi(s_1), \ldots, \psi(s_m))$.

4.24 Note that if $\rho \in \widehat{X_m^+}$ then indeed $(\rho_S)_{S \in \mathbf{S}}$ is an m-ary implicit operation by Exercise 4.23.

Let π, π' be distinct elements of $\widehat{X_m^+}$. Then there is a continuous morphism $\psi : \widehat{X_m^+} \to S$ onto a finite semigroup S with $\psi(\pi) \neq \psi(\pi')$. For every $\rho \in \widehat{X_m^+}$, one has $\psi(\rho) = \psi(\rho_{\widehat{X_m^+}}(x_1, \ldots, x_m)) = \rho_S(\psi(x_1), \ldots, \psi(x_m))$, thus $\pi_S(\psi(x_1), \ldots, \psi(x_m)) \neq \pi'_S(\psi(x_1), \ldots, \psi(x_m))$. This shows that the correspondence $\rho \in \widehat{X_m^+} \mapsto (\rho_S)_{S \in \mathbf{S}}$ is injective.

We proceed to show that it is surjective. Let $(\rho_S)_{S \in \mathbf{S}}$ be an m-ary implicit operation. We use the construction of $\widehat{X_m^+}$ as the projective limit

$$\widehat{X^+} = \lim_{\substack{\leftarrow \\ \theta \in \mathcal{C}}} X_m^+/\theta,$$

with $\mathcal{C}$ the directed set of congruences of finite index of X_m^+. For each $\theta \in \mathcal{C}$, let $v_\theta = \rho_{X_m^+/\theta}([x_1]_\theta, \ldots, [x_m]_\theta)$. If $\vartheta, \theta \in \mathcal{C}$ are such that $\vartheta \subseteq \theta$, then, for the connecting morphism $q_{\vartheta,\theta} : X_m^+/\vartheta \to X_m^+/\theta$, one has $q_{\vartheta,\theta}(v_\vartheta) = v_\theta$, since ρ is an implicit operation. Therefore, $v = (v_\theta)_{\theta \in \mathcal{C}}$ is an element of the projective limit $\widehat{X_m^+}$. Let S be an arbitrary finite semigroup. Take any element $\alpha = (s_1, \ldots, s_m)$ of S^m. Let θ be the kernel of the restriction of $\hat{\alpha}$ to X_m^+. Consider the canonical morphism $\psi : X_m^+/\theta \to S$, satisfying $\psi([x_i]_\theta) = \hat{\alpha}(x_i) = s_i$ for all $x_i \in X_m$. Observe that

$\psi \circ \hat{q}_\theta = \hat{\alpha}$, where $\hat{q}_\theta$ is the projection $\widehat{X_m^+} \to \widehat{X_m^+/\theta}$, extending the canonical morphism $q_\theta : X_m^+ \to X_m^+/\theta$. Then, we have the chain of equalities

$$v_S(s_1, \ldots, s_m) = \hat{\alpha}(v) = \psi(\hat{q}_\theta(v)) = \psi(v_\theta) = \rho_S(\psi([x_1]_\theta), \ldots, \psi([x_m]_\theta))$$

$$= \rho_S(s_1, \ldots, s_m).$$

This shows that $v_S = \rho_S$ for every finite semigroup S, concluding the proof that every m-ary implicit operation is in the image of the map $\rho \in \widehat{X_m^+} \mapsto (\rho_S)_{S \in \mathsf{S}}$.

4.25

(i) This is immediate by Proposition 4.5.5 and Example 4.5.1.
(ii) Let $M = [0, 1]$ be the compact monoid for the usual topology and multiplication of real numbers. The space is not totally disconnected, and so M is not profinite. For each positive integer n, let $x_n = 1 - \frac{1}{n}$. Then $x_n^\omega = 0$ and $\lim x_n = 1$, whence $\lim x_n^\omega \neq (\lim x_n)^\omega$.

4.11.5 Section 4.6

4.26

1. Assume that X is recognized by a morphism $\varphi : FG(A) \to G$ from $FG(A)$ into a finite group G. By the universal property of $\widehat{FG(A)}$, there is a unique continuous morphism $\hat{\varphi} : \widehat{FG(A)} \to G$ extending φ. We claim that the equality $\overline{X} = \hat{\varphi}^{-1}(\varphi(X))$ holds. We begin to note that $\hat{\varphi}^{-1}(\varphi(X))$ is closed by continuity of $\hat{\varphi}$, which implies $\overline{X} \subseteq \hat{\varphi}^{-1}(\varphi(X))$. On the other hand, given $y \in \hat{\varphi}^{-1}(\varphi(X))$, and since A^* is dense in $\widehat{FG(A)}$, there is a net $(y_i)_{i \in I}$ of elements of $FG(A)$ converging to y. It then follows that there is $i_0 \in I$ such that $i \geqslant i_0$ implies $\varphi(y_i) = \hat{\varphi}(y) \in \varphi(X)$, by continuity of $\hat{\varphi}$. Since $X = \varphi^{-1}(\varphi(X))$, we then get $y_i \in X$ whenever $i \geqslant i_0$, which implies $y \in \overline{X}$. This establishes the claimed equality $\overline{X} = \hat{\varphi}^{-1}(\varphi(X))$. Hence, $\overline{X}$ is open, by continuity of $\hat{\varphi}$. Moreover, because φ is the restriction of $\hat{\varphi}$ to $FG(A)$, one has $\hat{\varphi}^{-1}(\varphi(X)) \cap FG(A) = \varphi^{-1}(\varphi(X))$, and so the equality $\overline{X} \cap FG(A) = X$ holds.

2. Suppose that K is a clopen subset of $\widehat{FG(A)}$ such that $X = K \cap FG(A)$. By Proposition 3.9.8, there exists a continuous morphism $\psi : \widehat{FG(A)} \to G$ into a finite group G which recognizes K. Let φ be the restriction of ψ to $F(A)$. Then $X = FG(A) \cap K = FG(A) \cap \psi^{-1}(\psi(K)) = \varphi^{-1}(\psi(K))$ and so X is recognizable. For the converse, we may take $K = \overline{X}$, as seen in the first item of the exercise.

3. Let $X = FG(A) \setminus \{1\}$. Then we have $1 = \lim u^{n!} \in \overline{X}$ for every $u \in FG(A)$, whence $\overline{X} = \widehat{FG(A)}$ is clopen, $X \neq \overline{X} \cap FG(A)$ and X is not recognizable.

4.27 Since H is a recognizable subset of $FG(A)$ (cf. Lemma 4.6.2), we know that $\overline{H}$ is clopen in $\widehat{FG(A)}$, as seen in Exercise 4.26. Let $h \in \overline{H}$. Because A^* is dense in $FG(A)$, there is a sequence $(h_n)_n$ of elements of A^* converging to h. As $\overline{H}$ is clopen, we have $h_n \in \overline{H} \cap A^*$ for all large enough n. Since $H \cap A^* \subseteq \overline{H} \cap FG(A) = H$ by Exercise 4.26, we conclude that $h \in \overline{H \cap A^*}$.

For $A = \{a, b\}$, consider the subgroup H of $FG(A)$ generated by ba^{-1}. Then $H \cap A^* = \{\varepsilon\}$, whence $\overline{H \cap A^*} \neq \overline{H}$.

4.28 Since, by Lemma 4.6.2, the open subgroups of $F(A)$ are closed, an intersection of such subgroups is also closed. Conversely, let H be a closed subgroup of $FG(A)$. If $H = FG(A)$ then we are done. Hence, we may assume that H is a proper subgroup. Given x in the complement of H, there is an open normal subgroup K such that $Kx \cap H = \emptyset$. It follows that x does not belong to the open subgroup KH, which contains H. This shows that H is the intersection of all open subgroups that contain it.

4.29 Note that a reduction in a product of generators and their inverses never allows to remove the middle letters of each generator unless the whole generator is removed. Hence, H does not contain the letters a and b. On the other hand, we claim that H cannot be contained in a proper subgroup U of finite index in $FG(A)$ and so H cannot be closed by Exercise 4.28 and Lemma 4.6.2. To prove the claim, suppose U is a subgroup of finite index containing H. Since U has finite index, there are m, n such $m > n$ and $a^m \in a^n U$ and, therefore, there is $i \geqslant 1$ such that $a^i \in U$. Since U contains the element $a^i ba^i$ of H, we deduce that $b \in U$. By symmetry, we also have $a \in U$. Hence, U is not a proper subgroup, which establishes the claim.

4.30 The argument is the same as for finitely generated free profinite monoids (see Proposition 3.11.1 and Theorem 4.4.20).

4.11.6 Section 4.7

4.31 Let S be a finite semigroup and let φ be a morphism from A^+ onto S which is injective on A. For each $s \in S$, let $w_s \in A^+$ be a word of minimum length such that $\varphi(w_s) = s$. Let R be the finite set of all pairs $(w_s a, w_t)$ with $s \in S$, $a \in A$, and $t = s\varphi(a)$. Then S has the presentation $\langle A \mid R \rangle$. Indeed all relations (w_s, w) for $s \in S$ and $\varphi(w) = s$ are consequences of R.

4.32 Let $R\#$ be the congruence of A^+ generated by R and let $R\alpha$ be the admissible congruence on $\widehat{A^+}$ generated by R. Let $\varphi : \widehat{A^+} \to T$ be a continuous morphism from $\widehat{A^+}$ onto a finite semigroup T such that $\varphi(u) = \varphi(v)$ for every pair $(u, v) \in R$. Then we have a continuous morphism φ' from the profinite completion C of $S = A^+/R\#$ onto T having the same restriction to A as φ.

Conversely, a morphism $\psi : C \to T$ to a finite semigroup composed with the natural morphisms $A^+ \to A^+/R\# \to C$ induces a continuous morphism $\theta : \widehat{A^+} \to T$ such that $\theta(u) = \theta(v)$ for every pair $(u, v) \in R$ and thus a continuous morphism $\widehat{A^+}/R\alpha \to T$, again with the property that the restriction to A is the same as that of ψ. We conclude that the profinite semigroups C and $\widehat{A^+}/R\alpha$ are isomorphic by an isomorphism fixing the elements of A.

4.33 In any finite monoid, the relation $uv = 1$ implies $vu = 1$ because the $\mathcal{J}$-class of 1 is a group (Exercise 3.65). Thus this monoid is a group and therefore it is the group $\widehat{\mathbb{Z}}$.

4.34 If S is projective, it has clearly the lifting property for every extension. Conversely, assume that S has the lifting property for every extension and consider $f : S \to T$ and $g : U \to T$ with g surjective. Let $P = \{(u, s) \in U \times S \mid g(u) = f(s)\}$. Then P is a profinite semigroup and the projection π_2 from P to the second component is a surjective morphism from P to S. We have therefore a morphism $h' : S \to P$ such that $\pi_2 \circ h'$ is the identity on S. Then $h = \pi_1 \circ h'$ is such that $g \circ h = f$.

4.35 If a profinite group is projective in the category of profinite semigroups, then it is immediate that it is projective in the category of profinite groups, since a continuous semigroup morphism between profinite groups is a continuous group morphism.

Conversely, suppose that S is a profinite group which is projective in the category of profinite groups. Let $f : S \to T$ be a continuous morphism from S into a profinite semigroup, and let $g : U \to T$ be a continuous morphism from the profinite semigroup U onto T. Take $R = f(S)$ and let $i : R \to T$ be the inclusion, and denote by f' the continuous morphism $S \to R$ such that $i \circ f' = f$. Then R is a profinite group. By Exercise 3.64, there is a closed subgroup V of U such that $g(V) = R$. Consider the restriction $g' : V \to R$ of g, and denote by j the inclusion morphism $V \to U$, so that $g \circ j = i \circ g'$. See Diagram 4.14. Since S is a projective group in the category of profinite groups, there is a continuous morphism of groups $h : S \to V$ such that $g' \circ h = f'$.

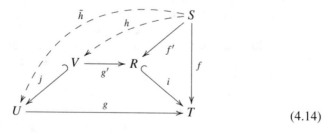

$$(4.14)$$

Noting that Diagram 4.14 is commutative, and taking $\tilde{h} = j \circ h$, we get $g \circ \tilde{h} = f$, and so S is a projective semigroup in the category of profinite semigroups.

4.11.7   Section 4.8

4.36 It is easy to verify by induction on $|w|$ that every $w \in S$ is in X^* and thus that X generates S. If Y generates S, then $X \subseteq Y$ and thus X is minimal. If S is recognizable, then X is recognizable since the family of recognizable sets is a boolean algebra which is closed under product by Kleene's Theorem.

4.11.8   Section 4.9

4.37 Let $\varphi : S \to T$ be a morphism of finite semigroups from S onto T. Any subgroup of T is the image of a subgroup of S (Exercise 3.64). Thus, if S is aperiodic, then so is T. The case of a subsemigroup is clear. Finally, the subgroups of the direct product $S \times T$ are the direct products of subgroups of S and T. Thus, if S, T are aperiodic, so is $S \times T$.

4.38 Let $\varphi : S \to T$ be a morphism of finite semigroups from S onto T. If 0_S is a zero of S, then $0_T = \varphi(0_S)$ is a zero of T. Let $n \geqslant 1$ be such that $S^n = \{0_S\}$. Then $T^n = \varphi(S^n) = \{0_T\}$, proving that the morphic image of a nilpotent semigroup is nilpotent. Moreover, if R is a subsemigroup of T, then $R^n \subseteq T^n = \{0_T\}$, thus $R^n = \{0_T\}$ and R is nilpotent. Also, if S and T are finite nilpotent semigroups, with absorbing elements 0_S and 0_T such that $S^n = \{0_S\}$ and $T^m = \{0_T\}$, respectively, then, for $k = \max\{m, n\}$, the pair $(0_S, 0_T)$ is a zero of $S \times T$ and $(S \times T)^k = \{(0_S, 0_T)\}$, showing that $S \times T$ is also nilpotent.

4.39 Note that L is recognized by a morphism if and only if the complement of L is recognized by the same morphism, and so it suffices to consider the case where L is finite. Since $\emptyset$ is recognized by the trivial semigroup, we may also suppose that $L \neq \emptyset$ and thus consider the greatest possible length m of an element of the finite set L. Let I be the ideal of A^+ formed by the words over A with length at least $m + 1$. Then the Rees quotient $q : A^+ \to A^+/I$ is such that $L = q^{-1}(L)$, and $S = A^+/I$ is a finite nilpotent semigroup.

Conversely, suppose that $L \subseteq A^+$ is such that $L = \varphi^{-1}\varphi(L)$ for some morphism $\varphi : A^+ \to S$ onto a finite nilpotent semigroup with zero 0, and let $n \geqslant 1$ be such that $S^n = \{0\}$. Note that $I = \varphi^{-1}(0)$ is an ideal of A^+. Moreover, the language $A^+ \setminus I$ is finite, consisting of some (possibly all) words in A^+ with length less than n. Therefore, the image of the Rees quotient $q : A^+ \to A^+/I$ is a finite nilpotent semigroup. Note that $q(u) = q(v)$ implies $\varphi(u) = \varphi(v)$, and so there is a factorization $\varphi = \psi \circ q$ for a unique morphism $\psi : A^+/I \to S$. Then, we have $L = q^{-1}(\psi^{-1}(\varphi(L)))$. This shows that L is finite or cofinite: indeed, $q^{-1}(P)$ is finite or cofinite for every subset P of A^+/I.

The family of finite and cofinite languages of A^+ is clearly closed under Boolean operations and taking residues. Finally, take a morphism $\varphi : B^+ \to A^+$. Since, for every $u \in B^+$, the length of u is not bigger than that of $\varphi(u)$, clearly $\varphi^{-1}(L)$ is finite when $L \subseteq A^+$ is finite. Moreover, because the equality $B^+ \setminus \varphi^{-1}(L) = \varphi^{-1}(A^+ \setminus L)$ always holds, if L is cofinite then $\varphi^{-1}(L)$ is cofinite.

4.40 The singletons of A^+ are recognized by finite Rees quotients of A^+, which are nilpotent semigroups. Therefore, adapting the arguments used in the proof of Proposition 4.4.3 and those discussed in Example 4.4.14, we conclude straightforwardly that A^+ embeds in $\widehat{A^+}/V$ as a dense subset whose induced topology is discrete.

Let u be an element of $\widehat{A^+}/N$ not in A^+, and let $\varphi : \widehat{A^+}/N \to S$ be a continuous morphism onto a finite nilpotent semigroup S such that $S^k = \{0\}$. Let (u_n) be a sequence of elements of A^+ converging to u. As the subspace A^+ is discrete, one has $|u_n| \geqslant k$ for all large enough n, thus $\varphi(u) = \lim \varphi(u_n) = 0$. Therefore, since $\widehat{A^+}/N$ is a projective limit of finite nilpotent semigroups, there is only one element in $\widehat{A^+}/N$ which is not in A^+, and this element is absorbing.

4.12 Notes

A detailed presentation of formal languages, including a proof of Kleene's Theorem can be found in any textbook on this subject, for example Pin (1986), Lallement (1979), or Eilenberg (1974).

The notion of recognizable language and of recognizable series introduced in Chap. 2 are two particular cases of the same notion. Indeed, let K be a commutative semiring and A a finite alphabet. A K-subset of the free monoid A^* is a map X from A^* into K associating to a word w its multiplicity (X, w). A power series $\sum s_n z^n$ with coefficients in $\mathbb{Q}$ is a K-subset of a^* for $K = \mathbb{Q}$ considering the map $a^n \mapsto s_n$. A language $L \subseteq A^*$ is a K-subset of A^* for K being the Boolean semiring. A K-subset X is *recognizable* if there is a morphism φ from A^* into the monoid of $n \times n$-matrices with coefficients in K a row vector $\lambda \in K^n$ and a column vector $\gamma \in K^n$ such that $(X, w) = \lambda \varphi(w) \gamma$ for every $w \in A^*$. One recovers both the notions of recognizability for series and for languages. See Eilenberg (1974) for a systematic presentation.

For a general introduction to free groups, see Lyndon and Schupp (2001) or Magnus et al. (2004). The notion of Stallings automaton is originally from Stallings (1983). See also Margolis et al. (2001) or Kapovich and Myasnikov (2002). Theorem 4.3.4 is Theorem 5.1 in Hall (1949) (see also Lyndon and Schupp (2001, Proposition I.3.10)).

For an expanded introduction to the case of free profinite semigroups on infinite alphabets, see Almeida and Steinberg (2009).

Theorem 4.4.9 is a special case of a general result from Almeida (1994, Section 3.6). The Stone duality given by Corollary 4.4.13 is an immediate consequence. It was discovered independently by Pippenger (1997). Stone duality only identifies the topological structure of free profinite semigroups. Gehrke et al. (2008) (see also Gehrke (2016)) further showed how the algebraic structure of the free profinite semigroup on a finite set A appears in the dualization of the Boolean algebra of recognizable languages over the alphabet A with the additional operations of quotients (as introduced in Sect. 4.9).

Proposition 4.4.17 is stated and shown in Almeida et al. (2019, Section 3). It is also embedded in the proof of Henckell et al. (2010, Lemma 4.3). Equidivisible monoids were introduced in McKnight and Storey (1969). See also Almeida and Costa (2017).

The natural metric, Proposition 3.11.1 and Theorem 4.4.20 may be extended to abstract profinite algebras (Almeida 2002b).

The profinite topology of the free group has been introduced by Hall (1950). For that reason, some authors call it the Hall topology (Pin and Reutenauer 1991; Pin 1996). The pro-G topology on A^* was introduced by Reutenauer (1979). Proposition 4.6.3 is from Hall (1950, p. 131).

Theorem 4.6.7 is from Coulbois et al. (2003, Corollary 2.2).

The group analogue of Proposition 4.7.8, asserting that any finitely generated projective profinite group is finitely presented, is due to Lubotzky (2001, Proposition 1.1). Theorem 4.7.9 was shown by Rhodes and Steinberg (2008, Theorem 1). Before that, it was already well known that every closed subgroup of a free profinite group is projective (Ribes and Zalesskii 2010, Lemma 7.6.3).

For a detailed introduction to codes, see Berstel et al. (2009). The prefix transducer is described in Berstel et al. (2009, Chapter 4).

Theorem 4.8.3 is from Margolis et al. (1998). This has been extended to rational codes in Almeida and Steinberg (2009): Theorem 4.8.5 is from Almeida and Steinberg (2009, Corollary 5.7).

The approach on finite semigroups using pseudovarieties is borrowed from Birkhoff's theory of varieties of algebras, for which (Burris and Sankappanavar 1981) is a good reference. In fact, pseudovarieties of semigroups are sometimes called *varieties of finite semigroups* (Pin 1986), which can cause some confusion, as pseudovarieties are not, in general, varieties in Birkhoff's sense (the latter are closed under arbitrary products, and therefore, may contain algebras with infinitely many elements). One of the main reasons for the predominance of this "varietal" approach to the study of finite semigroups stems from Eilenberg's theorem, relating pseudovarieties of semigroups to *varieties of +-languages* (Eilenberg 1976). Eilenberg's correspondence theorem proved to be a good and influential framework for finite semigroup theory. For instance, Schützenberger's characterization of the star-free ∗-languages as being the ones which are recognized by finite aperiodic monoids is an instance of Eilenberg's theorem (we remark that Schützenberger's result precedes in time Eilenberg's theorem). For an introduction to this framework, see for example Pin (1986).

Since the 1980s, relatively free profinite semigroups (and monoids) have been used as effective means to study pseudovarieties of semigroups (respectively, monoids), overcoming the difficulty that V in general does not contain a free object (for example, for each natural n, there is no finite n-generated semigroup whose morphic images are all the finite n-generated semigroups, as the latter can have arbitrarily large cardinal). The interpretation of pseudowords as implicit operations (cf. Exercise 4.24), allowing an equational theory that suits pseudovarieties, plays a key role in this context. This approach, motivated by the seminal works of Reiterman (1982) and Banaschewski (1983), is presented in depth in the books Almeida (1994) and Rhodes and Steinberg (2009). The text Almeida (2005a) is an alternative short introduction to the subject.

Various attempts have been made to understand the "middle" of an infinite pseudoword. Examples can be found in Almeida and Weil (1997) and Moura (2011), where pseudowords over certain pseudovarieties are represented by certain labeled linear orders. Recently, this perspective was successfully extended in two distinct descriptions, respectively made in the papers van Gool and Steinberg (2019) and Almeida et al. (2019), of the elements of free profinite aperiodic semigroups.

Chapter 5
Shift Spaces

5.1 Introduction

In this chapter, we introduce basic definitions of symbolic dynamics like shift spaces and minimal shifts. We study the topological closures, in the free profinite monoid, of the set of factors of a minimal shift and relate them to uniformly recurrent pseudowords.

The first section (Sect. 5.2) presents factorial, recurrent and uniformly recurrent sets.

The second section (Sect. 5.3) contains the definitions of shift spaces (also called subshifts), irreducible and minimal subshifts. We define shift spaces by their sets of factors and we prove that they are the closed and invariant subsets of the full shift.

Section 5.4 introduces the natural notion of morphism between shift spaces which turns out to be the sliding block codes.

In Sect. 5.5, we put special emphasis on shift spaces obtained by iterating a substitution. We define primitive substitutions and relate them to primitive matrices. We state without proof the Perron-Frobenius Theorem (Theorem 5.5.13). We develop results allowing us to prove the decidability of the periodicity of a substitution shift. We quote without proof Mossé's Theorem concerning the recognizability of morphisms on one-sided fixed points (Theorem 5.5.19). Using that theorem, we then prove a variant concerning two-sided infinite words (Theorem 5.5.22) which will be used in Chap. 7.

In Sect. 5.6, we come to the study of the topological closure, in the free profinite monoid, of a uniformly recurrent set. The main result (Theorem 5.6.7) is that an infinite pseudoword is uniformly recurrent if and only if it is $\mathcal{J}$-maximal as an infinite pseudoword. This result establishes a natural bijection between the minimal subshifts and the $\mathcal{J}$-classes of $\mathcal{J}$-maximal infinite pseudowords. In Sect. 5.7 we briefly discuss some aspects of a generalization of this correspondence to recurrent sets, which is not necessary for the following chapters.

© Springer Nature Switzerland AG 2020
J. Almeida et al., *Profinite Semigroups and Symbolic Dynamics*, Lecture Notes in Mathematics 2274, https://doi.org/10.1007/978-3-030-55215-2_5

5.2   Factorial Sets

We follow the convention of working only with finite alphabets. We begin with some background on factorial sets of free monoids. Let A be an alphabet. A nonempty set of finite words on A is *factorial* if it contains all the factors of its elements.

A factorial set F is *recurrent* if F contains some nonempty word and for any $x, z \in F$ there is some $y \in F$ such that $xyz \in F$. A factorial subset F of A^* is *biextendable* if, for every $x \in F$, one has $AxA \cap F \neq \emptyset$. Note that we are including the property of being factorial in the definitions of recurrent set and of biextendable set. Note also that every recurrent set is biextendable, and that every biextendable set is infinite.

Example 5.2.1   The set of words on $A = \{a, b\}$ without the factor bb is recurrent.

An infinite factorial set F of finite words is said to be *uniformly recurrent* or *minimal* if for any $x \in F$ there is an integer $n \geqslant 1$ such that x is a factor of every word in F of length n. A uniformly recurrent set is obviously recurrent.

We present an elementary example of a uniformly recurrent set. More interesting examples will be given in Sect. 5.5.

Example 5.2.2   The set of factors of the powers w^n of a word $w \in A^+$ is uniformly recurrent. On the contrary, the set of words over $\{a, b\}$ without the factor bb is not uniformly recurrent since, for all $n \geqslant 1$, the word a^n does not contain any factor b.

The following proposition explains the use of the alternative nomenclature "minimal".

Proposition 5.2.3   *Let F be an infinite factorial subset of A^*. Then F is uniformly recurrent if and only if it is minimal for inclusion among the infinite factorial subsets of A^*.*

Proof   Assume first that F is uniformly recurrent. Let $F' \subseteq F$ be an infinite factorial set. Let $x \in F$ and let n be such that x is a factor of every word of F of length n. Since F' is factorial infinite, it contains a word y of length n. Since x is a factor of y, it is in F'. Thus F is minimal.

Conversely, suppose that F is minimal among the infinite factorial subsets of A^*. Consider $x \in F$ and let T be the set of words in F which do not contain x as a factor. Then T is factorial. Since F is minimal among infinite factorial sets, T is finite. Thus there is an n such that x is a factor of all words of F of length n.   ■

5.3   Shift Spaces

A *two-sided infinite word* (or *biinfinite word*) on A is an element $x = (x_n)_{n \in \mathbb{Z}}$ of $A^{\mathbb{Z}}$. For $x \in A^{\mathbb{Z}}$ and $i < j$, we denote $x[i, j] = x_i x_{i+1} \cdots x_j$. The set $A^{\mathbb{Z}}$ is called the *full shift* on A.

We consider on $A^{\mathbb{Z}}$ the product topology, after endowing A with the discrete topology. In this way, $A^{\mathbb{Z}}$ is a compact topological space. It is actually a Cantor space for $\text{Card}(A) \geqslant 1$.

Note that the product topology on $A^{\mathbb{Z}}$ is generated by the metric d such that, for $x, y \in A^{\mathbb{Z}}$ with $x \neq y$, if k is the least possible value for $|i|$, with $i \in \mathbb{Z}$ and $x_i \neq y_i$, then $d(x, y) = 2^{-k}$.

The *shift map* $\sigma_A : A^{\mathbb{Z}} \to A^{\mathbb{Z}}$ is defined by $y = \sigma_A(x)$ if $y_n = x_{n+1}$ for each $n \in \mathbb{Z}$. The map σ_A is continuous and so is its inverse. We will frequently drop the subscript A, denoting σ_A by σ.

By a *factor* of a biinfinite word $x \in A^{\mathbb{Z}}$ we mean a word of the form $x_n x_{n+1} \cdots x_{n+m}$ for some integers n, m, with $m \geqslant 0$.

Given a biextendable subset F of A^*, the set of two-sided infinite words with all their factors in F, denoted $X(F)$, is called a *shift space* of $A^{\mathbb{Z}}$ (or *symbolic dynamical system*). If $X \subseteq Y$ are shift spaces, we say that X is a *subshift* of Y. In particular, every shift space on the alphabet A is a subshift of the full shift on A.

Example 5.3.1 The *golden mean shift* is the shift $X = X(F)$ on the alphabet $\{a, b\}$ where F is the set words without factor bb. It is also the set of labels of biinfinite paths in the graph of Fig. 5.1.

A shift space $X = X(F)$ is closed for the product topology. Indeed, if $x = \lim x^{(n)}$ with $x^{(n)} \in X$, then a factor of x is a factor of every $x^{(n)}$ for n large enough and thus $x \in X$. It is also shift invariant, that is such that $\sigma_A(X) = X$. Indeed, the factors of x and $\sigma_A(x)$ are the same.

Proposition 5.3.2 *The shift spaces are the closed and shift invariant nonempty subsets of $A^{\mathbb{Z}}$.*

Proof We have already seen that a shift space is closed and shift invariant. Conversely, let X be a nonempty closed and shift invariant subset of $A^{\mathbb{Z}}$. Let F be the set of factors of the elements of X. Clearly F is biextendable. Let us show that $X = X(F)$. The inclusion $X \subseteq X(F)$ holds by definition of F. Conversely, let $x \in X(F)$. For every $n \geqslant 0$, there is a word $y^{(n)} \in X$ such that $x[-n, n]$ is a factor of $y^{(n)}$. Since X is shift invariant, we may assume that $x[-n, n] = y^{(n)}[-n, n]$. Then $x = \lim y^{(n)}$ and, since X is closed, we conclude that $x \in X$. ∎

For each $x \in A^{\mathbb{Z}}$, the set $\mathcal{O}(x) = \{\sigma^n(x) \mid n \in \mathbb{Z}\}$ is the *orbit* of x. Therefore, in Proposition 5.3.2 it is being said that the subshifts of $A^{\mathbb{Z}}$ are the nonempty closed subspaces of $A^{\mathbb{Z}}$ which are a union of orbits of elements of $A^{\mathbb{Z}}$. Note also that if $x \in A^{\mathbb{Z}}$, then $\overline{\mathcal{O}(x)}$ is a shift space, by Proposition 5.3.2.

A shift space is a particular case of a topological dynamical system, a notion already met in Chap. 2. Recall from Chap. 2 that a topological dynamical system

Fig. 5.1 The golden mean shift space

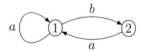

is a pair (X, T) consisting of a compact topological space and a homeomorphism $T : X \to X$. A subshift X of $A^{\mathbb{Z}}$ is a compact space (since it is a closed subset of the compact space $A^{\mathbb{Z}}$). Moreover the shift defines a homeomorphism from X to itself.

Note that whenever F and F' are infinite biextendable sets, one has $X(F) \subseteq X(F')$ if and only if $F \subseteq F'$, and so in particular each subshift is completely determined by a unique biextendable set.

Given a shift space X, we denote by $F(X)$ the set of words of the form $x[i, j]$, with $x \in X$, where $x[i, j]$ is the empty word when $i > j$. A word of this form is sometimes called a *block* of X. Note that the correspondence $X \mapsto F(X)$ between the subshifts of $A^{\mathbb{Z}}$ and the infinite biextendable sets of A^* is the inverse of the correspondence $F \mapsto X(F)$ already introduced. Moreover, one has $X \subseteq X'$ if and only if $F(X) \subseteq F(X')$, for all subshifts X, X'.

Dynamical properties of X may translate into combinatorial properties of $F(X)$. For instance, X is said to be *transitive* (also said *irreducible*) when it has a dense forward orbit, that is, when for some $x \in X$ the subshift X is the topological closure in $A^{\mathbb{Z}}$ of $\{\sigma^n(x) \mid n \in \mathbb{N}\}$; it turns out that X is transitive if and only if $F(X)$ is recurrent (Exercise 5.1).

An element x of $A^{\mathbb{Z}}$ is said to be *periodic* if there is a positive integer n such that $\sigma^n(x) = x$. Note that $x \in A^{\mathbb{Z}}$ is periodic if and only if its orbit $\mathcal{O}(x)$ is finite, in which case $\mathcal{O}(x)$ becomes a subshift of $A^{\mathbb{Z}}$. A subshift of $A^{\mathbb{Z}}$ is said to be a *periodic subshift* when it is the orbit of a periodic element of $A^{\mathbb{Z}}$. A positive integer n such that $\sigma^n(x) = x$ is said to be a *period* of x. If x is a periodic element of $A^{\mathbb{Z}}$, then the smallest period of x is called the *least period* of x. Note that the least period of a periodic element x of $A^{\mathbb{Z}}$ divides all the periods of x, and is equal to the size of the orbit of x. A factorial subset F of A^* is *periodic* if it is of the form $F = F(X)$ for some periodic subshift $X \subseteq A^{\mathbb{Z}}$. Note that a subset F of A^* is factorial if and only if it is the set of factors of the powers of a nonempty finite word w. One may always assume w to be a *primitive* word, which means that w is not a power of another word. The length of such primitive word is the least period of $X(F)$.

A subshift X of $A^{\mathbb{Z}}$ is called *minimal* if it does not contain subshifts other than itself. It turns out that the minimal shift spaces are those of the form $X(F)$ for some uniformly recurrent set F (Exercise 5.2). Therefore, a minimal subshift is a transitive subshift. Note that the periodic shift spaces are minimal. The periodic shift spaces are the unique finite transitive shift spaces, and every other transitive subshift is uncountable (since it is a Cantor space, see Exercise 5.4).

If $\mathrm{Card}(A) = 1$, then A^* is the only infinite factorial subset of A^*. But if $\mathrm{Card}(A) \geqslant 2$, then A^* contains $2^{\aleph_0}$ uniformly recurrent sets (see the Notes Section for references), and there are only $\aleph_0$ periodic subsets of A^*, as there are only $\aleph_0$ recognizable languages of A^*. Later in this chapter, we shall see in Proposition 5.5.4 an easy way of defining examples of nonperiodic uniformly recurrent sets.

As a variant, one may use infinite words instead of two-sided infinite words. The *one-sided shift map* is the map $\sigma : A^{\mathbb{N}} \to A^{\mathbb{N}}$ defined by $y = \sigma(x)$ if $y_n = x_{n+1}$ for all $n \geqslant 0$. Note that the one-sided shift is not one-to-one (except if $\mathrm{Card}(A) = 1$).

The set $A^{\mathbb{N}}$ of (one-sided) infinite words on A is again a topological space for the product topology, which is defined by the distance $d(x, y) = 2^{-k}$ where k is the length of the longest common prefix of x, y.

A *one-sided shift space* is a closed invariant subset of $A^{\mathbb{N}}$.

Let X be a (two-sided) subshift of $A^{\mathbb{Z}}$. Let θ be the natural projection $\theta : A^{\mathbb{Z}} \to A^{\mathbb{N}}$. The image by θ of X is a one-sided shift space called the one-sided subshift *associated* to X.

5.4   Block Maps and Conjugacy

Let X be a shift space on the alphabet A. For each positive integer n, let $F_n(X) \subseteq F(X)$ be the set of blocks of X with length n. Given an alphabet B and integers $k, \ell \geqslant 0$, a map $f : F_{k+\ell+1}(X) \to B$ is called a $k + \ell + 1$-*block map*. It defines a map $\gamma : X \to B^{\mathbb{Z}}$ by $\gamma(x) = y$ if, for every $n \in \mathbb{Z}$, the equality

$$y_n = f(x_{n-k} \cdots x_{n+\ell})$$

holds (see Fig. 5.2).

The map γ is called the *sliding block code* induced by f with memory k and anticipation ℓ. We denote $\gamma = f^{[-k,\ell]}$. If Y is a shift space on the alphabet B such that $\gamma(X) \subseteq Y$, we say that γ is a sliding block code from X into Y.

The simplest case occurs when $k = \ell = 0$, in which case γ is a simple change of alphabet. In this case, we say that γ is a 1-*block code*.

The following result shows that sliding block codes are the natural maps between shift spaces.

Proposition 5.4.1 *Let X, Y be shift spaces, with $X \subseteq A^{\mathbb{Z}}$ and $Y \subseteq B^{\mathbb{Z}}$. A map $\gamma : X \to Y$ is a sliding block code if and only if it is continuous and commutes with the shift map in the sense that the following diagram commutes:*

$$
\begin{array}{ccc}
X & \xrightarrow{\sigma_A} & X \\
\gamma \downarrow & & \downarrow \gamma \\
Y & \xrightarrow{\sigma_B} & Y.
\end{array}
$$

The proof of this proposition is left as an exercise (Exercise 5.5).

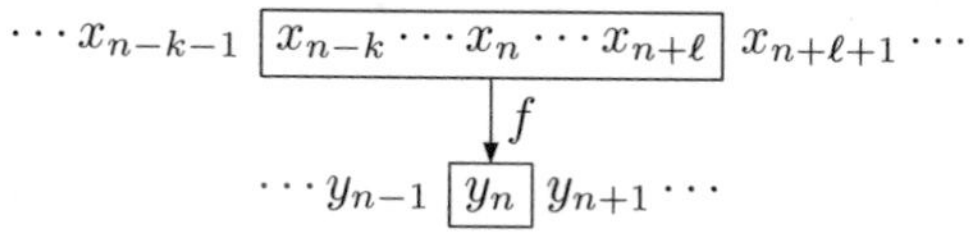

Fig. 5.2 The sliding block code

A sliding block code $\gamma : X \to Y$ is a *conjugacy* if it is a bijection from X onto Y. In this case, since a shift space is compact, the inverse of γ is also a sliding block code, in view of Proposition 5.4.1.

As a particular case, if $f : F_m(X) \to B$ with $m \geqslant 1$ is a bijection, the corresponding sliding block code $\beta_m = f^{[0,m-1]}$ is called the *m*th *higher block code* and the shift space $Y = \gamma(X)$ is called the *m*th *higher block shift space* or *higher block presentation* of X. The *m*th higher block code is clearly a conjugacy. Its inverse is induced by the 1-block map sending $b \in B$ to the first letter of $f^{-1}(b)$.

Example 5.4.2 Let X be the golden mean shift space on $A = \{a, b\}$ and let $B = \{a, b, c\}$. Set $k = 0$ and $\ell = 1$. The block map $f : aa \mapsto a, ab \to b, ba \to c$ induces the 2nd higher block presentation of X. The effect of the sliding block map is illustrated below.

$$\cdots abaababa \cdots$$

$$\cdots bcabcbc \cdots$$

Proposition 5.4.3 *Let $\gamma : X \to Y$ be a sliding block code. There exists a higher block shift space $\tilde{X}$, a conjugacy $\psi : X \to \tilde{X}$ whose inverse is a 1-block code and a 1-block code $\tilde{\gamma} : \tilde{X} \to Y$ such that the following diagram commutes:*

Proof Assume that $X \subseteq A^{\mathbb{Z}}$ and $Y \subseteq B^{\mathbb{Z}}$. Suppose that γ is induced by a block map f with memory k and anticipation ℓ. Let $g : F_{k+\ell+1}(X) \to C$ be a bijection and let $\psi : X \to C^{\mathbb{Z}}$ be defined by $y = \psi(x)$ if $y_i = g(x[i - k, i + \ell])$. Then $\psi = \sigma_B^k \circ \beta_{k+\ell+1}$ where $\beta_{k+\ell+1}$ is the $(k + \ell + 1)$th higher block code. Thus $\tilde{X} = \psi(X) = \beta_{k+\ell+1}(X)$ is a higher block shift space of X and ψ is a conjugacy whose inverse is a 1-block code. The map $\tilde{\gamma} = \gamma \circ \psi^{-1}$ is a 1-block code. ∎

We will use in Chap. 7 the following corollary.

Corollary 5.4.4 *Every conjugacy is a composition of a conjugacy which is a 1-block code with a conjugacy which is the inverse of a 1-block code.*

Proof Let X, Y be shift spaces and let $\gamma : X \to Y$ be a conjugacy. By Proposition 5.4.3, there is a conjugacy ψ whose inverse is a 1-block code and a 1-block code $\tilde{\gamma}$ such that $\gamma = \tilde{\gamma} \circ \psi$. Since ψ and γ are conjugacies, $\tilde{\gamma}$ is a conjugacy. ∎

5.5 Substitutive Shift Spaces

Let A be a finite alphabet. An endomorphism of A^* is frequently called a *substitution* over A. A substitution is *nonerasing* if $\varphi(a)$ is nonempty for every $a \in A$. Thus a nonerasing substitution is essentially an endomorphism of A^+.

A nonerasing substitution $\varphi : A^+ \to A^+$ extends to a map from $A^{\mathbb{N}}$ to $A^{\mathbb{N}}$. If $x \in A^{\mathbb{N}}$, then $\varphi(x)$ is the unique infinite word in $A^{\mathbb{N}}$ such that, for every finite prefix u of x, the word $\varphi(u)$ is a finite prefix of $\varphi(x)$ (note that $\varphi(x)$ is infinite because φ is nonerasing).

Remark 5.5.1 Although the following information is not necessary for the sequel, we remark that the aforementioned extension is the restriction to $A^{\mathbb{N}}$ of the unique continuous endomorphism of the relatively free profinite semigroup $\widehat{A^+}/K = A^+ \cup A^{\mathbb{N}}$ that extends $\varphi : A^+ \to A^+$ (see Example 4.9.8).

The nonerasing substitution $\varphi : A^+ \to A^+$ also extends to a map from $A^{\mathbb{Z}}$ to $A^{\mathbb{Z}}$. If $x = \cdots x_{-1}x_0x_1 \cdots$ is in $A^{\mathbb{Z}}$, then $\varphi(x) = \cdots y_{-1}y_0y_1 \cdots$ with

$$\cdots y_{-2}y_{-1} = \cdots \varphi(x_{-2})\varphi(x_{-1})$$

$$y_0y_1 \cdots = \varphi(x_0)\varphi(x_1) \cdots$$

If $\varphi : A^* \to A^*$ is a substitution, then φ^n is again a substitution for all $n \geqslant 1$. The *language* of φ, denoted $F(\varphi)$, is the set of factors of the words of the form $\varphi^n(a)$, with $n \geqslant 1$ and $a \in A$.

A nonerasing substitution is *biextendable* if for every $a \in A$ there are $b, c \in A$ such that $bac \in F(\varphi)$.

The language of a substitution φ is factorial. When φ is nonerasing, it is biextendable if and only if φ is biextendable (Exercise 5.8).

For a biextendable substitution φ, the shift space X formed of the biinfinite words having all their factors in $F(\varphi)$ is denoted $X(\varphi)$ and called a *substitutive shift space*.

5.5.1 Primitive Substitutions

Suppose that the finite alphabet A has at least two letters. A substitution $\varphi : A^+ \to A^+$ is said to be *primitive* if there is an integer k such that, for every letter a, all letters appear in $\varphi^k(a)$. When $\mathrm{Card}(A) = 1$, we also consider as a primitive substitution every endomorphism $A^+ \to A^+$ distinct of the identity map.

For an arbitrary finite alphabet A, say that a substitution $\varphi : A^+ \to A^+$ is *expansive* if for all large enough natural n the length of $\varphi^{n+1}(a)$ is greater than that of $\varphi^n(a)$ whenever $a \in A$. All primitive substitutions are expansive (Exercise 5.9).

If φ is primitive, then φ is biextendable (Exercise 5.10), and so we may consider the associated substitutive shift space $X(\varphi)$ and the corresponding language of blocks $F(\varphi)$.

Example 5.5.2 The *Fibonacci morphism* $\varphi : a \to ab, b \to a$ is primitive. The corresponding shift space $X(\varphi)$ is called the *Fibonacci shift space*.

Example 5.5.3 The *Thue-Morse morphism* $\tau : a \to ab, b \to ba$ is primitive. The shift space $X(\tau)$ is called the *Thue-Morse shift space*.

The proof of the following proposition is left as an exercise (Exercise 5.11).

Proposition 5.5.4 *If φ is a primitive substitution, then the set $F(\varphi)$ is uniformly recurrent.*

The primitive substitution φ is said to be *periodic* when $F(\varphi)$ is periodic. Given φ, one can decide whether φ is periodic or not (see Exercise 5.15).

A *periodic point* of a substitution φ is an infinite word $x \in A^{\mathbb{N}}$ such that $\varphi^n(x) = x$ for some positive integer n (one should not make the confusion with a periodic point of the shift map). It is called a *fixed point* when $n = 1$.

Proposition 5.5.5 *Every expansive substitution (and therefore, every primitive substitution) has a periodic point.*

Proof Let $\varphi : A^+ \to A^+$ be an expansive substitution. Take $b \in A$. Since A is finite, there are natural numbers i, j with $i < j$ and such that $\varphi^i(b)$ and $\varphi^j(b)$ start with the same letter a. Let $n = j - i \geqslant 1$. From the equality $\varphi^n(\varphi^i(b)) = \varphi^j(b)$ we deduce that $\varphi^n(a)$ begins with a. Set $\psi = \varphi^n$. Then $\psi^{m+1}(a)$ begins with $\psi^m(a)$ for every $m \geqslant 1$. Since φ is expansive, the length of $\psi^m(a)$ tends to infinity. Then the infinite word in $A^{\mathbb{N}}$ which has all $\psi^m(a)$ as prefixes is a fixed point of ψ and thus a periodic point of φ. ∎

Example 5.5.6 The Fibonacci morphism φ has a unique fixed point $x = abaababa\cdots$ called the *Fibonacci word*.

Example 5.5.7 The Thue-Morse substitution $\tau : a \mapsto ab, b \mapsto ba$ has two fixed points $x = abbabaab\cdots$, which is called the *Thue-Morse word*, and $y = baababba\cdots$.

Example 5.5.8 Let $\varphi : a \to ba, b \to a$. Then $\psi = \varphi^2$ has two fixed points with prefixes the $\psi^n(a)$ and $\psi^n(b)$ respectively.

One may also consider two-sided fixed points, that is two-sided infinite words x such that $\varphi(x) = x$.

For this, we introduce the notion of a *connection* of a morphism $\varphi : A^* \to A^*$. It is a pair (a, b) of letters such that $ba \in F(\varphi)$, that $\varphi^\omega(a)$ begins with a and $\varphi^\omega(b)$ ends with b.

Remark 5.5.9 Let the unique continuous endomorphism of $\widehat{A^*}$ extending the substitution $\varphi : A^* \to A^*$ be also denoted by φ. Let $a \in A$. In accordance with Theorem 3.12.1 and the definition of the pointwise topology, the pseudoword $\varphi^\omega(a)$ is the limit in $\widehat{A^*}$ of the words $\varphi^{n!}(a)$ and thus $\varphi^\omega(a)$ begins with a if (and only if) there is a positive integer k such that $\varphi^k(a)$ begins with a, since indeed $\varphi^n(a)$ starts with a for all multiple n of such integer k. A dual remark holds if we replace "first letter" by "last letter".

Proposition 5.5.10 *Every nonerasing substitution has a connection.*

Proof Let $a, b \in A$ be such that $ba \in F(\varphi)$. Since the alphabet is finite, there are integers $i_0 \geqslant 0$ and $p \geqslant 1$ such that the first letter of $\varphi^{i_0}(a)$ is equal to the first letter of $\varphi^{i_0+p}(a)$, and so the first letter of $\varphi^{\ell}(\varphi^{i_0+p}(a))$ is the first letter of $\varphi^{\ell}(\varphi^{i_0}(a))$ whenever $\ell \geqslant 0$. In other words, the first letter of $\varphi^i(a)$ is the first letter of $\varphi^{i+p}(a)$ for all $i \geqslant i_0$. Similarly, there are $j_0 \geqslant 0$ and $q \geqslant 1$ such that, for every $j \geqslant j_0$, the last letter of $\varphi^j(b)$ is equal to the last letter of $\varphi^{j+q}(b)$. Set $k = \max\{i_0, j_0\}$ and let c be the first letter of $\varphi^k(a)$ and d be the last letter of $\varphi^k(b)$. Then dc is a factor of $\varphi^k(ba)$ and thus $dc \in F(\varphi)$. Moreover $\varphi^p(c)$ begins by c and thus $\varphi^{\omega}(c) \in c\widehat{A^*}$. Similarly we have $\varphi^{\omega}(d) \in \widehat{A^*}d$. This shows that dc is a connection of φ. ∎

Let (a, b) be a connection of φ. A *connective power* of φ, with respect to the connection (a, b), is a finite power $\tilde{\varphi}$ of φ such that the first letter of $\tilde{\varphi}(a)$ is a and the last letter of $\tilde{\varphi}(b)$ is b. It follows from Proposition 5.5.10 that every primitive morphism has a connective power.

Whenever φ is primitive, it is clear that (a, b) is a connection of φ if and only if there is a two-sided infinite word $x \in X(\varphi)$ such that $x_{-1} = b$ and $x_0 = a$ and $\tilde{\varphi}(x) = x$. If (a, b) is a connection, such a two-sided infinite word is unique and is called an *admissible two-sided fixed point* of φ.

Example 5.5.11 Let $\tau : a \mapsto ab, b \mapsto ba$ be the Thue-Morse morphism. The word aa is a connection for τ and $\tilde{\tau} = \tau^2$ is a connective power of τ.

5.5.2 Matrix of a Substitution

Let φ be a substitution on A. The *matrix associated* to φ is the integer $A \times A$-matrix $M(\varphi)$ with entries

$$M_{a,b} = |\varphi(a)|_b,$$

where $|v|_b$ denotes the number of times the letter b appears in the word v. It is easy to verify that $M(\varphi^n) = M(\varphi)^n$ for every $n \geqslant 1$.

A real square matrix M is said to be *primitive* when, for some positive integer n, all entries in M^n are positive. Note that φ is primitive if and only if $M(\varphi)$ is primitive.

Example 5.5.12 The Fibonacci morphism $\varphi : a \mapsto ab, b \mapsto a$ is primitive. Indeed, both letters a, b appear in each of the words $\varphi^2(a)$ and $\varphi^2(b)$. One has

$$M(\varphi) = \begin{pmatrix} 1 & 1 \\ 1 & 0 \end{pmatrix}, \quad M(\varphi^2) = \begin{pmatrix} 2 & 1 \\ 1 & 1 \end{pmatrix}.$$

We will use the following classical result, known as the *Perron-Frobenius Theorem* (see the Notes Section for a reference).

Theorem 5.5.13 *Let M be an $n \times n$-matrix with real nonnegative coefficients. If M is primitive, then it has a positive eigenvalue ρ such that $|\lambda| < \rho$ for every other eigenvalue λ of M. Moreover, there corresponds to ρ an eigenvector with strictly positive coefficients.*

For a substitution φ on A, we denote

$$|\varphi| = \min\{|\varphi(a)| \mid a \in A\}, \quad \|\varphi\| = \max\{|\varphi(a)| \mid a \in A\}.$$

Note that for every $k \geqslant 1$,

$$|\varphi^k| \leqslant \|\varphi\| \, |\varphi^{k-1}|, \text{ and } \|\varphi^k\| \leqslant \|\varphi\| \, \|\varphi^{k-1}\|. \tag{5.1}$$

We deduce from Theorem 5.5.13 the following result which we will use in Chap. 6.

Proposition 5.5.14 *Let φ be a primitive morphism. There is a $k \geqslant 0$ such that $\|\varphi^n\| \leqslant k|\varphi^n|$ for all $n \geqslant 1$.*

Proof Set $M = M(\varphi)$. By Theorem 5.5.13 there is a real number ρ and a strictly positive vector v such that $Mv = \rho v$. Let $\alpha = \min\{v_a \mid a \in A\}$ and $\beta = \max\{v_a \mid a \in A\}$. Then for every $a \in A$,

$$\frac{\alpha}{\beta}\rho^n \leqslant \frac{\rho^n v_a}{\beta} = \sum_{b \in A} \frac{(M^n)_{a,b} v_b}{\beta} \leqslant \sum_{b \in A} (M^n)_{a,b} \leqslant \sum_{b \in A} \frac{(M^n)_{a,b} v_b}{\alpha} = \frac{\rho^n v_a}{\alpha} \leqslant \frac{\beta}{\alpha}\rho^n.$$

Since for every $a \in A$, $|\varphi^n(a)| = \sum_{b \in A}(M^n)_{a,b}$, this shows that

$$\frac{\alpha}{\beta}\rho^n \leqslant |\varphi^n| \leqslant \|\varphi^n\| \leqslant \frac{\beta}{\alpha}\rho^n$$

and, thus,

$$\|\varphi^n\| \leqslant \frac{\beta}{\alpha}\rho^n \leqslant \frac{\beta^2}{\alpha^2}|\varphi^n|.$$

This proves the statement with $k = \beta^2/\alpha^2$.   ∎

We will also need later the following property of substitutive shift spaces. The *exponent of periodicity* of a factorial set F is the minimal value of $n \geqslant 1$ such that F does not contain words of the form w^n with w nonempty. It can be finite or infinite.

Theorem 5.5.15 *Let φ be a primitive morphism such that $F(\varphi)$ is not periodic. The exponent of periodicity of φ is finite.*

The proof relies on the following lemma.

Lemma 5.5.16 *Let φ be a primitive substitution. If there exists a primitive word v and integers $n, p \geqslant 1$ such that*

(i) *for every $a, b \in A$ with $ab \in F(\varphi)$, $\varphi^p(ab)$ is a factor of v^n.*
(ii) *$2|v| \leqslant |\varphi^p|$.*

Then $F(\varphi)$ is periodic.

Proof For every $a \in A$, since $\varphi^p(a)$ is a factor of v^n, there is an integer n_a, a proper prefix w_a of v and a proper suffix v_a of v such that $\varphi^p(a) = v_a v^{n_a} w_a$. If $ab \in F(\varphi)$, then $v^{n_a} w_a v_b v^{n_b} \in F(\varphi)$ and is a factor of v^n. Since v is primitive, this forces $w_a v_b = v$ or $w_a v_b = \varepsilon$. This implies that for every word w and $a, b \in A$ such that $awb \in F(\varphi)$, one has $\varphi^p(awb) \in v_a v^* w_b$. Since all these words are factors of v^*, it shows that $F(\varphi)$ is periodic. ∎

Proof of Theorem 5.5.15 Since φ is primitive, $F(\varphi)$ is uniformly recurrent. Let $r \geqslant 1$ be such that every word $ab \in F(\varphi)$ of length 2 is a factor of every word of length r in $F(\varphi)$.

Assume that $w^n \in F(\varphi)$ for some nonempty $w \in F(\varphi)$. Let $p \geqslant 1$ be such that $|\varphi^{p-1}| \leqslant 2|w| < |\varphi^p|$. For every $a, b \in A$ with $ab \in F(\varphi)$, there is an occurrence of $\varphi^p(ab)$ in each word of $F(\varphi)$ of length $r\|\varphi^p\|$. Indeed, if $u \in F(\varphi)$ is of length $(r+2)\|\varphi^p\|$, there is a word $x \in F(\varphi)$ of length at least r such that $u = u_1\varphi^p(x)u_2$. By the choice of r, ab is a factor of x and thus there is an occurrence of $\varphi^p(ab)$ in u. Hence, by Lemma 5.5.16, we have

$$|w^n| < 2r\|\varphi^p\| \tag{5.2}$$

since otherwise every $\varphi^p(ab)$ would be factor of w^n with $2|w| < |\varphi^p|$ and thus $F(\varphi)$ would be periodic.

Since $|w^n| = n|w|$, we deduce from (5.2) the inequality

$$n < \frac{2r\|\varphi^p\|}{\frac{1}{2}|\varphi^{p-1}|}$$

$$< 4r\frac{\|\varphi^p\|}{|\varphi^{p-1}|} \leqslant 4r\|\varphi\|\frac{\|\varphi^{p-1}\|}{|\varphi^{p-1}|} \leqslant 4r\|\varphi\|k$$

where k is the constant of Proposition 5.5.14. This shows that the exponent of periodicity of $F(\varphi)$ is bounded by $4r\|\varphi\|k$. ∎

5.5.3   Recognizable Substitutions

Let $\varphi : A^* \to A^*$ be a substitution. Let $u \in A^{\mathbb{N}}$ be a one-sided fixed point of φ. Define a map $\alpha_u : \mathbb{N} \to \mathbb{N}$ by

$$\alpha_u(n) = \left| \varphi\big(u[0, n[\big) \right|$$

where $u[0, n[= u_0 \cdots u_{n-1}$. Thus

$$\varphi\big(u[0, n[\big) = u[0, \alpha_u(n)[.$$

Note that $\alpha_u(0) = 0$.

Let $E_u(\varphi)$ be the set

$$E_u(\varphi) = \{\alpha_u(n) \mid n \geqslant 0\}.$$

For every $n \geqslant 0$, we define an equivalence relation on the set of integers at least equal to n by

$$i \sim_{u,n} j \text{ if and only if } u[i - n, i + n] = u[j - n, j + n].$$

The morphism φ is said to be *recognizable* on u if there is an integer $n \geqslant 0$ such that for every $i, j \geqslant n$, whenever $i \sim_{u,n} j$ we have $i \in E_u(\varphi)$ if and only if $j \in E_u(\varphi)$.

The least integer n such that the above condition holds is called the *recognizability constant* of φ on u.

Example 5.5.17 Let $\varphi : a \mapsto ab, b \mapsto a$ be the Fibonacci morphism. It is recognizable on its fixed point $u = abaab \cdots$ with $n = 0$. Indeed, one has $i \in E_u(\varphi)$ if and only if $u_i = a$.

Example 5.5.18 Let $\varphi : a \mapsto ab, b \mapsto ba$ be the Morse morphism. Let $u = abbabaab \cdots$ be the fixed point beginning with a. Then φ is recognizable on u with $n = 2$. Indeed, let x be a word of length 5 in $F(\varphi)$ and let $i \geqslant 2$ be such that $u[i - 2, i + 2] = x$. Thus, $x = u_{i-2}u_{i-1}u_iu_{i+1}u_{i+2}$. Since x has length 5, it has a factor aa or bb. If such factor is $u_{i-1}u_i$ or $u_{i+1}u_{i+2}$, then $i \in E_u(\varphi)$ and otherwise $i \notin E_u(\varphi)$.

We quote without proof the following result (see the Notes Section).

Theorem 5.5.19 (Mossé) *Every primitive nonperiodic substitution is recognizable on its one-sided fixed points.*

As a complement, we mention that sometimes in the literature the following condition, called *asymptotic injectivity*, is attached to the definition of recognizability: there is an integer n such that whenever $\alpha_u(k) \sim_{u,n} \alpha_u(\ell)$ then $u_k = u_\ell$. Theorem 5.5.19 still holds with this additional condition. We will not need this stronger form of Mossé's Theorem.

Example 5.5.20 The Fibonacci morphism does not satisfy the asymptotic injectivity
with $n = 0$ on its fixed-point u. Indeed, $\alpha_u(1) = 2$ and $\alpha_u(2) = 3$ with $u_2 = u_3 = a$
although $u_1 = b$ and $u_2 = a$. It is satisfied however with $n = 1$. Indeed, $u_k = a$ if
and only if $u_{\alpha(k)+1} = b$.

For a two-sided fixed point $w \in A^{\mathbb{Z}}$ of φ, let $\beta_w : \mathbb{Z} \to \mathbb{Z}$ be the map

$$\beta_w(n) = \begin{cases} |\varphi(w[0, n[)| & \text{if } n \geqslant 0 \\ -|\varphi(w[0, n[)| & \text{otherwise} \end{cases}$$

Thus for $n \geqslant 0$, one has $\beta_w = \alpha_u$ where $u = w_0 w_1 \cdots$. Note that if $i < j$, then

$$\beta_w(j) - \beta_w(i) = |\varphi(w[i, j[)| \tag{5.3}$$

Indeed, this is true if i, j have the same sign. If $i < 0 < j$, then $\beta_w(j) - \beta_w(i) =$
$|\varphi(w[0, j[)| + |\varphi(w[i, 0[)| = |\varphi(w[i, j[)|$. Similarly, if $i < j$, then

$$\varphi(w[i, j[) = w[\beta(i), \beta(j)[\tag{5.4}$$

Let $E_w(\varphi)$ be the set

$$E_w(\varphi) = \{\beta_w(n) \mid n \in \mathbb{Z}\}.$$

For every $n \geqslant 1$, we define an equivalence relation on $\mathbb{Z}$ by

$$i \sim_{w,n} j \text{ if and only if } w[i - n, i + n] = w[j - n, j + n].$$

We say that φ is *recognizable* on w if there is an integer $n \geqslant 0$ such that for every
$i, j \in \mathbb{Z}$, whenever $i \sim_{w,n} j$ we have $i \in E_w(\varphi)$ if and only if $j \in E_w(\varphi)$.

There is also a bilateral version of the notion of asymptotic injectivity, defined
by: there is an integer n such that $\beta_w(k) \sim_{w,n} \beta_w(\ell)$ implies $w_k = w_\ell$.

Example 5.5.21 Let $\varphi : a \mapsto ab, b \mapsto a$ be the Fibonacci morphism. Then (a, b)
is a connection. Let $w = \cdots abab \cdot abaab \cdots$ be the fixed point of φ^2 such that
$w_{-1} w_0 = ba$. Then φ^2 is recognizable on w with $n = 1$. Indeed, one has $i \in E_w(\varphi)$
if and only if $w[i - 1, i + 1] = aab$ or bab. The asymptotic injectivity requires
$n = 3$. Indeed $w_k = a$ if and only if $w_{\beta_w(k)+3} = a$.

We will use the following variant of Mossé's Theorem.

Theorem 5.5.22 *A primitive nonperiodic substitution φ is recognizable on each of
its two-sided admissible fixed-points.*

Proof Let φ be primitive and nonperiodic. Let n be the recognizability constant of
φ on the one-sided infinite word $u = w_0 w_1 \cdots$. We show that φ is recognizable on
w with the same constant.

Let $i, j \in \mathbb{Z}$ be such that $i \sim_{w,n} j$. We may suppose $i < j$. Let $k < \ell$ be such that $\beta_w(k) \leqslant i - n < j + n < \beta_w(\ell)$ (see Fig. 5.3).

Since w is admissible, there is some $t > 0$ such that $k+t \geqslant 0$ and $w[k+t, \ell+t[= w[k, \ell[$.

Suppose that $i \in E_w(\varphi)$. Let $m \in \mathbb{Z}$ be such that $\beta_w(m) = i$. Then

$$\beta_w(m + t) \sim_{w,n} \beta_w(m) = i$$

$$\sim_{w,n} j$$

$$\sim_{w,n} j - i + \beta(m + t)$$

where the first and the last equivalence follow from the fact that $w[k + t, \ell + t[= w[k, \ell[$ and thus, using (5.3), $w[\beta_w(k + t), \beta_w(\ell + t)[= w[\beta_w(k), \beta_w(\ell)[$ (see Fig. 5.4).

Since φ is recognizable on u with constant n, we infer that $\beta_w(m + t) + j - i \in E_u(\varphi)$. Let $q \geqslant 0$ be such that $\beta_w(q) = \alpha_u(q) = \beta_w(m+t) + j - i$. Set $p = q - t$. Then, using (5.3),

$$\beta_w(p + t) - \beta_w(m + t) = \left| \varphi(w[m + t, p + t[) \right|$$

$$= \left| \varphi(w[m, p[) \right| = \beta_w(p) - \beta_w(m).$$

Since $\beta_w(m) = i$, we conclude that $\beta_w(p) = \beta_w(m) + j - i = j$ and, thus, that $j \in E_w(\varphi)$. The proof that $j \in E_w(\varphi)$ implies $i \in E_w(\varphi)$ is entirely analogous. ∎

Theorem 5.5.22 also holds if we add to the definition of bilateral recognizability the aforementioned asymptotic injectivity condition for biinfinite words. One may

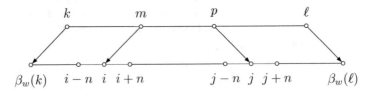

Fig. 5.3 Locating k, m, p, ℓ

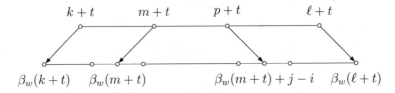

Fig. 5.4 Locating $k + t, m + t, p + t, \ell + t$

deduce from that stronger form of Theorem 5.5.22, that for every primitive nonperiodic substitution φ, the map $\varphi : X(\varphi) \to X(\varphi)$ is injective (Exercise 5.17).

5.6   The Topological Closure of a Uniformly Recurrent Set

In the following pages, we introduce a connection between symbolic dynamics and free profinite semigroups that is based on the simple idea of considering the topological closure in $\widehat{A^*}$ of a (uniformly) recurrent subset of A^*.

We begin to see how the properties of being factorial, biextendable or recurrent translate into the topological closure. For that purpose, we introduce the following extensions to $\widehat{A^*}$ of these definitions. Say that a nonempty subset K of $\widehat{A^*}$ is *factorial* if it contains all factors of elements of K; that K is *recurrent* if it is factorial, contains some pseudoword different from the empty word, and for all $x, z \in K$ there is some $y \in K$ such that $xyz \in K$; and that K is *biextendable* if it is factorial and, for every $x \in K$, one has $AxA \cap K \neq \emptyset$. Because $\widehat{A^*} \setminus A^*$ is an ideal of $\widehat{A^*}$, a subset of A^* is factorial (respectively, recurrent) in A^* if and only if it is factorial (respectively, recurrent) in $\widehat{A^*}$. It is also clear that the definition of biextendable subset of $\widehat{A^*}$ is consistent with the one given for subsets of A^*.

Proposition 5.6.1 *Let F be a factorial subset of A^*. Then $\overline{F}$ is a factorial subset of $\widehat{A^*}$. Moreover, if F is a recurrent (respectively, biextendable) subset of A^*, then $\overline{F}$ is a recurrent (respectively, biextendable) subset of $\widehat{A^*}$.*

Proof To say that $\overline{F}$ is factorial can be translated into saying that $uv \in \overline{F}$ implies $u, v \in \overline{F}$. Assuming that $uv \in \overline{F}$, let (w_n) be a sequence of elements of F converging to uv. By Proposition 4.4.17, there are sequences (u_n) and (v_n) such that $\lim u_n = u$, $\lim v_n = v$ and $w_n = u_n v_n$. The set $\widehat{A^*} \setminus A^*$ is an ideal of $\widehat{A^*}$, thus $u_n, v_n \in A^*$ in particular. Actually, we have $u_n, v_n \in F$, as we are assuming that F is factorial, which yields $u, v \in \overline{F}$.

Suppose moreover that F is a recurrent subset of A^*. Given $u, v \in \overline{F}$, let (u_n) and (v_n) be sequences of elements of F respectively converging to u and v. Since F is recurrent, for each n there is $w_n \in A^*$ such that $u_n w_n v_n \in F$. By the compactness of $\widehat{A^*}$, we may consider an accumulation point w of the sequence (w_n). Thanks to the continuity of the multiplication in $\widehat{A^*}$, we then have $uwv \in \overline{F}$, thus establishing that $\overline{F}$ is recurrent. The proof that $\overline{F}$ is biextendable when F is biextendable is similar.   ∎

5.6.1   *Uniformly Recurrent Pseudowords*

For an infinite pseudoword w, we denote by $F(w)$ the set of finite factors of w. It is an infinite factorial set. An infinite pseudoword w is *uniformly recurrent* if $F(w)$ is uniformly recurrent. Every uniformly recurrent set is of this form, as next shown.

Proposition 5.6.2 *If F is a uniformly recurrent subset of A^*, then there is an infinite pseudoword u such that $F(u) = F$. More precisely, for every infinite pseudoword u in $\overline{F} \subseteq \widehat{A^*}$ we have $F(u) = F$.*

Proof First notice that, since F is infinite, there is indeed some element u in $\overline{F} \cap \widehat{A^*} \setminus A^*$, namely any accumulation point of a sequence of elements of A^* with strictly increasing lengths. By Proposition 5.6.1, we have $F(u) \subseteq \overline{F}$. In fact, because the elements of A^* are isolated points of $\widehat{A^*}$ and thus $\overline{F} \cap A^* = F$, we actually have $F(u) \subseteq F$. Since F is minimal among the infinite factorial subsets of A^* (cf. Proposition 5.2.3), the equality $F(u) = F$ holds. ∎

If u is a word, then u^ω is a uniformly recurrent word, as $F(u^\omega)$ is a periodic set (the finite factors of u^ω are the finite factors of some finite power of u). More interesting examples are provided by primitive substitutions. Let φ be a substitution over A. Then φ extends to a unique continuous endomorphism of $\widehat{A^*}$, which we still denote φ. In the following lines we again use the information gathered in Sect. 3.12, namely that the topological monoid $\mathrm{End}(\widehat{A^*})$ is profinite (Theorem 3.12.1).

Proposition 5.6.3 *Let φ be a primitive substitution over a finite alphabet A. For every $a \in A$, the pseudoword $\varphi^\omega(a)$ is uniformly recurrent. More precisely, one has $F(\varphi^\omega(a)) = F(\varphi)$, for every $a \in A$.*

Proof Let u be a finite factor of $\varphi^\omega(a)$. Since $\mathrm{End}(\widehat{A^*})$ is endowed with the pointwise topology, we have $\varphi^\omega(a) = \lim \varphi^{n!}(a)$. Therefore, u is a finite factor of $\varphi^{n!}(a)$ for all sufficiently large n (recall that the set $\widehat{A^*}u\widehat{A^*}$ is open), establishing that $F(\varphi^\omega(a)) \subseteq F(\varphi)$. Because φ is expansive, the pseudoword $\varphi^\omega(a)$ is infinite, thus $F(\varphi^\omega(a))$ is an infinite factorial set. It follows that $F(\varphi^\omega(a)) = F(\varphi)$, as $F(\varphi)$ is a minimal infinite factorial set (cf. Propositions 5.5.4 and 5.2.3). ∎

The following result characterizes uniformly recurrent pseudowords.

Proposition 5.6.4 *An infinite pseudoword is uniformly recurrent if and only if all its infinite factors have the same finite factors.*

Proof Let w be an infinite pseudoword. Assume first that w is uniformly recurrent. Let u be an infinite factor of w. Let us show that $F(u) = F(w)$. The inclusion $F(u) \subseteq F(w)$ is clear. Conversely, let $x \in F(w)$. Since w is uniformly recurrent, x is a factor of every long enough finite factor of w. In particular, x is a factor of every long enough finite factor of u. Thus $x \in F(u)$.

Conversely, assume that $F(w) = F(u)$ for all infinite factors u of w. Let v be a finite factor of w. Arguing by contradiction, assume that there are arbitrary long factors of w which do not have v as a factor. This infinite set contains, by Proposition 3.3.15, a subsequence converging to some infinite pseudoword u which is also a factor of w, and such that $v \notin F(u)$ (as $\widehat{A^*}v\widehat{A^*}$ is open), a contradiction. ∎

We briefly look at the special case of periodic pseudowords, that is, pseudowords whose set of finite factors is periodic. If the pseudoword is periodic, then its $\mathcal{J}$-class has a very simple structure, with a finite number of $\mathcal{H}$-classes, and its

Fig. 5.5 The $\mathcal{J}$-class of
$(ab)^\omega$

$(ab)^\omega$	$(ab)^\omega a$
$b(ab)^\omega$	$(ba)^\omega$

maximal subgroups are isomorphic with the free profinite group on one generator (Exercise 5.20).

Example 5.6.5 The $\mathcal{J}$-class of a^ω in $\widehat{A^*}$ is made of one uncountable $\mathcal{H}$-class. The $\mathcal{J}$-class of $(ab)^\omega$ has four uncountable $\mathcal{H}$-classes. It is represented in Fig. 5.5 where only representatives of each $\mathcal{H}$-class are depicted.

Before we proceed to consider arbitrary uniformly recurrent pseudowords, we register the following simple observation (recall Proposition 5.6.3).

Proposition 5.6.6 *Let φ be a primitive substitution over a finite alphabet A. The pseudowords $\varphi^\omega(a)$, with $a \in A$, are all $\mathcal{J}$-equivalent.*

Proof Let $a, b \in A$. Since φ is primitive, for every sufficiently large n, every element b of A is a factor of $\varphi^n(a)$. By continuity of the multiplication we conclude that b is a factor of $\varphi^\omega(a)$, whence $\varphi^\omega(a) \leqslant_\mathcal{J} b$. As φ^ω is an idempotent endomorphism, it follows that $\varphi^\omega(a) = \varphi^\omega(\varphi^\omega(a)) \leqslant_\mathcal{J} \varphi^\omega(b)$. By symmetry, we conclude that $\varphi^\omega(a)\mathcal{J}\varphi^\omega(b)$.   ∎

The next result gives an algebraic characterization of uniform recurrence in the free profinite monoid. By an *infinite $\mathcal{J}$-maximal pseudoword* we mean a infinite pseudoword u all of whose factors which are not $\mathcal{J}$-equivalent to u must be words, that is, finite pseudowords.

Theorem 5.6.7 *An infinite pseudoword is uniformly recurrent if and only if it is an infinite $\mathcal{J}$-maximal pseudoword.*

Before we present a proof of Theorem 5.6.7, we extract a couple of straightforward consequences.

Corollary 5.6.8 *Every uniformly recurrent pseudoword is regular.*

Proof Let w be a uniformly recurrent pseudoword. Then w has an infinite idempotent pseudoword e as a factor (cf. Exercise 4.22). Since, by Theorem 5.6.7, the pseudoword w is $\mathcal{J}$-maximal among infinite pseudowords, it follows that $w\mathcal{J}e$, thus w is regular.   ∎

In view of Proposition 5.6.3, one should consider Proposition 5.6.6 as a special case of the following corollary of Theorem 5.6.7.

Corollary 5.6.9 *Let w be a uniformly recurrent pseudoword. For every infinite pseudoword u, we have $u\mathcal{J}w$ if and only if $F(u) = F(w)$, if and only if $F(u) \subseteq F(w)$.*

Proof Note that $F(u) \subseteq F(w)$ implies $F(u) = F(w)$ by Proposition 5.2.3, and so we only have to show that $F(u) = F(w)$ implies that $u \mathcal{J} w$. Let u_n be a factor of u with length at least n, and let u' be an accumulation point of the sequence (u_n). Since $F(u) = F(w)$, the infinite pseudoword u' is a factor of both u and w. As u and w are uniformly recurrent, we have $u \mathcal{J} u' \mathcal{J} w$ by Theorem 5.6.7. ∎

Corollary 5.6.9 is remarkable in the sense that in general the $\mathcal{J}$-equivalence with a pseudoword is not determined by the set of its finite factors. For example, the pseudowords $(b^{\omega}ab^{\omega})^{\omega}$ and $b^{\omega}ab^{\omega}$ have the same set of finite factors (namely, $b^*ab^* \cup b^*$), but they are not $\mathcal{J}$-equivalent: the first is idempotent, while the second is not even regular, because a regular pseudoword having a as a factor must belong to the clopen set $\widehat{A^*a\widehat{A}^*a}\,\widehat{A^*}$.

To prove Theorem 5.6.7, we first show the next lemma.

Lemma 5.6.10 *Let u be a uniformly recurrent pseudoword and suppose v is a pseudoword such that uv is still uniformly recurrent. Then u and uv are $\mathcal{R}$-equivalent.*

Proof Suppose first that v is finite. For each positive integer n, let u_n be a suffix of u of length at least n. By compactness, we may assume that (u_n) converges to a pseudoword u'. Since u is an infinite factor of the uniformly recurrent pseudoword uv, we have $F(u) = F(uv)$ by Proposition 5.6.4. Therefore, for every n, the words $u_n v$ and u_n are elements of the uniformly recurrent set $F(u)$. We conclude that there is a strictly increasing sequence m_n of positive integers for which there is a factorization $u_{m_n} = x_n u_n v y_n$, for some (possibly empty) words x_n, y_n. By compactness, we may consider an accumulation point (x, y) of the sequence (x_n, y_n). Then, by continuity of multiplication, we get $u' = x u' v y$, whence $u' v \mathcal{J} u'$ and thus $u' \mathcal{R} u' v$ by stability. Finally, since u' is a suffix of u, we conclude that $u \mathcal{R} u v$ as the $\mathcal{R}$-equivalence is a left congruence.

Assume next that v is infinite. We assume by contradiction that $u >_{\mathcal{R}} uv$. By a standard compactness argument (see Exercise 5.19), it follows that there is a continuous morphism $\varphi : \widehat{A^*} \to M$ into a finite monoid M such that $\varphi(u) >_{\mathcal{R}} \varphi(uv)$, where A denotes the alphabet under consideration. Let v_n be a sequence of finite words converging to v. Taking a subsequence, we may assume that $\varphi(v_n)$ is constant and so $\varphi(u) >_{\mathcal{R}} \varphi(uv_n)$ for all n. Thus, for each n, we have a factorization $v_n = x_n a_n y_n$ with $a_n \in A$ such that $\varphi(u)\mathcal{R}\varphi(ux_n) >_{\mathcal{R}} \varphi(ux_n a_n) \geqslant_{\mathcal{R}} \varphi(uv_n)$. Since the alphabet is finite and $\widehat{A^*}$ is compact, we may, up to taking a subsequence, assume that the letter sequence a_n is constant and the sequences x_n, y_n converge to x and y respectively. Thus we have $v = xay$ with $a \in A$ and $\varphi(u)\mathcal{R}\varphi(ux) >_{\mathcal{R}} \varphi(uxa) \geqslant_{\mathcal{R}} \varphi(uv)$ so that, in particular, we get $ux >_{\mathcal{R}} uxa \geqslant_{\mathcal{R}} uv$. On the other hand, since ux and uxa are infinite factors of uv, they are both uniformly recurrent by Proposition 5.6.4. By the first part, we have $ux \mathcal{R} uxa$, a contradiction. ∎

A dual result holds for the $\mathcal{L}$-order.

Proof of Theorem 5.6.7 Suppose first that w is $\mathcal{J}$-maximal as an infinite pseudoword. If v is an infinite factor of w, then it is $\mathcal{J}$-equivalent to w. Hence v, w
have the same factors. Therefore, by Proposition 5.6.4, w is uniformly recurrent.

Conversely, suppose that $u, w \in \widehat{A^*} \setminus A^*$ are such that $u \geqslant_{\mathcal{J}} w$ with w uniformly
recurrent. Set $w = puq$ with $p, q \in \widehat{A^*}$. Notice that u and pu are also uniformly
recurrent pseudowords, by Proposition 5.6.4. By the dual of Lemma 5.6.10, we have
$pu \mathcal{L} u$. And by Lemma 5.6.10, we have $pu \mathcal{R} puq$. Thus u and w are $\mathcal{J}$-equivalent.

5.6.2   The $\mathcal{J}$-Class of a Uniformly Recurrent Set

Let F be a uniformly recurrent subset of A^*. By Proposition 5.6.2 there is a uniformly recurrent pseudoword w such that $F = F(w)$. Moreover, by Corollary 5.6.9,
for every uniformly recurrent pseudoword w' such that $F = F(w')$, we have $w \mathcal{J} w'$.
Therefore, the $\mathcal{J}$-class of any pseudoword w such that $F = F(w)$, which we denote
by $J(F)$, is a $\mathcal{J}$-class that depends only on F and not on the particular w.

The following observation is basically a reformulation of Corollary 5.6.9.

Remark 5.6.11 For every uniformly recurrent set $F \subseteq A^*$ and every infinite
pseudoword $u \in \widehat{A^*}$, one has $u \in J(F)$ if and only if $F(u) = F$, if and only if
$F(u) \subseteq F$.

The proof of the following proposition is left as an exercise (Exercise 5.21).

Proposition 5.6.12 *The mapping $F \mapsto J(F)$ is a bijection from the set of
uniformly recurrent subsets of A^* to the set of $\mathcal{J}$-classes of $\mathcal{J}$-maximal infinite
pseudowords over A.*

In particular, the assignment $F \mapsto J(F)$ turns out to be a bijective correspondence between the set of uniformly recurrent subsets of A^* and the set of $\mathcal{J}$-classes
of $\mathcal{J}$-maximal infinite pseudowords. Since there are $2^{\aleph_0}$ uniformly recurrent sets of
A^* when A has more than one letter, we immediately deduce that $\widehat{A^*}$ is very "large"
with respect to the $\mathcal{J}$-order, in the following precise sense.

Proposition 5.6.13 *If $\operatorname{Card}(A) > 1$, then, for the $\leqslant_{\mathcal{J}}$-order, there is an anti-chain
of $2^{\aleph_0}$ elements in $\widehat{A^*}$. Moreover, the elements can be chosen to be maximal regular.*

Next is another perspective of the dependence of $J(F)$ on F only.

Proposition 5.6.14 *The equality $J(F) = \overline{F} \setminus A^*$ holds for every uniformly
recurrent subset F of A^*.*

The proof is left as an exercise (Exercise 5.22).

For $w \in \widehat{A^*}$, denote by $\overrightarrow{w}$ the right infinite word in $A^{\mathbb{N}}$ whose finite prefixes are
those of w. Symmetrically, $\overleftarrow{w}$ is the left infinite word in $A^{\mathbb{Z}_-}$ whose finite suffixes
are those of w, where $\mathbb{Z}_-$ denotes the set of negative integers.

Proposition 5.6.15 *For a uniformly recurrent set F, two pseudowords $u, v \in J(F)$ are $\mathcal{R}$-equivalent if and only if $\overrightarrow{u} = \overrightarrow{v}$, and they are $\mathcal{L}$-equivalent if and only if $\overleftarrow{u} = \overleftarrow{v}$.*

Proof Assume that $\overrightarrow{u} = \overrightarrow{v}$. Let w be an accumulation point of the sequence of prefixes of length n of u. Then w is a prefix of u. It is also in $\overline{F}$ and thus in $J(F)$ by Proposition 5.6.14. Since $\widehat{A^*}$ is stable, this implies that $w\mathcal{R}u$. On the other hand, since u and v have the same set of finite prefixes, the same argument applies to v and thus u, v, w belong to the same $\mathcal{R}$-class of $J(F)$. ∎

Over the finite alphabet A, a right infinite word $x \in A^{\mathbb{N}}$ and a left infinite word $y \in A^{\mathbb{Z}_-}$ can be concatenated in a natural way to produce a two-sided infinite word $x \cdot y \in A^{\mathbb{Z}}$. According to Proposition 5.6.15, the $\mathcal{H}$-class of a uniformly recurrent pseudoword w is completely determined by $\overleftarrow{w} \cdot \overrightarrow{w}$. Previous observations lead to the following remark concerning maximal subgroups that are $\mathcal{H}$-classes of uniformly recurrent pseudowords.

Remark 5.6.16 Let $w \in J(F)$. Recall that it follows from basic properties of Green's relations that $w^2 \in J(F)$ if and only if w belongs to a subgroup of $J(F)$ (cf. Exercise 3.60). Since $w^2 \in J(F)$ is equivalent to $F(w^2) \subseteq F$ (cf. Remark 5.6.11), it follows that w belongs to a subgroup if and only if the two-sided infinite word $\overleftarrow{w} \cdot \overrightarrow{w}$ has all its factors in F. Indeed, the finite factors of w^2 are those of w plus the products uv where u is a finite suffix of w and v is a finite prefix of w (cf. Example 4.4.19). Thus the maximal subgroups of $J(F)$ are in natural bijection with the elements of the subshift $X(F)$ (cf. Exercise 5.23).

Example 5.6.17 Let F be the Fibonacci set and let $w = \varphi^{\omega}(a)$. The right infinite word $\overrightarrow{w}$ is the Fibonacci word. The left infinite word $\overleftarrow{w}$ is the word with suffixes $\varphi^{2n}(a)$. The two sided infinite word $\overleftarrow{w} \cdot \overrightarrow{w}$ is a fixed point of φ^2.

Example 5.6.18 let $A = \{a, b\}$ and let $\varphi : A^* \to A^*$ be defined by $\varphi(a) = ab$ and $\varphi(b) = a^3b$. Let F be the set of finite factors of $\varphi^{\omega}(a)$. Since φ is primitive, F is uniformly recurrent. The pseudowords $\varphi^{\omega}(a)$ and $\varphi^{\omega}(b)$ belong to the same $\mathcal{H}$-class of $J(F)$. Indeed, we have $\overrightarrow{\varphi^{\omega}(a)} = \overrightarrow{\varphi^{\omega}(b)}$ and $\overleftarrow{\varphi^{\omega}(a)} = \overleftarrow{\varphi^{\omega}(b)}$.

5.7 Generalization to Recurrent Sets

Using a standard compactness argument, one may easily see that if F is a recurrent subset of A^*, then there is an element w of $\overline{F}$ whose set of factors is $\overline{F}$, and, moreover, if w' is another element of $\overline{F}$ whose set of factors is $\overline{F}$, then $w'\mathcal{J}w$ (Exercise 5.25). Therefore, the $\mathcal{J}$-class of such w depends only on F. This $\mathcal{J}$-class is denoted $J(F)$, a notation consistent with the one already used when F is uniformly recurrent (cf. Proposition 5.6.14).

For recurrent sets, the correspondence $F \mapsto J(F)$ is one-to-one (Exercise 5.26). Moreover, $F \subseteq F'$ if and only if the elements of $J(F)$ are factors of the elements

Fig. 5.6 The even shift space

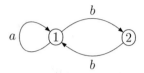

of $J(F')$. We now combine this fact with the following: if the alphabet A has more than one letter, then there is a chain, for the inclusion, consisting of $2^{\aleph_0}$ recurrent subsets of A^* (a proof of this fact can be done with the help of β-shifts, see the Notes Section). The result of the combination is the following proposition, which can be considered as the "height" counterpart of Proposition 5.6.13.

Proposition 5.7.1 *If* $\mathrm{Card}(A) > 1$, *then, for the* $\leqslant_{\mathcal{J}}$*-order, there is a chain of* $2^{\aleph_0}$ *elements in* $\widehat{A^*}$. *Moreover, the elements can be chosen to be regular.*

Several properties that hold for the topological closure of uniformly recurrent sets are not valid for arbitrary recurrent sets.

Example 5.7.2 Let F be the set of words that label paths in the graph of Fig. 5.6.

The set F is known as the *even shift space*, because separating two consecutive occurrences of a in a word belonging to F we always have an even power of b, a fact that extends to pseudowords in $\overline{F}$. The set F is a rational recurrent set which is not uniformly recurrent. The properties expressed in Remark 5.6.11 do not hold here. For example, every finite factor of the pseudoword $ab^{\omega+1}a$ belongs to F, but $ab^{\omega+1}a \notin \overline{F}$, as $b^{\omega+1}$ has odd length. Also, the pseudowords $ab^\omega a$ and $(ab^\omega a)^\omega$ are not $\mathcal{J}$-equivalent (the first one is not even regular), but both belong to $\overline{F}$.

It is an easy exercise to show that, for every recurrent subset F of A, the $\mathcal{J}$-class $J(F)$ is regular (Exercise 5.26). We denote by $G(F)$ the Schützenberger group of $J(F)$.

We quote without proof the following result in which one uses a variation in the definition of a free profinite group. For a set X, the *free profinite group with basis X converging to the identity* is the projective limit of all X-generated finite groups G such that all but finitely many elements of X are mapped to the identity of G. The cardinal of the set X determines, up to isomorphism of topological groups, this group, and the cardinal of X is the *rank* of the group. Note that, if X is finite, then we are dealing with the free profinite group generated by X, as introduced in Chap. 4.6.

Theorem 5.7.3 *If* F *is a nonperiodic rational recurrent set, then* $G(F)$ *is a free profinite group with basis converging to the identity, with countable infinite rank.*

For a reference to the proof see the Notes Section.

Remark 5.7.4 If F is a periodic rational set then $G(F)$ is a free profinite group of rank 1, a fact whose proof is much simpler than that of Theorem 5.7.3 (see Exercise 5.20).

The symbolic dynamical systems corresponding to rational biextendable sets are known as *sofic* shift spaces. The minimal and the sofic shift spaces are the most studied classes of subshifts of $A^{\mathbb{Z}}$ (in the latter case, the subclass of shift spaces of finite type deserves special attention). These two classes share only the periodic subshifts, and are known to be very different under several aspects. For uniformly recurrent sets, there is more diversity concerning what $G(F)$ may be. This is analyzed in Chap. 7.

5.8 Exercises

5.8.1 Section 5.2

5.1 Show that a shift space X is transitive if and only if $F(X)$ is recurrent.

5.2 Show that a shift space X is minimal if and only if $F(X)$ is uniformly recurrent.

5.3 Show that the set of words commuting with a word in A^+ is of the form w^* for some unique primitive word w in A^+.

5.8.2 Section 5.3

5.4 Show that an infinite transitive subshift is a Cantor space.

5.5 Prove Proposition 5.4.1.

5.6 Show that every sliding block code is induced by some block map with memory equal to r and anticipation also equal to r, for some natural number r.

5.8.3 Section 5.5

5.7 Suppose that $\mathrm{Card}(A) > 1$. Show that the substitution $\varphi : A^+ \to A^+$ is primitive if and only if there is $k \geqslant 1$ such that, for all $n \geqslant k$, and for all $a \in A$, all letters appear in $\varphi^n(a)$.

5.8 Prove that the language of a nonerasing substitution φ is biextendable if and only if φ is biextendable.

5.9 Show that every primitive substitution is expansive.

5.10 Show that if φ is primitive, then φ is biextendable.

5.11 Prove Proposition 5.5.4.

5.12 A submonoid M of A^* is *stable* if it satisfies

$$u, uv, vw, w \in M \Rightarrow v \in M \tag{5.5}$$

for all $u, v, w \in A^*$ (note that this notion is unrelated with the notion of stable semigroup defined in Chap. 3). Prove that a submonoid M of A^* is free if and only if it is stable.

Conclude that an intersection of free submonoids is free.

5.13 A substitution $\varphi : A^* \to A^*$ is *elementary* if it cannot be written as $\varphi = \alpha \circ \beta$ with $\beta : A^* \to B^*$, $\alpha : B^* \to A^*$ and $\text{Card}(B) < \text{Card}(A)$. Prove that an elementary substitution is injective as a map from $A^{\mathbb{N}}$ to $A^{\mathbb{N}}$. Hint: consider the intersection of all free submonoids containing $\varphi(A^*)$.

5.14 Prove that if x is a periodic fixed point of a primitive elementary substitution, then every letter $a \in A$ can be followed by at most one letter in x and thus that the least period of x is at most $\text{Card}(A)$.

5.15 Prove that if a fixed point x of a primitive substitution φ is periodic, then the least period of x is at most $\|\varphi\|^{\text{Card}(A)-1}$ where $\|\varphi\| = \max\{|\varphi(a)| \mid a \in A\}$ (Hint: Use Exercise 5.14 and the fact that if $\varphi = \alpha \circ \beta$ and if x is a periodic fixed point of φ, then $y = \beta(x)$ is a periodic fixed point of $\psi = \beta \circ \alpha$ and that the least period of x is at most the least period of y times $\|\alpha\| \leqslant \|\varphi\|$).

5.16 Show that the exponent of periodicity of the Fibonacci set is 3.

5.17 Let $\varphi : A^* \to A^*$ be a primitive nonperiodic substitution. Show that $\varphi : X(\varphi) \to X(\varphi)$ is injective.

5.8.4 Section 5.6

5.18 Let $F \subseteq A^*$ be a factorial set, and let $L \subseteq A^*$ be a recognizable language. Show that the equality $\overline{F \cap L} = \overline{F} \cap \overline{L}$, between subsets of $\widehat{A^*}$, holds.

5.19 Let $\mathcal{K} \in \{\mathcal{R}, \mathcal{L}, \mathcal{J}\}$. Let S be a profinite semigroup. Show that $s \geqslant_{\mathcal{K}} t$ if and only if $\varphi(s) \geqslant_{\mathcal{K}} \varphi(t)$ for all continuous morphisms $\varphi : S \to T$ with T finite.

5.20 Let $u \in A^+$. Show that the $\mathcal{H}$-class of u^ω is a free profinite monogenic group and that the $\mathcal{J}$-class of u^ω has a finite number of $\mathcal{H}$-classes.

5.21 Prove Proposition 5.6.12.

5.22 Prove Proposition 5.6.14.

5.23 For a subshift X of $A^{\mathbb{Z}}$, denote by $\overrightarrow{X}$ the subset of $A^{\mathbb{N}}$ whose finite blocks belong to $F(X)$. Suppose that F is a uniformly recurrent set, and consider the minimal subshift $X = X(F)$. Let $\rho(F)$ be the set of $\mathcal{R}$-classes of $J(F)$. For each

$R \in \rho(F)$, denote by $\vec{R}$ the right infinite word $\vec{w}$, where $w \in \rho(F)$. Show that the mapping $R \in \rho(F) \mapsto \vec{R} \in \vec{X}$ is a well-defined bijection from $\rho(F)$ onto $\vec{X}$. (Note that a dual result holds for the set of $\mathcal{L}$-classes of $J(F)$.)

5.24 Show that every infinite pseudoword of $\widehat{A^*}$ has a factor which is an infinite $\mathcal{J}$-maximal pseudoword. (Hint: use Zorn's Lemma.)

Deduce the following basic fact: every subshift of $A^{\mathbb{Z}}$ contains a minimal subshift.

5.8.5 Section 5.7

5.25 Let F is a recurrent subset of A^*. Show that there is $w \in \overline{F}$ whose set of factors is $\overline{F}$, and every $w' \in \overline{F}$ whose set of factors is $\overline{F}$ belongs to the $\mathcal{J}$-class of w, denoted $J(F)$.

5.26 Show that if F_1 and F_2 are recurrent sets such that $J(F_1) = J(F_2)$, then $F_1 = F_2$.

5.27 Show that if F is any recurrent subset of A^*, then $J(F)$ is a regular $\mathcal{J}$-class of $\widehat{A^*}$.

5.28 Show that if u belongs to the recurrent set $F \subseteq A^*$, then u is the prefix of some idempotent $e \in J(F)$.

5.9 Solutions

5.9.1 Section 5.2

5.1 Assume first that the set $F \subseteq A^*$ is recurrent. Let (u_n) be an enumeration of the words of F. Build another sequence (v_n) of words of F as follows. Set $v_0 = \varepsilon$. Assuming v_n already defined, let $w_n \in F$ be such that $v_{n+1} = v_n w_n u_{n+1} \in F$. Let $y \in A^{\mathbb{N}}$ be the infinite word having all v_n as prefixes. Let $x \in X(F)$ be such that $x_0 x_1 \cdots = y$. Then x has a dense forward orbit.

Conversely, let $x \in X$ be such that the set $\{\sigma^n(x) \mid n \in \mathbb{N}\}$, the forward orbit of x, is dense in X. Let $u, v \in F(X)$, with $|u| = r$ and $|v| = s$. Consider the open sets $U = \{y \in A^{\mathbb{Z}} \mid y_0 \ldots y_{r-1} = u\}$ and $V = \{y \in A^{\mathbb{Z}} \mid y_0 \ldots y_{s-1} = v\}$. By the density of the forward orbit of x, there is $n \in \mathbb{N}$ such that $\sigma^n(x) \in U$. Since $\sigma^{-(n+r)}(V)$ is open, there is also $m \in \mathbb{N}$ such that $\sigma^m(x) \in \sigma^{-(n+r)}(V)$, that is to say $\sigma^{n+r+m}(x) \in V$. Therefore, the word $x[n, n+r+m+s-1]$ is an element of $F(X) \cap u A^* v$, thus showing that $F(X)$ is recurrent.

5.2 This is a translation of Proposition 5.2.3.

5.3 Let u, v be nonempty words such that $uv = vu$. Then, for every $k \geq 1$, we have $(uv)^k = u^k v^k = v^k u^k$. This implies that $uu \cdots = vv \cdots$ and thus that u, v are powers of a common primitive word, which is obviously unique.

5.9.2 Section 5.3

5.4 Let $X \subseteq A^{\mathbb{Z}}$ be a transitive subshift. Since X is a closed subspace of a compact space, it is compact. And being a subspace of the totally disconnected space $A^{\mathbb{Z}}$, it is itself totally disconnected. To show that X is a Cantor space, it remains to check that it has no isolated point. Let $y \in X$. Let U be a neighborhood of y. Since X is transitive, there is $x \in X$ such that $X = \overline{\{\sigma^n(x) \mid n \in \mathbb{N}\}}$. Therefore, there is $k \in \mathbb{N}$ such that $\sigma^k(x) \in U$. Since $X = \sigma(X) = \sigma^{k+1}(X)$ and σ is continuous, one has $X = \overline{\{\sigma^n(x) \mid n > k\}}$, and so there is $\ell > k$ such that $\sigma^\ell(x) \in U$. If $\sigma^\ell(x) = \sigma^k(x)$, then x has period $\ell - k$ and so $X = \mathcal{O}(x)$ is finite, contradicting our hypothesis. Hence $\sigma^\ell(x) \neq \sigma^k(x)$, thus $U \setminus \{y\}$ is nonempty. Therefore y is not an isolated point of X.

5.5 It is clear that a sliding block code is continuous and commutes with the shift map. Conversely, let $\gamma : X \to Y$ be a continuous shift commuting map from a shift space X on the alphabet A to a shift space Y on the alphabet B. For each $b \in B$, set $[b] = \{y \mid y_0 = b\}$. Since γ is continuous, the set $\gamma^{-1}([b])$ is a clopen set. Thus there is an integer $n \geq 1$ and a set $W_b \subseteq F_{2n+1}(X)$ such that $\gamma(x) \in [b]$ if and only if $x[-n, n] \in W_b$. Since B is finite, we may assume that the integer n is the same for every $b \in B$. Let $f : F_{2n+1}(X) \to B$ be defined by $f(w) = b$ if $w \in W_b$. Then $\gamma = f^{[-n,n]}$. Indeed, let $y = \gamma(x)$ and let $i \in \mathbb{Z}$. Let $x' = \sigma^i(x)$ and $y' = \sigma^i(y)$. Then, since γ commutes with the shift map,

$$y' = \sigma^i(y) = \sigma^i(\gamma(x)) = \gamma(\sigma^i(x))$$
$$= \gamma(x').$$

We conclude that

$$y_i = b \Leftrightarrow y'_0 = b \Leftrightarrow x'[-n, n] \in W_b$$
$$f(x'[-n, n]) = b \Leftrightarrow f(x[i - n, i + n]) = b.$$

5.6 Let $\gamma : X \subseteq A^{\mathbb{Z}} \to Y \subseteq B^{\mathbb{Z}}$ be a sliding block code. By Proposition 5.4.1, there is a block map f with $\gamma = f^{[-k,\ell]}$. Take $r \geq k, \ell$. Let $g : F_{2r+1}(X) \to B$ be such that $g(a_{-r}a_{-r+1} \cdots a_{r-1}a_r) = f(a_{-k}a_{-k+1} \cdots a_{\ell-1}a_\ell)$ whenever the letters $a_i \in A$, with $-r \leq i \leq r$, are such that $a_{-r}a_{-r+1} \cdots a_{r-1}a_r \in F(X)$. Then we have $\gamma = g^{[-r,r]}$.

5.9.3   Section 5.5

5.7 Suppose that φ is primitive. Then there is $k \geqslant 1$ such that all letters of A appear in $\varphi^k(a)$, for every $a \in A$. Since $\varphi^{n+k}(a)$ belongs to the image of φ^k, all letters of A appear in $\varphi^{n+k}(a)$, whenever $a \in A$. The converse follows trivially from the definition.

5.8 If $F(\varphi)$ is biextendable, then φ is clearly biextendable. Conversely, let $w \in F(\varphi)$. Let $a \in A$ and $n \geqslant 1$ be such that w is a factor of $\varphi^n(a)$. Since φ is biextendable there are $b, c \in A$ such that $bac \in F(\varphi)$. Since φ is nonerasing, this implies that $AwA \cap F(\varphi) \neq \emptyset$.

5.9 Let $\varphi : A^+ \to A^+$ be a primitive substitution. If $A = \{a\}$ then $\varphi(a) = a^m$ for some $m > 1$, and $\varphi^k(a) = a^{m^k}$ has length m^k, strictly increasing with k since $m > 1$. Suppose that $\mathrm{Card}(A) > 1$. There is a natural k such that the set of letters appearing in $\varphi^k(a)$ is A. Hence, $\varphi(A) \not\subseteq A$, otherwise we would get $\varphi^n(A) \subseteq A$ for all $n \geqslant 1$. Therefore, if the set of letters appearing in the word w is A, then $|\varphi(w)| > |w|$. In particular, $|\varphi(\varphi^n(a))| \geqslant |\varphi^n(a)|$ for all $n \geqslant k$.

5.10 Let $w = \varphi^n(a)$ with $a \in A$. Since φ is primitive, there is a letter b and an integer k such that $\varphi^k(b) = uav$ with u, v nonempty. Thus $\varphi^{n+k}(b) = \varphi^n(u)w\varphi^n(v)$ showing that $AwA \cap F(\varphi) \neq \emptyset$ since φ is nonerasing.

5.11 Let $\varphi : A^* \to A^*$ be a primitive substitution and let x be a fixed point of φ. Replacing φ by some power, we may assume that every letter appears in the image of every letter. Then every letter appears at bounded distances in x and, thus, so does every factor of x.

5.12 Assume first that M is free with basis U. Then, the unique factorisation of $u(vw) = (uv)w$ forces $v \in M$ whenever $u, uv, vw, w \in M$. Conversely, let U be the set of nonempty words in M which cannot be written as a product of strictly shorter words in M. Take $x_1 x_2 \cdots x_n = y_1 y_2 \cdots y_m$ with $x_i, y_j \in U$, and $n, m \geqslant 0$ minimal. By definition of U, we have $n, m \geqslant 1$. Assume $|x_1| \leqslant |y_1|$ and set $y_1 = x_1 v$. Then, $x_2 \cdots x_n = vy_1 \cdots y_m$. By (5.5), we have $v \in M$, which forces $v = \varepsilon$ and $x_1 = y_1$, a contradiction with the minimality of n, m.

The intersection of stable submonoids is clearly stable.

5.13 Let $\varphi : A^* \to A^*$ be a morphism which is not injective as a map from $A^{\mathbb{N}}$ to itself. Set $X = \varphi(A)$. Let Y be the basis of the intersection of all free submonoids containing X^* and let $\beta : B \to Y$ be a bijection from an alphabet B with Y.

For every $a \in A$ there are unique $y_1, y_2, \ldots, y_k \in Y$ such that $\varphi(a) = y_1 y_2 \cdots y_k$. Set $\alpha(a) = b_1 b_2 \cdots b_k$ where $b_i \in B$ is, for $1 \leqslant i \leqslant k$ such that $\beta(b_i)y_i$. Extend α to a morphism from A^* to B^*. Then α is such that $\varphi = \alpha \circ \beta$. For a word $x \in X$, let $\lambda(x) \in Y$ be the first symbol in its decomposition in words of Y, that is $x \in \lambda(x)Y^*$. If some $y \in Y$ does not appear as an initial symbol in the words of X, set $Z = (Y \setminus y)y^*$, which is a suffix code. Then, Z^* is free and $X^* \subseteq Z^* \subseteq Y^*$. Thus, $Y = Z$, a contradiction. Since φ is not injective on $A^{\mathbb{N}}$, the map λ is not injective and thus $\mathrm{Card}(Y) < \mathrm{Card}(X)$, showing that φ is not elementary.

5.14 Let $p(n)$ be the number of factors of length n of x. Let us show that if $p(n) < p(n+1)$, then there is an $m > n$ such that $p(m) < p(m+1)$. Indeed if $p(n) < p(n+1)$ there is a factor u of length n of x which is right-special, that is, there are two distinct letters a, b such that ua, ub are factors of x. Since φ is elementary, it is injective on $A^{\mathbb{N}}$ (Exercise 5.13). Thus there are some v, w such that $\varphi(av) \neq \varphi(bw)$. If $|\varphi(u)| > |u|$ or if $\varphi(a), \varphi(b)$ begin by the same letter, the longest common prefix of $\varphi(uav), \varphi(ubw)$ is a right-special word of length $m > n$. Otherwise we may replace u by $\varphi(u)$. Since φ is primitive, the second case can happen only a bounded number of times. If x is periodic, then $p(n)$ is bounded and by the preceding argument, no letter can be right-special, which implies that the least period of x is at most $\mathrm{Card}(A)$.

5.15 We use an induction on $\mathrm{Card}(A)$. If $\mathrm{Card}(A) = 1$, the result is true since the least period is 1. Otherwise, assume first that φ is elementary and, thus, injective by Exercise 5.13. Then, by Exercise 5.14, the least period of x is at most equal to $\mathrm{Card}(A)$. Finally, assume that $\varphi = \alpha \circ \beta$ with $\beta : A^* \to B^*$ and $\alpha : B^* \to A^*$ and $\mathrm{Card}(B) < \mathrm{Card}(A)$. Set $y = \beta(x)$ and $\psi = \beta \circ \alpha$. Then $\psi(y) = \beta \circ \alpha \circ \beta(x) = \beta(x) = y$. Thus y is a fixed point of ψ. Since ψ is primitive, we may apply the induction hypothesis. Since the least period of y is at most $\|\psi\|^{\mathrm{Card}(B)-1}$, the least period of x is at most $\|\alpha\| \, \|\psi\|^{\mathrm{Card}(B)-1} \leqslant \|\varphi\|^{\mathrm{Card}(A)-1}$.

5.16 Let $F = F(\varphi)$ be the Fibonacci set. Note that $a^3, b^2 \notin F$ as one may verify easily. We use an induction on the length of w to prove that $w^3 \notin F$. The statement is true if $|w| = 1$ since $a^3, b^2 \notin F$. Next assume that $w^3 \in F$. If w ends with b, then it begins with a (since $b^2 \notin F$). Thus $w = \varphi(u)$ and $u^3 \in F$ because if w^3 is a factor of $\varphi(v)$, then $v = v_1 u^3 v_2$, which is impossible by induction hypothesis. If w ends and begins with a, then $w = \varphi(u)$ again and $u^3 \in F$, a contradiction. Finally, if w ends with a and begins with b, set $w = bua$. Then $aw \in F$ and $abu = \varphi(v)$ with $v^3 \in F$ again.

5.17 Let $x, y \in X(\varphi)$ be such that $\varphi(x) = \varphi(y)$. Let n be the recognizability constant of φ (with respect to the asymptotic injectivity). Let w be an admissible fixed point of φ. For every $i \in \mathbb{Z}$, there exist $k, \ell \in \mathbb{Z}$ such that $w[k-n, k+n] = x[i-n, i+n]$ and $y[i-n, i+n] = w[\ell-n, \ell+n]$. Then, by asymptotic injectivity, we have $w_\ell = w_k$ and thus $x_i = y_i$.

5.9.4 Section 5.6

5.18 Since the intersection of closed sets is closed, the inclusion $\overline{F \cap L} \subseteq \overline{F} \cap \overline{L}$ is trivial. Conversely, let $x \in \overline{F} \cap \overline{L}$. Then $x = \lim x_n$ for some sequence (x_n) of elements of F. Because L is recognizable, the closure $\overline{L}$ is a neighborhood of x, by Theorem 4.4.9, whence $x_n \in \overline{L}$ for all sufficiently large n. Since the elements of A^* are isolated in $\widehat{A^*}$, this gives $x_n \in F \cap L$ for all sufficiently large n, thus $x \in \overline{F \cap L}$.

5.19 It suffices to show for $\mathcal{K} = \mathcal{J}$, as the other cases are similar. We may consider a projective limit $S = \varprojlim_{i \in I} S_i$, with S_i finite for all i, and the continuous projections $\varphi_i : S \to S_i$ all onto. By hypothesis, we may take $x_i, y_i \in S^1$ such that $\varphi_i(t) = \varphi_i(x_i s y_i)$, for all i. Let (x, y) be an accumulation point of $(x_i, y_i)_{i \in I}$. Note that $\varphi_i(t) = \varphi_i(x_j s y_j)$ for all $j \geqslant i$, thus $\varphi_i(t) = \varphi_i(xsy)$ by continuity of φ_i. Since i is arbitrary, it follows that $t = xsy$.

5.20 We may suppose that u is primitive. Let k be the length of u. Consider the element x of $A^{\mathbb{Z}}$ with period k such that $x[0, k - 1] = u$. As u is primitive, k is the least period of x (that is, the least integer p such that $x_{n+p} = x_n$ for all $n \in \mathbb{Z}$). Consider the periodic shift space $X = \{\sigma^n(x) \mid n \in \mathbb{N}\}$, and let $F = F(X)$. Then we have $u^\omega \in J(F)$.

Note that if $v \in \widehat{\mathbb{N}} \setminus \mathbb{N}$, then $u^v \mathcal{H} u^\omega$, a fact that may be justified by a simple direct argument or by invoking Proposition 5.6.15. We claim that the converse holds: if $h \mathcal{H} u^\omega$, then $h = u^v$ for some $v \in \widehat{\mathbb{N}}$. Let us then take h in the $\mathcal{H}$-class of u^ω. Since $h \in \overline{F} \cap u\widehat{A^*} \cap \widehat{A^*}u$, one has $h = \lim v_n$ for some sequence $v_n = x[i_n, j_n[$, with $i_n + k \leqslant j_n$, $x[i_n, i_n + k[= u$ and $x[j_n - k, j_n[= u$. Let $i_n = q_n k + r_n$ with $q_n \in \mathbb{Z}$ and $r_n \in \{0, \ldots, k - 1\}$. Suppose that $r_n \neq 0$. Consider the word $z = x[q_n k, q_n k + r_n[= x[(q_n + 1)k, (q_n + 1)k + r_n[$. Then we have $x[q_n k, (q_n + 1)k + r_n[= zu = uz$. By Exercise 5.3, it follows that z and u are powers of a primitive word w, contradicting that u is a primitive word of length greater than that of z. To avoid the contradiction, we must have $r_n = 0$, thus i_n is a multiple of k. With the same kind of reasoning, one concludes that j_n is also a multiple of k. Therefore, v_n is a power of u, for every n, and so h is of the form u^v for some $v \in \widehat{\mathbb{N}} \setminus \mathbb{N}$.

Because $\{u\}$ is a code, the mapping $v \mapsto u^v$ is a bijection (cf. Theorem 4.8.3) and, therefore, a continuous isomorphism between $\widehat{\mathbb{N}} \setminus \mathbb{N} = \widehat{\mathbb{Z}}$ and the $\mathcal{H}$-class of u^ω.

Finally, since the shift space X is finite, it follows from Proposition 5.6.15 that the number of $\mathcal{R}$-classes and of $\mathcal{L}$-classes (and hence, of $\mathcal{H}$-classes) in the $\mathcal{J}$-class of u^ω is finite.

5.21 If $w \in \widehat{A^*}$ is a $\mathcal{J}$-maximal infinite pseudoword, then $F(w)$ is uniformly recurrent by Theorem 5.6.7. The $\mathcal{J}$-class $J(F(w))$ is the $\mathcal{J}$-class of w, by definition, which shows that $F \mapsto J(F)$ is surjective.

On the other hand, suppose that F_1 and F_2 are uniformly recurrent sets such that $J(F_1) = J(F_2)$. We may take infinite pseudowords u_1, u_2 such that $F_i = F(u_i)$ (Proposition 5.6.2). By the definition of $J(F_i)$, one has $u_i \in J(F_i)$. As $u_1 \mathcal{J} u_2$, we conclude that $F_1 = F(u_1) = F(u_2) = F_2$, establishing that $F \mapsto J(F)$ is injective.

5.22 Let $u \in \overline{F} \setminus A^*$ and let $v \in J(F)$. By Proposition 5.6.2 we have $F = F(u)$. Also, $F = F(v)$ and $u \mathcal{J} v$ by Corollary 5.6.9. Therefore, we have $\overline{F} \setminus A^* \subseteq J(F)$. As $\overline{F}$ is factorial (cf. Proposition 5.6.1), we actually have $\overline{F} \setminus A^* = J(F)$.

5.23 By Proposition 5.6.15, the map $R \in \rho(F) \mapsto \overrightarrow{R} \in \overrightarrow{X}$ is a well-defined injective function. Let $x = (x_i)_{i \in \mathbb{N}}$ be in $\overrightarrow{X}$ and let $u_n = x_0 x_1 \ldots x_n$. Take an accumulation point u in $\widehat{A^*}$ of the sequence $(u_n)_n$. Note that u_n is a prefix of u

for every n, that is, $\overrightarrow{w} \in \overrightarrow{X}$. Then, $u \in \overline{F} \setminus A^*$ and so we have $u \in J(F)$ by Proposition 5.6.14. This shows that $R \in \rho(F) \mapsto \overrightarrow{R} \in \overrightarrow{X}$ is surjective.

5.24 Let $u \in \widehat{A^*} \setminus A^*$. Let C be a nonempty chain of elements of $\widehat{A^*} \setminus A^*$, ordered by the $\mathcal{J}$-order $\leqslant_\mathcal{J}$, and such that $u \leqslant_\mathcal{J} z$ for every $z \in C$. Consider the net $(v_z)_{z \in C}$ in which $v_z = z$ for every $z \in C$. Let $(v_{z_i})_{i \in I}$ be a subnet of $(v_z)_{z \in C}$ converging in the compact space $\widehat{A^*} \setminus A^*$ to v. Given $z \in C$, there is $i_0 \in I$ such that $i_0 \leqslant i$ implies $z \leqslant_\mathcal{J} z_i$. Since $\leqslant_\mathcal{J}$ is a closed relation, it follows that $z \leqslant_\mathcal{J} \lim z_i = \lim v_{z_i} = v$. This shows that v is an upper bound of C for the relation $\leqslant_\mathcal{J}$. Therefore, by Zorn's Lemma, the set $\{v \in \widehat{A^*} \setminus A^* \mid u \leqslant_\mathcal{J} v\}$ has some maximal element w for the $\mathcal{J}$-order.

Let X be a subshift of $A^{\mathbb{Z}}$. Then $\overline{F(X)}$ contains some infinite pseudoword u. By the preceding paragraph, there is an infinite $\mathcal{J}$-maximal pseudoword w such that $u \leqslant w$. By Theorem 5.6.7, the pseudoword w is uniformly recurrent. Therefore, we may consider the unique minimal subshift Y of $A^{\mathbb{Z}}$ such that $F(w) = F(Y)$. We then have $F(Y) \subseteq F(u) \subseteq \overline{F(X)}$, the last inclusion holding in accordance with the fact that $\overline{F(X)}$ is factorial (cf. Proposition 5.6.1). Because the elements of A^* are isolated in $\widehat{A^*}$, we have $\overline{F(X)} \cap A^* = F(X)$, whence $F(Y) \subseteq F(X)$. This inclusion gives $Y \subseteq X$.

5.9.5 Section 5.7

5.25 Let x be an element of $X(F)$ with dense forward orbit. For each $n \in \mathbb{N}$, let $w_n = x[0, n]$. The sequence (w_n) has an accumulation point w in $\overline{F} \setminus A^*$. Let $v \in F$. By the choice of x, one has $v \in F(w_n)$ for all sufficiently large n. We conclude that $v \in F(w)$, and $\overline{F} \subseteq F(w)$, since the set of factors of any pseudoword is closed.

Suppose that $w' \in \overline{F}$ is such that all elements of $\overline{F}$ are factors of w'. Since $w' \in \overline{F}$, we get $w' \geqslant_\mathcal{J} w$ by the previous paragraph. On the other hand, since $w \in \overline{F}$, we have $w \geqslant_\mathcal{J} w'$ by the choice of w'.

5.26 We show something stronger: if F_1, F_2 are recurrent sets and the pseudowords $w_1 \in J(F_1)$ and $w_2 \in J(F_2)$ are such that $w_1 \geqslant_\mathcal{J} w_2$, then $F_1 \subseteq F_2$. Let $v \in F_1$. Then $v \geqslant_\mathcal{J} w_1$ by the definition of $J(F_1)$, thus $v \geqslant_\mathcal{J} w_2$. Since, A^* being discrete, the intersection of $\overline{F_2}$ with A^* is F_2, we deduce that $v \in F_2$.

5.27 Let $u \in J(F)$. Since $\overline{F}$ is recurrent by Proposition 5.6.1, there is $v \in \widehat{A^*}$ such that $uvu \in \overline{F}$. By the definition of $J(F)$, we have $uvu \mathcal{J} u$, thus $u = xuvuy$ for some $x, y \in \widehat{A^*}$. It follows that $u = x^k u(vuy)^k$ for all $k \in \mathbb{N}$, thus $u = x^\omega u(vuy)^\omega$ and $u \mathcal{J} (vuy)^\omega$, establishing that the $\mathcal{J}$-class of u is regular.

5.28 By the definition of $J(F)$, and since it is a regular $\mathcal{J}$-class, there is an idempotent $f \in J(F)$ such that $f = xuy$. Let $e = (uyx)^2$. Then e is an idempotent that is $\mathcal{J}$-equivalent to e.

5.10   Notes

We refer to Lind and Marcus (1995) for a more detailed introduction to symbolic dynamics.

In Lothaire (2002, Chapter 2) one finds a proof of the existence of $2^{\aleph_0}$ nonperiodic uniformly recurrent sets. More precisely, it is shown that there are $2^{\aleph_0}$ *Sturmian sets* over A when $\text{Card}(A) = 2$, where a Sturmian set is a factorial set $F \subseteq A^*$ such that $\text{Card}(F \cap A^n) = n+1$ for every n (cf. Lothaire (2002, Proposition 2.1.18)). Sturmian sets are also studied in this book, in Chap. 6.

A proof of the Perron-Frobenius Theorem (Theorem 5.5.13) can be found in many textbooks (see Gantmacher (1959) for example).

Proposition 5.5.14 appears in Queffélec (2010) while Theorem 5.5.15 is originally from Mossé (1992).

Theorem 5.5.19 is an important result originally due to Mossé (1992) (with the complement concerning asymptotic injectivity in Mossé (1996)). It is the final version of a first result announced by Martin (1973) with an inaccurate statement and proof. A more recent proof can be found in Kurka (2003). The variant stated as Theorem 5.5.22 appears in Fogg (2002, Theorem 7.2.2) (without mention of the admissibility of the fixed-point). It also appears in Durand et al. (1999, Theorem 11). Recognizability of substitutions plays an important role in several problems concerning substitutive shift spaces. See for example Durand et al. (1999) where it is used to build a decomposition in Kakutani-Rokhlin towers of a substitutive shift space. It will be used here in Chap. 7.

The connection between symbolic dynamics and free profinite semigroups was explored in several papers (Almeida 2005b; Costa 2006; Almeida and Costa 2009; Costa and Steinberg 2011; Almeida and Costa 2012, 2013, 2016).

The characterization of uniformly recurrent pseudowords of Proposition 5.6.4 is from Almeida (2005b, Lemma 2.2). Theorem 5.6.7 is from Almeida (2005b, Theorem 2.6).

Proposition 5.6.15 is from Almeida and Costa (2009, Lemma 6.6). Several aspects concerning this proposition are developed in Almeida and Costa (2012).

A reference concerning β-shifts is Lothaire (2002, Chapter 7) or Ito and Takahashi (1974, Chapter 4). The weaker versions of Propositions 5.6.13 and 5.7.1 that do not mention regularity were first proved in Costa (2001). A stronger result, with the same statement of Proposition 5.7.1 except that we replace the $\leqslant_{\mathcal{J}}$-order by the $\leqslant_{\mathcal{R}}$-order, is proved in the last section of Almeida et al. (2019), applying ideas similar to those used in the proof of Proposition 5.7.1, together with a sophisticated use of the axiom of choice.

Theorem 5.7.3 is a difficult result from Costa and Steinberg (2011).

The decidability of the periodicity of the fixed-point of a morphism (Exercise 5.15) is from Pansiot (1986) and Harju and Linna (1986).

Chapter 6
Sturmian Sets and Tree Sets

6.1 Introduction

In this chapter, we present notions concerning factorial sets of words such as neutral and Sturmian sets, episturmian sets and tree sets.

In Sect. 6.2, we introduce return words. We prove that a substitutive shift defined by a primitive morphism is bounded in the sense that the number of return words to a given word is bounded (Theorem 6.2.4).

In Sect. 6.3, we define the factor complexity of a factorial set. We define neutral words by a property of the numbers of left, right and two sided extensions of a word by a letter. We define neutral sets as sets containing only neutral words. We define Sturmian words and Sturmian sets as those with minimal nontrivial complexity on a binary alphabet. We prove that the number of return words to a given word w in a uniformly recurrent neutral set is independent of w (Theorem 6.3.6).

In Sect. 6.4, we define episturmian words as the natural generalization of Sturmian words on arbitrary alphabets. We state without proof Justin's Formula and the characterization of standard episturmian words using the iterated palindromic closure.

In Sect. 6.5, we introduce the extension graph of a word with respect to a factorial set of words. This graph represents the possible two-sided extensions of a word by letters. We define tree sets as those for which the extension graphs are always trees. Next we prove, as a main result of this chapter, the Return Theorem asserting that in a tree set, the set of return words to a given word is a basis of the free group on the alphabet (Theorem 6.5.5).

In Sects. 6.6 and 6.7 we show how to associate to a uniformly recurrent pseudoword a sequence of return words. We derive from this a set of generators of maximal subgroups in the $\mathcal{J}$-class of a bounded set (Theorem 6.7.3).

© Springer Nature Switzerland AG 2020
J. Almeida et al., *Profinite Semigroups and Symbolic Dynamics*, Lecture Notes in Mathematics 2274, https://doi.org/10.1007/978-3-030-55215-2_6

6.2 Return Words

In this section, we introduce return words. We begin with the classical notion of left and right return words in factorial sets. We prove a result on the cardinality of return sets in substitutive shifts (Theorem 6.2.4) which is instrumental to deduce conclusions about the maximal subgroups of $J(F)$ when F is the set of factors of a substitutive shift.

6.2.1 Left and Right Return Words

Let F be a factorial set. A *return word* to $x \in F$ is a nonempty word $w \in F$ such that xw begins and ends by x and the factor x occurs in xw only as a prefix and a suffix of xw. We denote by $\mathcal{R}_F(x)$ the set of return words to x, which is a nonempty set if F is recurrent. Note that the $\mathcal{R}_F(x)$ is a prefix code, provided it is nonempty.

For $x \in F$, we denote

$$\Gamma_F(x) = \{w \in F \mid xw \in F \cap A^*x\}.$$

Thus $\mathcal{R}_F(x)$ is the set of nonempty words in $\Gamma_F(x)$ without any proper prefix in $\Gamma_F(x)$.

A subset N of a monoid M is called *right unitary* if for every $u, v \in M$, $u, uv \in N$ implies $v \in N$. It is well known that a nontrivial submonoid of A^* is right unitary if and only if it is generated by a prefix code (in this chapter, we shall not need this fact, which we shall revisit in Chap. 8, and whose proof is proposed there in Exercise 8.2). Note that $\Gamma_F(x)$ a right unitary subset of F, and that the prefix code $\mathcal{R}_F(x)$ is the set of words in $\Gamma_F(x)$ without any proper prefix in $\Gamma_F(x)$.

A return word may also be called a *right return word*. One also defines a *left return word* to $x \in F$ as a nonempty word w such that wx begins and ends with x and the factor x occurs in xw only as a prefix and as a suffix. We denote by $\mathcal{R}'_F(x)$ the set of left return words to x. One has obviously the equality $\mathcal{R}'_F(x) = x\,\mathcal{R}_F(x)x^{-1}$, where, for $Y \subseteq A^*$ and $v \in A^*$, one uses the notation Yv^{-1} for the set $\{u \in A^* \mid uv \in Y\}$ (cf. Exercise 6.1).

Example 6.2.1 Let F be the Fibonacci set. The sets of right and left return words to a are $\mathcal{R}_F(a) = \{a, ba\}$ and $\mathcal{R}'_F(a) = \{a, ab\}$. Similarly, one has $\mathcal{R}_F(b) = \{ab, aab\}$ and $\mathcal{R}'_F(b) = \{ba, baa\}$.

Example 6.2.2 Let F be a periodic set. Let w be a primitive word of length n such that $F = F(w^*)$. Then, for any word $x \in F$ of length at least n, the set $\mathcal{R}_F(w)$ is reduced to one word of length n.

Proposition 6.2.3 *Let F be a factorial set. For any $x \in F$, one has $\Gamma_F(x) = \mathcal{R}_F(x)^* \cap x^{-1}F$.*

Proof If a nonempty word w is in $\Gamma_F(x)$ and it is not in $\mathcal{R}_F(x)$, then $w = uv$ with $u \in \Gamma_F(x)$ and v nonempty. Since $\Gamma_F(x)$ is right unitary, we have $v \in \Gamma_F(x)$, whence the conclusion $w \in \mathcal{R}_F(x)^*$ by induction on the length of w. Moreover, one has $xw \in F$ and thus $w \in x^{-1}F$.

Conversely, assume that w is a nonempty word in $\mathcal{R}_F(x)^* \cap x^{-1}F$. Set $w = uv$ with $u \in \mathcal{R}_F(x)$ and $v \in \mathcal{R}_F(x)^*$. Then $xw = xuv \in A^*xv \subseteq A^*x$ and $xw \in F$. Thus $w \in \Gamma_F(x)$.   ∎

Note that, as a consequence, for $x, y \in F$ such that $xy \in F$, we have

$$. \quad \mathcal{R}_F(xy) \subseteq \mathcal{R}_F(y)^*. \tag{6.1}$$

Indeed, if $w \in \mathcal{R}_F(xy)$, then $xyw \in F \cap A^*xy$ implies $yw \in F \cap A^*y$ and thus the result follows since $\Gamma_F(y) \subseteq \mathcal{R}_F(y)^*$ by Proposition 6.2.3.

The dual of Proposition 6.2.3 and that of Eq. (6.1) hold for left return words.

Let $k \geqslant 1$ be an integer. A recurrent set F is *k-bounded* if every $x \in F$ has at most k return words. The set F is *bounded* if it is k-bounded for some k. We shall encounter along this chapter several important classes of bounded sets (cf. Theorems 6.2.4 and 6.3.6). Note that a biextendable set is uniformly recurrent if and only if, for every $u \in F$, the set $\mathcal{R}_F(u)$ is finite and nonempty. In particular, every bounded set is uniformly recurrent. There exist uniformly recurrent sets which are not bounded (see the Notes Section).

Theorem 6.2.4 *Let* $X = X(\varphi)$ *be a substitutive shift with* φ *primitive. If* X *is not periodic then there are integers* k, ℓ, m *such that for every* $u \in F(\varphi)$,

(i) *for every* $v \in \mathcal{R}(u)$, *one has* $|u| \leqslant \ell|v| \leqslant m|u|$,
(ii) *one has* $\mathrm{Card}(\mathcal{R}(u)) \leqslant k$ *and thus* X *is* k-*bounded.*

Proof

(i) By Proposition 5.5.14 there is an integer q such that $\|\varphi^p\| \leqslant q|\varphi^p|$ for all $p \geqslant 1$. Let r be the maximal length of a word in $ab\,\mathcal{R}(ab)$ for $a, b \in A$.

Consider now $u \in F(\varphi)$. Let p be the smallest integer such that $|u| \leqslant |\varphi^p|$. By definition of p, we have $|\varphi^{p-1}| \leqslant |u|$ and there are letters $a, b \in A$ such that u is a factor of $\varphi^p(ab)$.

Let ℓ be the exponent of periodicity of X (Theorem 5.5.15). Then $|u| \leqslant \ell|v|$ since otherwise, v being a period of u, the word v^ℓ is a suffix of u (see Fig. 6.1). Next, we have

$$|v| \leqslant r\|\varphi^p\| \leqslant rq|\varphi^p| \leqslant rq\|\varphi\|\,|\varphi^{p-1}| \leqslant rq\|\varphi\|\,|u|$$

so that we can set $m = \ell rq\|\varphi\|$.

Fig. 6.1 The words u and v

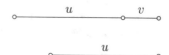

(ii) Let n be minimal such that $|\varphi^n| \geqslant (m+1)|u|$. For all $v \in \mathcal{R}(u)$, we have $|uv| \leqslant (m+1)|u|$ since $|v| \leqslant m|u|$ by assertion (i). Hence there are letters $a, b \in A$ such that uv is a factor of $\varphi^n(ab)$. The word $\varphi^n(ab)$ contains at most $s = |\varphi^n(a)|\ell|u|^{-1}$ occurrences of v since v has length at least $\ell^{-1}|u|$ by assertion (i). On the other hand, by Proposition 5.5.14 again, there is a $q \geqslant 1$ such that $\|\varphi^n\| \leqslant n|\varphi^n|$ for all $n \geqslant 1$. Hence, we have

$$|\varphi^n(ab)| \leqslant 2\|\varphi^n\| \leqslant 2\|\varphi\| \, \|\varphi^{n-1}\| \leqslant 2\|\varphi\| \, q|\varphi^{n-1}| \leqslant 2\|\varphi\|q(m+1)|u|.$$

It follows that $s \leqslant 2\|\varphi\| q(m+1)\ell$. Since every $v \in \mathcal{R}(u)$ appears in one of the at most $\mathrm{Card}(A)^2$ words $\varphi^n(ab)$, we conclude that

$$\mathrm{Card}(\mathcal{R}(u)) \leqslant s \, \mathrm{Card}(A) \leqslant 2\|\varphi\|q(m+1)\, \mathrm{Card}(A)^2\ell$$

whence the conclusion with $k = 2\|\varphi\|q(m+1)\, \mathrm{Card}(A)^2\ell$.   ∎

6.3   Neutral Sets

Let F be a factorial set on the alphabet A. For $w \in F$, we denote

$$L_F(w) = \{a \in A \mid aw \in F\},$$
$$R_F(w) = \{a \in A \mid wa \in F\},$$
$$E_F(w) = \{(a, b) \in A \times A \mid awb \in F\}$$

and further

$$\ell_F(w) = \mathrm{Card}(L_F(w)), \quad r_F(w) = \mathrm{Card}(R_F(w)), \quad e_F(w) = \mathrm{Card}(E_F(w)).$$

For $w \in F$, we denote

$$m_F(w) = e_F(w) - \ell_F(w) - r_F(w) + 1.$$

If $e_F(w) \geqslant 1$, then the word w is *biextendable in F*. We already saw in the previous chapter that a factorial set F is said to be *biextendable* if every word $w \in F$ is biextendable in F. The set $F(x)$ of factors of a right infinite recurrent word x is obviously biextendable (this is false without recurrency, as the example $baaaaa \cdots$ shows).

A word w is *right-special* (resp. *left-special*) if $r_F(w) \geqslant 2$ (resp. $\ell_F(w) \geqslant 2$). It is *bispecial* if it is right-special and left-special. A right-special (resp. left-special) word w is *strict* if $\ell_F(w) = \mathrm{Card}(A)$ (resp. $r_F(w) = \mathrm{Card}(A)$). In the case of a two-letter alphabet, all special words are strict.

A word w is called *neutral* if $m_F(w) = 0$, *strong* if $m(w) > 0$ and *weak* if $m(w) < 0$.

A factorial set F is *neutral* if every word in F is neutral.

Clearly, a biextendable word w which is not both left-special and right-special is neutral. Indeed, assume for instance that $\ell_F(w) = 1$. Then $e_F(w) = r_F(w)$ and thus $m_F(w) = 0$.

We now introduce an important class of words. A *Sturmian word* is an infinite word x on a binary alphabet such that $F(x)$ has $n + 1$ elements of length n for each $n \geqslant 1$. A *Sturmian set* is the set of factors of a Sturmian word.

Note that, in the set of factors of a Sturmian word, there is exactly one left-special (and exactly one right-special) word of each length. It is classical that the Fibonacci word is Sturmian (see Exercise 6.8).

Example 6.3.1 The set of factors of a Sturmian word is neutral. Let indeed x be a Sturmian word on the alphabet $A = \{a, b\}$ and let $F = F(x)$. Let $w \in F$ be a bispecial word. Then wa and wb cannot both be left special. Assume that wa is left special and that $awb \in F$. Then $E_F(w) = \{(a, a), (a, b), (b, a)\}$ and $m_F(w) = 0$. Thus w is neutral.

Example 6.3.2 The set of factors T of the Thue-Morse word is not neutral. Indeed, since $A^2 \subseteq T$, one has $m_T(\varepsilon) = 1$.

Example 6.3.3 Let $\varphi : a \mapsto ab$, $b \mapsto a^3b$ be as in Example 5.6.18. Let F be the set of factors of $\varphi^\omega(a)$. It is not neutral since $m(a) = 1$ and $m(aa) = -1$.

The *complexity* of a factorial set F containing the alphabet A is the sequence $p(n) = \operatorname{Card}(F \cap A^n)$. A neutral set containing the alphabet A has complexity $kn + 1$ where $k = \operatorname{Card}(A) - 1$ (see Exercise 6.9).

We now give an example of a set of complexity $2n + 1$ on an alphabet with three letters which is not neutral.

Example 6.3.4 Let $A = \{a, b, c\}$. The *Chacon word* on three letters is the fixpoint x starting with a of the morphism f from A^* into itself defined by $f(a) = aabc$, $f(b) = bc$ and $f(c) = abc$. Thus, we have $x = aabcaabcbcabc \cdots$. The *Chacon set* is the set F of factors of x. It is of complexity $2n + 1$ (see the Notes Section for a reference).

It contains strong, neutral and weak words. Indeed, $F \cap A^2 = \{aa, ab, bc, ca, cb\}$ and thus $m(\varepsilon) = 0$ showing that the empty word is neutral. Next $m(abc) = 1$ and $m(bca) = -1$, showing that abc is strong while bca is weak.

An important property of neutral sets is that the numbers $\rho(w) = r(w) - 1$ are left additive, that is, they satisfy the equation

$$\rho(w) = \sum_{a \in L(w)} \rho(aw). \tag{6.2}$$

Indeed

$$\sum_{a \in L(w)} \rho(aw) = \sum_{a \in L(w)} (r(aw) - 1)$$

$$= e(w) - \ell(w) = r(w) - 1 = \rho(w).$$

Thus ρ is almost a left probability, except for its value on the empty word which is $\mathrm{Card}(A \cap F) - 1$.

This implies in particular the following very useful property. An *F-maximal suffix code* is a suffix code $X \subseteq F$ which is not properly included in any suffix code $Y \subseteq F$. Denoting $\rho(X) = \sum_{x \in X} \rho(x)$, we have the following useful result.

Proposition 6.3.5 *For any neutral set F and any finite F-maximal suffix code $X \subseteq F$,*

$$\rho(X) = \rho(\varepsilon) \tag{6.3}$$

Proof Let us use an induction on the sum of the lengths of the words of X. The result is true by Eq. (6.2) if all words have length 1 since then $X = A \cap F$. Otherwise, let $x \in X$ be of maximum length. Then $|x| \geqslant 2$ and thus $x = ay$ with $a \in A$ and $y \in F$ nonempty. We have $L(y)y \subseteq X$ and the set $Y = (X \setminus L(y)y) \cup \{y\}$ is a finite F-maximal suffix code. Note that

$$\rho(Y) = \rho(X) - \sum_{a \in L(y)} \rho(ay) + \rho(y) = \rho(X).$$

By induction hypothesis, we have $\rho(Y) = \rho(\varepsilon)$ and thus $\rho(X) = \rho(\varepsilon)$. ∎

This applies in particular when $X = F \cap A^n$ and gives an alternative way to show the linearity of the complexity of a neutral set.

The following theorem shows in particular that in a uniformly recurrent neutral set the number of return words to a given word is constant.

Theorem 6.3.6 *If F is a uniformly recurrent neutral set containing the alphabet A, then the equality $\mathrm{Card}(\mathcal{R}_F(x)) = \mathrm{Card}(A)$ holds.*

Proof Consider the set Y formed of the proper prefixes of the set $w \mathcal{R}_F(w)$ which are not proper prefixes of w. We show that Y is an F-maximal suffix code. Assume that $y, y' \in Y$. If y' is a suffix of y, then $y = uy' = uwv$ which implies that u is empty and thus $y = y'$. This shows that Y is a suffix code.

Next, to show that Y is an F-maximal suffix code, we need to show that every long enough word in F has a suffix in Y. Since F is uniformly recurrent, every long enough word $u \in F$ has a factor w. Set $u = pwv$ with w unioccurrent in wv. Then wv is a proper prefix of a word in $w \mathcal{R}_F(w)$ and thus $wv \in Y$, showing that u has a suffix in Y. In any finite tree, we have

$$\text{Card}(L) = 1 + \sum_{i \in I}(d(i) - 1)$$

where L is the set of leaves, I the set of internal nodes and $d(i)$ is the number of children of $i \in I$. Since $w\,\mathcal{R}_F(w)$ is a prefix code, the set $w\,\mathcal{R}_F(w)$ is the set of leaves of a tree with Y as its set of internal nodes. Thus

$$\text{Card}(w\,\mathcal{R}_F(w)) = 1 + \rho(Y)$$

On the other hand, since Y is an F-maximal suffix code, we have by Eq. (6.3), $\rho(Y) = \rho(\varepsilon)$ and since $A \subseteq F$, we have $\rho(\varepsilon) = \text{Card}(A) - 1$. Thus

$$\text{Card}(\mathcal{R}_F(w)) = \text{Card}(w\,\mathcal{R}_F(w))$$
$$= 1 + \rho(Y) = \text{Card}(A)$$

which completes the proof. ∎

Example 6.3.7 Let $\varphi : a \mapsto ab,\ b \mapsto a^3b$ be the substitution in Example 5.6.18. Let F be the set of finite factors of $\varphi^\omega(a)$. One has $\mathcal{R}_F(a) = \{a, ba\}$ but $\mathcal{R}_F(aa) = \{a, babaa, bababaa\}$. Thus the number of return words may not be constant in a uniformly recurrent set which is not neutral.

6.4 Episturmian Words

We recall here some notions concerning episturmian words, which are a generalization of Sturmian words on alphabets with more than two letters. We will not give the proofs, which are difficult, and do not contribute to the understanding of the other results because we will mostly use episturmian words to give examples.

The *reversal* of a word $w = a_1a_2 \cdots a_n$ is the word $\tilde{w} = a_n \cdots a_2a_1$. A set of words F is closed under reversal if $w \in F$ implies $\tilde{w} \in F$.

By definition, a right infinite word x is *episturmian* if $F(x)$ is closed under reversal and if $F(x)$ contains, for each $n \geqslant 1$, at most one word of length n which is right-special.

Since $F(x)$ is closed under reversal, the reversal of a right-special factor of length n is left-special, and it is the only left-special factor of length n of x. A suffix of a right-special factor is again right-special. Symmetrically, a prefix of a left-special factor is again left-special.

Note that the assumption of closure under reversal guarantees that the episturmian words are biextendable. It can be shown that an episturmian word is uniformly recurrent (cf. Remark 6.4.6).

As a particular case, a *strict* episturmian word is an episturmian word x with the two following properties: x has exactly one right-special factor of each length and moreover each right-special factor u of x is strict, that is, it satisfies the inclusion $uA \subseteq F(x)$.

It is easy to see that for a strict episturmian word x on an alphabet A with k letters, $F(x) \cap A^n$ has $(k-1)n + 1$ elements for each n. Thus for a binary alphabet, the strict episturmian words are just the Sturmian words since a Sturmian word has one right-special factor for each length and its set of factors is closed under reversal (Exercise 6.7).

An episturmian word x is *standard* if all its left-special factors are prefixes of x. For any episturmian word x, there is a standard one y such that $F(x) = F(y)$.

Example 6.4.1 The *Tribonacci word* is the fixed point x of the morphism $\varphi : a \mapsto ab$, $b \mapsto ac$, $c \to a$, namely the infinite word $x = abacaba \cdots$. It is a strict standard episturmian word (Exercise 6.10).

An *Arnoux-Rauzy set* is the set of factors of a strict episturmian word. A shift space X is an *Arnoux-Rauzy shift* if $F(X)$ is an Arnoux-Rauzy set. We have chosen this name for Arnoux-Rauzy sets and shifts because it is more commonly used than episturmian. We reserve the term *Sturmian* to sets and shifts on binary alphabets.

For $a \in A$, denote by ψ_a the morphism of A^* into itself defined by

$$\psi_a(b) = \begin{cases} ab & \text{if } b \neq a \\ a & \text{otherwise} \end{cases}$$

Let $\psi : A^* \to \text{End}(A^*)$ be the morphism from A^* into the monoid of endomorphisms of A^* which maps each $a \in A$ to ψ_a. For $u \in A^*$, we denote by ψ_u the image of u by the morphism ψ. Thus, for three words u, v, w, we have $\psi_{uv}(w) = \psi_u(\psi_v(w))$. A *palindrome* is a word w which is equal to its reversal. Given a word w, we denote by $w^{(+)}$ the *palindromic closure* of w. It is, by definition, the shortest palindrome which has w as a prefix.

The *iterated palindromic closure* of a word w is the word $\text{Pal}(w)$ defined recursively as follows. One has $\text{Pal}(1) = 1$ and for $u \in A^*$ and $a \in A$, one has $\text{Pal}(ua) = (\text{Pal}(u)a)^{(+)}$. Since $\text{Pal}(u)$ is a proper prefix of $\text{Pal}(ua)$, it makes sense to define the iterated palindromic closure of an infinite word x as the infinite word which has all the iterated palindromic closures of the prefixes of x as prefixes.

Justin's Formula is the following: for all words u and v, one has

$$\text{Pal}(uv) = \psi_u(\text{Pal}(v)) \, \text{Pal}(u) \, .$$

This formula extends to infinite words: if u is a word and v is an infinite word, then

$$\text{Pal}(uv) = \psi_u(\text{Pal}(v)) \, . \tag{6.4}$$

There is a precise combinatorial description of standard episturmian words.

Theorem 6.4.2 *An infinite word s is a standard episturmian word if and only if there exists an infinite word $\Delta = a_0 a_1 \cdots$, where the a_n are letters, such that*

$$s = \lim_{n \to \infty} u_n \,,$$

where the sequence $(u_n)_{n \geqslant 0}$ is defined by $u_n = \mathrm{Pal}(a_0 a_1 \cdots a_{n-1})$. Moreover, the word s is episturmian strict if and only if every letter appears infinitely often in Δ.

See the Notes Section for a reference to a proof. The infinite word Δ is called the *directive word* of the standard word s. The description of the infinite word s can be rephrased by the equation

$$s = \mathrm{Pal}(\Delta) \,.$$

As a particular case of Justin's Formula, one has

$$u_{n+1} = \psi_{a_0 \cdots a_{n-1}}(a_n) u_n \,. \tag{6.5}$$

The words u_n are the only prefixes of s which are palindromes.

Example 6.4.3 The Fibonacci word is a standard episturmian word with directive word $\Delta = ababa \cdots$. Indeed, by Formula (6.5) one has $\mathrm{Pal}(\Delta) = \psi_{ab}(\mathrm{Pal}(\Delta))$. Since $\psi_{ab} = \varphi^2$ where φ is the Fibonacci morphism, we have $\mathrm{Pal}(\Delta) = \varphi^2(\mathrm{Pal}(\Delta))$. Since the Fibonacci word is the unique right infinite word fixed point of φ^2, this shows that $\mathrm{Pal}(\Delta)$ is the Fibonacci word.

We note that for $n \geqslant 1$, one has

$$|u_{n+1}| \leqslant 2|u_n|. \tag{6.6}$$

Indeed, set $u_n = v_n ba$. If $a_n = a$, the word $v_n baab\tilde{v}_n$ is a palindrome of length at most $2|u_n|$. If $a_n = b$, then $v_n bab\tilde{v}_n$ is a palindrome of length strictly less than $2|u_n|$.

Example 6.4.4 As a consequence of Eq. (6.5), when s is the Fibonacci word and φ the Fibonacci morphism, we have for every $n \geqslant 0$

$$u_{n+1} = \varphi^n(a) u_n. \tag{6.7}$$

In view of Eq. (6.5), we need to show that $\varphi^n(a) = \psi_{a_0 \cdots a_{n-1}}(a_n)$. By Example 6.4.3, the directive word of s is $ababa \cdots$. If n is even, then we have

$$\psi_{a_0 \cdots a_{n-1}}(a_n) = \psi_{(ab)^{n/2}}(a) = \varphi^n(a)$$

since $\psi_{ab} = \varphi^2$. If n is odd, then

$$\psi_{a_0 \cdots a_{n-1}}(a_n) = \psi_{(ab)^{(n-1)/2}a}(b) = \varphi^{n-1}(ab) = \varphi^n(a)$$

and the property is also true.

As a consequence, we have in the prefix ordering (that is, the partial order such that $x \leqslant y$ when x is a prefix of y) for every $n \geqslant 0$,

$$u_n < \varphi^{n+1}(a). \tag{6.8}$$

Indeed, both words are prefixes of the Fibonacci word and it is enough to compare their lengths. For $n = 0$, we have $|u_0| = 0$ and $|\varphi(a)| = 2$. Next, for $n \geqslant 1$, we have by (6.7), $u_n = \varphi^{n-1}(a)u_{n-1}$. Arguing by induction, we obtain $|u_n| < |\varphi^{n-1}(a)| + |\varphi^n(a)| = |\varphi^{n+1}(a)|$.

There is an explicit form for the return words in an Arnoux-Rauzy set.

Proposition 6.4.5 *Let s be a standard strict episturmian word over A, let $\Delta = a_0 a_1 \cdots$ be its directive word, and let $(u_n)_{n \geqslant 0}$ be its sequence of palindrome prefixes.*

(i) *The left return words to u_n are the words $\psi_{a_0 \cdots a_{n-1}}(a)$ for $a \in A$.*
(ii) *For each factor u of s, let n be the minimal integer such that u is a factor of u_n. There is a unique word z such that zu is a prefix of u_n and the left return words to u are the words $z^{-1}yz$, where y ranges over the left return words to u_n.*

See the Notes Section for a reference to a proof.

Remark 6.4.6 According to Proposition 6.4.5, and since the morphisms of the form ψ_a are injective, if k is the number of letters appearing in an Arnoux-Rauzy set F, then the cardinality of $\mathcal{R}_F(u)$ is k for every $u \in F$. In particular, the Arnoux-Rauzy sets are uniformly recurrent. More generally, every episturmian set is uniformly recurrent (see the Notes Section for a reference).

Example 6.4.7 Let φ be the Fibonacci morphism and let F be the Fibonacci set. We have for $n \geqslant 1$,

$$\mathcal{R}'_F(\varphi^{2n}(aa)) = \{\varphi^{2n}(a), \varphi^{2n+1}(a)\}. \tag{6.9}$$

For example, $\varphi^2(aa) = abaaba$ and $\mathcal{R}'_F(\varphi^2(aa)) = \{aba, abaab\}$. Note that Eq. (6.9) does not hold for $n = 0$. Indeed, $\mathcal{R}'_F(aa) = \{aab, aabab\} \neq \{a, ab\}$.

To show (6.9), we first observe that in the prefix order for $n \geqslant 1$,

$$u_{2n} < \varphi^{2n}(aa) \leqslant u_{2n+1}. \tag{6.10}$$

Indeed, one has by (6.7)

$$u_{2n+1} = \varphi^{2n}(a)u_{2n}, \tag{6.11}$$

and, since $n \geqslant 1$,

$$u_{2n} = \varphi^{2n-1}(a)u_{2n-1} = \varphi^{2n-1}(a)\varphi^{2n-2}(a)u_{2n-2} = \varphi^{2n}(a)u_{2n-2}. \tag{6.12}$$

Thus $u_{2n+1} = \varphi^{2n}(a)u_{2n} = \varphi^{2n}(aa)u_{2n-2}$. This proves the second inequality in (6.10). Since $|u_{2n}| \leqslant 2|u_{2n-1}|$ by (6.6), and $|u_{2n-1}| < |\varphi^{2n}(a)|$ by (6.8), we obtain $|u_{2n}| < 2|\varphi^{2n}(a)|$ and this proves the first inequality.

By (6.10), the minimal integer m such that $\varphi^{2n}(aa)$ is a factor of u_m is $m = 2n + 1$. Thus, by Proposition 6.4.5(ii), one has $\mathcal{R}'_F(\varphi^{2n}(aa)) = \mathcal{R}'_F(u_{2n+1})$. On the other hand, by Proposition 6.4.5(i), we have $\mathcal{R}'_F(u_{2n+1}) = \{\psi_{(ab)^n a}(a), \psi_{(ab)^n a}(b)\}$. Since $\psi_{(ab)^n a}(a) = \varphi^{2n}(a)$ and $\psi_{(ab)^n a}(b) = \varphi^{2n}(ab) = \varphi^{n+1}(a)$, this proves (6.9).

6.5   Tree Sets

Let F be a factorial set of words. For $w \in F$, we consider the set $E_F(w)$ as an undirected graph on the set of vertices which is the disjoint union of $L_F(w)$ and $R_F(w)$ with edges the pairs $(a, b) \in E_F(w)$. This graph is called the *extension graph* of w.

A biextendable set is a *tree set*, also called a *dendric set* if for every $w \in F$, the graph $E_F(w)$ is a tree. A tree set is neutral. A *tree shift*, also called *dendric shift* is a shift space X such that $F(X)$ is a tree set.

More generally one also defines a *connected set* (resp. *acyclic set*) as a biextendable set F such that, for every $w \in F$, the graph $E_F(w)$ is connected (resp. acyclic). Thus a biextendable set is a tree set if and only if it is both connected and acyclic.

Example 6.5.1   The Fibonacci set is a tree set. This follows from the fact that it is an Arnoux-Rauzy set and that every Arnoux-Rauzy set is a tree set (see Exercise 6.12).

Example 6.5.2   The *Tribonacci set* is the set of factors of the Tribonacci word (Example 6.4.1). It is a also a tree set, since it is Arnoux-Rauzy as we have seen. The graph $E(\varepsilon)$ is represented in Fig. 6.2.

Fig. 6.2   The extension graph of ε in the Tribonacci set

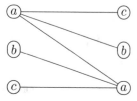

Example 6.5.3 Let $\varphi : a \mapsto ab$, $b \mapsto a^3b$ be as in Example 5.6.18. Let F be the set of factors of $\varphi^{\omega}(a)$. The graphs $E(a)$ and $E(aa)$ are shown in Fig. 6.3.

The first graph has a cycle of length 4 and the second one has two connected components. Thus F is neither an acyclic nor a connected set. In particular, it is not a tree set.

Example 6.5.4 Let T be the Thue-Morse set (see Example 5.5.7). The set T is not a tree set since $E_T(\varepsilon)$ is the complete bipartite graph $K_{2,2}$ on two sets with two elements.

The following result, called the Return Theorem, shows that, in a tree set, a property much stronger than Theorem 6.3.6 holds.

Theorem 6.5.5 *Let F be a uniformly recurrent tree set containing the alphabet A. For any $x \in F$, the set $\mathcal{R}_F(x)$ is a basis of $FG(A)$.*

Example 6.5.6 Let F be the Tribonacci set on $A = \{a, b, c\}$. Then $\mathcal{R}_F(a) = \{a, ba, ca\}$, which is easily seen to be a basis of $FG(A)$.

The proof uses Theorem 6.3.6 and the following result.

Theorem 6.5.7 *Let F be a uniformly recurrent connected set containing the alphabet A. For every $w \in F$, the set $\mathcal{R}_F(w)$ generates the free group $FG(A)$.*

To prove Theorem 6.5.7, we first introduce the notion of a Rauzy graph. Let F be a factorial set. The *Rauzy graph* of F of order $n \geqslant 0$ is the following labeled graph $G_n(F)$. Its vertices are the words in the set $F \cap A^n$. Its edges are the triples (x, a, y) for all $x, y \in F \cap A^n$ and $a \in A$ such that $xa \in F \cap Ay$.

Example 6.5.8 The Rauzy graphs of the Fibonacci set of orders 0, 1, 2 are shown in Fig. 6.4.

Let $u \in F \cap A^n$. The following properties follow easily from the definition of the Rauzy graph.

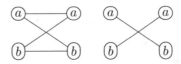

Fig. 6.3 The extension graphs $E(a)$ and $E(aa)$

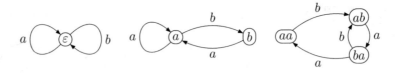

Fig. 6.4 The Rauzy graphs of order 0, 1, 2

(i) For every word w such that $uw \in F$, there is a path labeled w in $G_n(F)$ from u to the suffix of length n of uw.

(ii) Conversely, the label of every path of length at most $n + 1$ in $G_n(F)$ is in F.

When F is recurrent, all Rauzy graphs $G_n(F)$ are strongly connected. Indeed, let $u, w \in F \cap A^n$. Since F is recurrent, there is a $v \in F$ such that $uvw \in F$. Then there is a path in $G_n(F)$ from u to w labeled vw by property (i) above.

The Rauzy graph $G_n(F)$ of a recurrent set F with a distinguished vertex v can be considered as a finite automaton $\mathcal{A} = (Q, v, v)$ with set of states $Q = F \cap A^n$ (see Sect. 4.2).

Let G be a graph on a set Q of vertices with edges labeled by A. A *generalized path* in G is a sequence of edges or their inverses (the inverse of an edge being labeled by the inverse of the label). The label of the generalized path is the reduced word equivalent to the concatenation of its labels. The *group described* by G with respect to a vertex v is the set of labels of generalized paths from v to itself. We will prove the following statement.

Theorem 6.5.9 *Let F be a recurrent connected set containing the alphabet A. The group described by a Rauzy graph of F with respect to any vertex is the free group on A.*

Proof We consider on a Rauzy graph $G_n(F)$ the equivalence θ_n formed by the pairs of vertices (u, v) with $u = ax$, $v = bx$, $a, b \in L(x)$ such that there is a path from a to b in the extension graph $E(x)$ (and more precisely from the vertex corresponding to a to the vertex corresponding to b in the copy corresponding to $L(x)$ in the bipartite graph $E(x)$).

We claim that if F is connected then, for each $n \geqslant 1$, then the quotient of $G_n(F)$ by the equivalence θ_n is isomorphic to $G_{n-1}(F)$.

The map $\varphi : F \cap A^n \to F \cap A^{n-1}$ mapping a word of F of length n to its suffix of length $n - 1$ is clearly a graph morphism from $G_n(F)$ onto $G_{n-1}(F)$. If $u, v \in F \cap A^n$ are equivalent modulo θ_n, then $\varphi(u) = \varphi(v)$. Thus, there is a morphism ψ from $G_n(F)/\theta_n$ onto $G_{n-1}(F)$. It is defined for each word $u \in F \cap A^n$ by $\psi(\bar{u}) = \varphi(u)$ where $\bar{u}$ denotes the class of u modulo θ_n. But, since F is connected, the class modulo θ_n of a word ax of length n has $\ell(x)$ elements, which is the same as the number of elements of $\varphi^{-1}(x)$. This shows that ψ is a surjective map from a finite set onto a set of the same cardinality and thus that it is one-to-one. Thus ψ is an isomorphism.

Recall from Sect. 4.3 that a Stallings folding at vertex v relative to letter a in a graph G consists in identifying the edges coming into v labeled a and identifying their origins. A Stallings folding does not modify the group described by the graph with respect to some vertex. Indeed, if $p \xrightarrow{a} v$, $p \xrightarrow{b} r$ and $q \xrightarrow{a} v$ are three edges of G, then adding the edge $q \xrightarrow{b} r$ does not change the group described since the path $q \xrightarrow{a} v \xrightarrow{a^{-1}} p \xrightarrow{b} r$ has the same label. Thus merging p and q does not add new labels of generalized paths.

The quotient $G_n(F)/\theta_n$ can be obtained by a sequence of Stallings foldings from the graph $G_n(F)$. Indeed, a Stallings folding at vertex v identifies vertices which are equivalent modulo θ_n. Conversely, consider $u = ax$ and $v = bx$, with $u, v \in F \cap A^n$ and $a, b \in A$ such that a and b (considered as elements of $L(x)$), are connected by a path in $E(x)$. Let $a_0, \ldots, a_k$ and $b_1, \ldots, b_k$ with $a = a_0$ and $b = a_k$ be such that (a_i, b_{i+1}) for $0 \leqslant i \leqslant k - 1$ and (a_i, b_i) for $1 \leqslant i \leqslant k$ are in $E(x)$. The successive Stallings foldings at $xb_1, xb_2, \ldots, xb_k$ identify the vertices $u = a_0 x, a_1 x, \ldots, a_k x = v$. Indeed, since $a_i x b_{i+1}, a_{i+1} x b_{i+1} \in F$, there are two edges labeled b_{i+1} going out of $a_i x$ and $a_{i+1} x$ which end at $x b_{i+1}$. The Stallings folding identifies $a_i x$ and $a_{i+1} x$. The conclusion follows by induction.

Since the Stallings foldings do not modify the group described, we deduce from the claim that the group described by the Rauzy graph $G_n(F)$ is the same as the group described by the Rauzy graph $G_0(F)$. Since $G_0(F)$ is the graph with one vertex and with loops labeled by each of the letters, it describes the free group on A.   ■

Example 6.5.10 Let F be the tree set obtained by decoding the Fibonacci set S into blocks of length 2, that is the set $f^{-1}(S)$ on the alphabet $B = \{u, v, w\}$ with $f : u \to aa, v \to ab, w \to ba$. The graph $G_2(F)$ is represented on the left of Fig. 6.5.

The classes of θ_2 are $\{wv, vv\}$ $\{vu\}$ and $\{ww, uw\}$. The graph $G_1(F)$ is represented on the right.

The following example shows that Theorem 6.5.7 is false for sets which are not connected.

Example 6.5.11 Consider again the Chacon set (see Example 6.3.4).

The Rauzy graph $G_1(F)$ corresponding to the Chacon set is represented in Fig. 6.6 on the left. The graph $G_1(F)/\theta_1$ is represented on the right. It is not isomorphic to $G_0(F)$ since it has two vertices instead of one.

We are now ready to prove Theorem 6.5.5.

Proof of Theorem 6.5.5 We will prove that if F is connected, then for any $w \in F$, the set $\mathcal{R}_F(w)$ generates the free group on A. Since, by Theorem 6.3.6, the set $\mathcal{R}_F(w)$ has Card(A) elements, the result will follow.

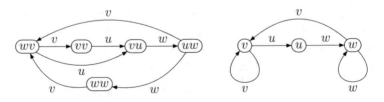

Fig. 6.5 The Rauzy graphs $G_2(F)$ and $G_1(F)$ for the decoding of the Fibonacci set into blocks of length 2

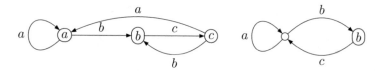

Fig. 6.6 The graphs $G_1(F)$ and $G_1(F)/\theta_1$

Since F is uniformly recurrent, the set $\mathcal{R}_F(w)$ is finite. Let n be the maximal length of the words in $w\,\mathcal{R}_F(w)$. In this way, every word in $F \cap A^n$ beginning with w has a prefix in $w\,\mathcal{R}_F(w)$. Moreover, recall from Property (ii) of Rauzy graphs, that the label of every path of length $n + 1$ in the Rauzy graph $G_n(F)$ is in F.

Let $x \in F$ be a word of length n ending with w. Let $\mathcal{A}$ be the finite automaton defined by $G_n(F)$ with initial and terminal state x. Let X be the prefix code generating the submonoid recognized by $\mathcal{A}$. The subgroup generated by X is equal to the group described by $\mathcal{A}$ since both groups are described by the Stallings reduction of $\mathcal{A}$.

We show that $X \subseteq \mathcal{R}_F(w)^*$. Indeed, let $y \in X$. Since y is the label of a path starting at x and ending in x, the word xy ends with x and thus the word wy ends with w. Let $\Gamma = \{z \in A^+ \mid wz \in A^*w\}$ and let $R = \Gamma \setminus \Gamma A^+$. Then R is a prefix code and $\Gamma \cup 1 = R^*$, as one may verify easily. Since $y \in \Gamma$, we can write $y = u_1 u_2 \cdots u_m$ where each word u_i is in R. Since F is recurrent and since $x \in F$, there is $v \in F \cap A^n$ such that $vx \in F$ and thus there is a path labeled x ending at the vertex x by Property (i) of Rauzy graphs. Thus, there is a path labeled xy in $G_n(F)$. This implies that for $1 \leqslant i \leqslant m$, there is a path in $G_n(F)$ labeled wu_i.

Assume that some u_i is such that $|wu_i| > n$. Then the prefix p of length n of wu_i is the label of a path in $G_n(F)$. This implies, by Property (ii) of Rauzy graphs, that p is in F and thus that p has a prefix in $w\,\mathcal{R}_F(w)$. But then wu_i has a proper prefix in $w\,\mathcal{R}_F(w)$, a contradiction. Thus we have $|wu_i| \leqslant n$ for all $i = 1, 2, \ldots, m$. But then the wu_i are in F by Property (i) again and thus the u_i are in $\mathcal{R}_F(w)$. This shows that $y \in \mathcal{R}_F(w)^*$.

Thus the group generated by $\mathcal{R}_F(w)$ contains the group generated by X. But, by Theorem 6.5.9, the group described by $\mathcal{A}$ is the free group on A. Thus $\mathcal{R}_F(w)$ generates the free group on A.   ■

We illustrate the proof in the following example.

Example 6.5.12 Let F be the Fibonacci set. We have $\mathcal{R}_F(aa) = \{baa, babaa\}$. The Rauzy graph $G_7(F)$ is represented in Fig. 6.7. The set recognized by the automaton obtained using $x = aababaa$ as initial and terminal state is X^* with $X = \{babaa, baababaa\}$. In agreement with the proof of Theorem 6.5.5, we have $X \subseteq \mathcal{R}_F(aa)^*$.

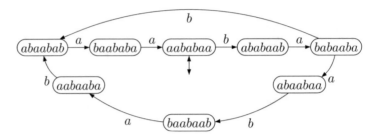

Fig. 6.7 The Rauzy graph $G_7(F)$

6.6 Sequences of Return Sets

In this section, we shall associate to a uniformly recurrent pseudoword a sequence of sets of return words.

Let F be a uniformly recurrent set. Consider a pseudoword x which belongs to a group in $J(F)$. Let $(r_n)_{n\geqslant 1}$ be a sequence of finite prefixes of x strictly increasing for the prefix order. Similarly, let $(\ell_n)_{n\geqslant 1}$ be a sequence of finite suffixes of x strictly increasing for the suffix order. Since $x^2 \in J(F)$, we have $\ell_n r_n \in F$ (cf. Remark 5.6.16). Let

$$\mathcal{R}_n = r_n\, \mathcal{R}_F(\ell_n r_n) r_n^{-1}.$$

We say that the sequence $(\ell_n, r_n)_{n\geqslant 1}$ is a *defining pair* of $H(x)$, a terminology justified by Proposition 5.6.15.

Recall the established convention according to which, if $L \subseteq A^+$, then L^+ denotes the subsemigroup of A^+ generated by L.

Lemma 6.6.1 *The inclusion*

$$\mathcal{R}_{n+1} \subseteq \mathcal{R}_n^+ \tag{6.13}$$

holds.

Proof Set $r_{n+1} = r_n s_n$. Since $\ell_{n+1} r_n$ is a prefix of $\ell_{n+1} r_{n+1}$, we have by the dual of (6.1) the inclusion $\mathcal{R}'_F(\ell_{n+1} r_{n+1}) \subseteq \mathcal{R}'_F(\ell_{n+1} r_n)^+$. Thus

$$\mathcal{R}_F(\ell_{n+1} r_{n+1}) = r_{n+1}^{-1} \ell_{n+1}^{-1} \mathcal{R}'_F(\ell_{n+1} r_{n+1}) \ell_{n+1} r_{n+1},$$
$$\subseteq r_{n+1}^{-1} \ell_{n+1}^{-1} \mathcal{R}'_F(\ell_{n+1} r_n)^+ \ell_{n+1} r_{n+1}$$
$$\subseteq s_n^{-1} \mathcal{R}_F(\ell_{n+1} r_n)^+ s_n.$$

Since $\ell_n r_n$ is a suffix of $\ell_{n+1} r_n$, we have by (6.1) the inclusion $\mathcal{R}_F(\ell_{n+1} r_n) \subseteq \mathcal{R}_F(\ell_n r_n)^+$. Thus we obtain

$$\mathcal{R}_{n+1} = r_{n+1} \mathcal{R}_F(\ell_{n+1} r_{n+1}) r_{n+1}^{-1},$$
$$\subseteq r_{n+1} s_n^{-1} \mathcal{R}_F(\ell_{n+1} r_n)^+ s_n r_{n+1}^{-1},$$
$$\subseteq r_n \mathcal{R}_F(\ell_n r_n)^+ r_n^{-1} = \mathcal{R}_n^+,$$

concluding the proof. ∎

Therefore, we have a chain

$$\cdots \subseteq \mathcal{R}_4^+ \subseteq \mathcal{R}_3^+ \subseteq \mathcal{R}_2^+ \subseteq \mathcal{R}_1^+. \tag{6.14}$$

Let $l(n)$ be the length of the shortest word in $\mathcal{R}_n$. Since $\mathcal{R}_{n+1} \subseteq \mathcal{R}_n^+$, we have $l(n) \leqslant l(n+1)$. If F is periodic, then the sequence $l(n)$ is bounded (cf. Example 6.2.2). In the opposite case, we have the following result.

Lemma 6.6.2 *If F is uniformly recurrent not periodic, then $\lim l(n) = \infty$.*

Proof Assume that $l(n) = k$ for all large enough n. Then, for n large enough, each $\mathcal{R}_n$ contains a word v_n of length k. For $n \geqslant k$, this implies that $\ell_n r_n$ is of period k. Indeed, let u_n be the prefix of length k of ℓ_n and w_n be the suffix of length k of r_n. Then (see Fig. 6.8) $u_n \ell_n = \ell_n v_n$ and $v_n r_n = r_n w_n$, showing that ℓ_n and r_n have period k. Thus all words of F have period k. ∎

Therefore, if F is uniformly recurrent non periodic, then the intersection of the chain (6.14) is empty. We proceed to see what happens if, in each element of the chain, we take the topological closure in $\widehat{A^*}$.

Proposition 6.6.3 *If the uniformly recurrent set F is not periodic, then we have*

$$H(x) = \bigcap_{n \geqslant 1} \overline{\mathcal{R}_n^+}.$$

Fig. 6.8 The periodicity of $\ell_n r_n$

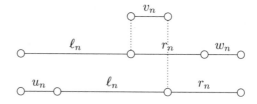

Irrespectively of the uniformly recurrent set F being periodic or not, the equality

$$H(x) = \bigcap_{n \geqslant 1} \overline{\mathcal{R}_n^+} \setminus A^*$$

holds.

Proof If F is not periodic, then the intersection of the sets $\mathcal{R}_n^+$ is empty by Lemma 6.6.2. Hence, from $\mathcal{R}_n^+ = \overline{\mathcal{R}_n^+} \cap A^*$ we get $\bigcap_{n \geqslant 1} \mathcal{R}_n^+ = \bigcap_{n \geqslant 1} \overline{\mathcal{R}_n^+} \setminus A^*$. Thus it suffices to show the second statement.

Let $y \in \bigcap_{n \geqslant 1} \overline{\mathcal{R}_n^+} \setminus A^*$. We wish to show that all finite factors of y belong to F.

Let $v \in A^*$ be a factor of y with length k, where k is an arbitrary positive integer. Then the set $\widehat{A^*}v\widehat{A^*}$ is an open neighborhood of y, since it is the topological closure of the recognizable set A^*vA^* (cf. Theorem 4.4.9). Therefore, we have $\mathcal{R}_n^+ \cap A^*vA^* \neq \emptyset$ for every $n \geqslant 1$.

If F is periodic, with least period p, then for all sufficiently large n there is a word w_n of length p such that $\mathcal{R}_n = \{w_n\}$, and F is the set of factors of $\mathcal{R}_n^*$ (cf. Example 6.2.2). Since $\mathcal{R}_n^+ \cap A^*vA^* \neq \emptyset$, this shows that $v \in F$.

Suppose that F is not periodic. Then, there is an integer m such that the length of r_m and of each element of $\mathcal{R}_m$ is at least k, by Lemma 6.6.2. Let $u_1, \ldots, u_q$ be elements of $\mathcal{R}_m$ such that the word $u_1 \cdots u_q$ belongs to A^*vA^*. We may suppose that $q > 1$. Since the words $u_1, \ldots, u_q$ have length at least k, there is $j \in \{1, \ldots, q-1\}$ such that v is a factor of $u_j u_{j+1}$. By the definition of $\mathcal{R}_m$, there is some word z such that $u_{j+1}r_m = r_m z$. Since v is a factor of $u_j u_{j+1} r_m = u_j r_m z$ and r_m has length at least k, the word v is a factor of $u_j r_m$ or of $r_m z = u_{j+1} r_m$. Again by the definition of $\mathcal{R}_m$, both words $u_j r_m$ and $u_{j+1} r_m$ belong to $r_m \mathcal{R}_F(\ell_m r_m)$, which in turn is contained in F. This shows that $v \in F$.

We have shown that $F(y) \subseteq F$, irrespectively of F being periodic or not. This establishes that $y \in \overline{F} \setminus A^*$, by Remark 5.6.11, whence $y \in J(F)$ by Proposition 5.6.14.

Since $\mathcal{R}_n \subseteq r_n A^+$, the infinite word $\overrightarrow{y}$ is the element of $A^\mathbb{N}$ having all r_n as prefixes and thus $\overrightarrow{y} = \overrightarrow{x}$. Similarly, since $\mathcal{R}_n \subseteq A^+\ell_n$, the equality $\overleftarrow{y} = \overleftarrow{x}$ also holds. Therefore, we have $y \in H(x)$ by Proposition 5.6.15.

Conversely, suppose that $y \in H(x)$. Let $n \geqslant 1$. Since the pseudoword xyx belongs to the group $H(x)$, its factor $l_n y r_n$ belongs to the factorial set $\overline{F}$. Therefore, as $l_n r_n$ is a prefix of $l_n y$ and a suffix of $y r_n$, there is, according to Exercise 6.3, some pseudoword $w \in \overline{\mathcal{R}_F(l_r r_n)^+}$ such that $l_n y r_n = l_n r_n w$. Because l_n is a finite pseudoword, we may cancel it in the last equality (cf. Exercise 4.20), thus obtaining $y r_n \in r_n \overline{\mathcal{R}_F(l_r r_n)^+}$, and so

$$y \in r_n \overline{\mathcal{R}_F(l_r r_n)^+} r_n^{-1}.$$

Since $\mathcal{R}_F(\ell_n r_n) \subseteq A^+ r_n$ and $\mathcal{R}_n = r_n \mathcal{R}_F(\ell_n r_n) r_n^{-1}$, one sees (cf. Exercise 6.2) that the equality

$$\overline{\mathcal{R}_n^+} = r_n \overline{\mathcal{R}_F(\ell_n r_n)^+} r_n^{-1}$$

holds. This shows that $y \in \overline{\mathcal{R}_n^+}$ for all $n \geqslant 1$, and so, as $J(F) \subseteq \widehat{A^*} \setminus A^*$, we have shown that $H(x)$ is indeed the intersection of the sets $\overline{\mathcal{R}_n^+} \setminus A^*$.   ∎

6.7   Limit Return Sets

Suppose that F is a k-bounded set. For an element x of a group contained in $J(F)$, consider a defining pair $(\ell_n, r_n)_{n \geqslant 1}$ of $H(x)$ and the associated sequence $(\mathcal{R}_n)_{n \geqslant 1}$ of return sets. Up to taking a subsequence we may choose a sequence $(\ell_n, r_n, \mathcal{R}_n)$ such that $\mathcal{R}_n$ has a fixed number q of elements $w_{n,1}, \ldots, w_{n,q}$ (recall that $q \leqslant k$) and that the sequence $(w_{n,1}, \ldots, w_{n,q})$ converges in $(\widehat{A^*})^q$ to a q-tuple $(w_1, \ldots, w_q)$. The set $\mathcal{R} = \{w_1, \ldots, w_q\}$ is called a *limit return set* to x.

Example 6.7.1 Let φ be the Fibonacci morphism and F be the Fibonacci set. We will show that $\{\varphi^\omega(a), \varphi^\omega(ba)\}$ is a limit return set to $x = \varphi^\omega(a)$. Note that the $\mathcal{H}$-class of the pseudoword $x = \varphi^\omega(a)$ is a group. Indeed, $x^2 = \lim \varphi^{n!}(aa)$ also belongs to $J(F)$, as $\varphi^k(aa)$ belongs to F for every $k \geqslant 0$ and $\overline{F} \setminus A^* = J(F)$ when F is uniformly recurrent.

Consider the sequence $\ell_n = r_n = \varphi^{2n}(a)$. The sequence is increasing both for the prefix and the suffix order and its terms are both prefixes and suffixes of x. The set $\mathcal{R}_n = r_n \mathcal{R}_F(\ell_n r_n) r_n^{-1} = \ell_n^{-1} \mathcal{R}'_F(\ell_n r_n) \ell_n$ is, by equality (6.9) in Example 6.4.7,

$$\mathcal{R}_n = \varphi^{2n}(a)^{-1} \{\varphi^{2n}(a), \varphi^{2n+1}(a)\} \varphi^{2n}(a) = \{\varphi^{2n}(a), \varphi^{2n}(b)\varphi^{2n}(a)\}$$

$$= \{\varphi^{2n}(a), \varphi^{2n}(ba)\}.$$

The pair $(\varphi^{n!}(a), \varphi^{n!}(ba))$ converges to $(\varphi^\omega(a), \varphi^\omega(ba))$, which proves the claim that $\{\varphi^\omega(a), \varphi^\omega(ba)\}$ is a limit return set to x.

The following example shows that in degenerated cases, a limit return set may contain finite words.

Example 6.7.2 Let F be a periodic set of least period n. By Example 6.2.2 a return set to any word of length larger than n is formed of one word of length n. Thus any limit return set to a pseudoword $x \in J(F)$ is formed of one word of length n.

The following result shows that a maximal group in $J(F)$ where F is nonperiodic and bounded is the closure of a finitely generated subsemigroup.

Theorem 6.7.3 *Let F be a nonperiodic bounded set. Let $x \in J(F)$ be such that $H(x)$ is a group and let $\mathcal{R}$ be a limit return set to x. Then $H(x)$ is the closure of the semigroup generated by $\mathcal{R}$.*

Proof Let $\mathcal{R} = \{w_1, \ldots, w_q\}$ be a limit return set to x. It is by definition associated to a defining pair $(\ell_n, r_n)_{n \geqslant 1}$ of $H(x)$ whose associated sequence of return sets $\mathcal{R}_n = \{w_{n,1}, \ldots, w_{n,q}\}$ is such that $w_{n,i} \to w_i$, for all $i \in \{1, \ldots, q\}$.

Let $i \in \{1, \ldots, q\}$. Choose an integer $k \geqslant 1$. By Lemma 6.6.1, for all $n \geqslant k$ we have $w_{n,i} \in \mathcal{R}_k^+$, thus $w_i \in \overline{\mathcal{R}_k^+}$. Since k is an arbitrary integer, and since F is not periodic, we deduce from Proposition 6.6.3 that $w_i \in H(x)$, and so $\mathcal{R} \subseteq H(x)$. Because $H(x)$ is a closed subsemigroup in our environment free profinite monoid, we conclude that indeed $\overline{\mathcal{R}^+} \subseteq H(x)$.

Conversely, let $y \in H(x)$. Then we have $y \in \overline{\mathcal{R}_n(x)^+}$ for all $n \geqslant 1$, by Proposition 6.6.3. Therefore, by Proposition 4.5.4, we know that for each $n \geqslant 1$ there is a pseudoword $\rho_n \in \widehat{X_q^+}$, where $X_q = \{x_1, \ldots, x_q\}$, such that

$$y = \rho_n(w_{n,1}, \ldots, w_{n,q}).$$

Since $\widehat{X_q^+}$ is compact, we may consider a convergent subsequence $(\rho_{n_k})_k$ of $(\rho_n)_n$. By the continuity of the evaluation on pseudowords (Proposition 4.5.5), one has

$$\rho_{n_k}(w_{n_k,1}, \ldots, w_{n_k,q}) \to \rho(w_1, \ldots, w_q)$$

thus $y = \rho(w_1, \ldots, w_q)$. Hence, again by Proposition 4.5.4, we have $y \in \overline{\mathcal{R}^+}$. ∎

Note that Theorem 6.7.3 does not hold without the hypothesis that F is nonperiodic (see Example 6.7.2).

Theorems 6.7.3 and 6.3.6, together, give the following.

Corollary 6.7.4 *If F is a uniformly recurrent neutral set on an alphabet of k letters, then any maximal subgroup of $J(F)$ is generated by k elements.*

Example 6.7.5 Let φ be the Fibonacci morphism and let F be the Fibonacci set. Take $x = \varphi^\omega(a)$. The group $H(x)$ is the closure of the semigroup generated by x and $y = \varphi^\omega(ba)$ (cf. Example 6.7.1). This implies that the image of $H(x)$ by the canonical projection $p_{\mathsf{G}} : \widehat{A^*} \to \widehat{FG(A)}$ is $\widehat{FG(A)}$, because the image of $\varphi^\omega(a)$ and $\varphi^\omega(ba)$ by p_{G} are respectively a and ba (cf. Example 4.6.10), and the subgroup of $FG(A)$ generated by $\{a, ba\}$ is $FG(A)$. We shall see in Chap. 7 that this projection of $H(x)$ onto $\widehat{FG(A)}$ is actually an isomorphism.

6.8   Exercises

6.8.1   Section 6.2

6.1 If $X \subseteq A^*$ and $v \in A^*$, then the set $\{u \in A^* \mid uv \in X\}$ is denoted Xv^{-1}, and is said to be a *right quotient* of X.

1. Give an example showing that this notation may conflict with the classic convention of denoting the set $\{y \in FG(A) \mid y = xv^{-1}$ for some $x \in X\}$ also by Xv^{-1}, when we view X as a subset of the free group $FG(A)$ and v as an element of $FG(A)$. In which cases this ambiguity does not occur? (We adhere to both these notations, since they are common in the literature, and any ambiguity is avoided by context.)
2. Give an example in which the right quotient $(uX)v^{-1}$ is distinct from $u(Xv^{-1}) \subseteq A^*$.
3. Why there is no ambiguity in the use of the expression $x \, \mathcal{R}_F(x)x^{-1}$, made in the main text of Sect. 6.2?
4. Why is the notation $Xv^{-1} = \{u \in A^* \mid uv \in X\}$ consistent with the one used in Exercise 4.21?

(Note: the previous discussion is mirrored in the obvious way when we consider the *left* quotients $v^{-1}X = \{u \in A^* \mid vu \in X\}$.)

6.2 Show that, for every nonempty subset X of A^+ and word w over A, if $wX \subseteq A^+w$ then $\emptyset \neq \overline{wXw^{-1}} \subseteq A^+$ and $(wXw^{-1})^+ = wX^+w^{-1}$. Deduce that the equality $\overline{(wXw^{-1})^+} = w\overline{X^+}w^{-1}$ holds in $\widehat{A^+}$. Finally, conclude that the equality

$$\overline{\mathcal{R}'_F(x)^*} = x\overline{\mathcal{R}_F(x)^*}x^{-1}$$

holds in $\widehat{A^*}$, for every x in a factorial set $F \subseteq A^*$.

6.3 In this exercise, we provide pseudoword companions for the set $\Gamma_F(x)$ and for Proposition 6.2.3: show that, for every factorial set $F \subseteq A^*$ and $x \in A^*$, and for the set

$$\Gamma_{\overline{F}}(x) = \{w \in \overline{F} \mid xw \in \overline{F} \cap \widehat{A^*}x\}$$

the chain of equalities $\Gamma_{\overline{F}}(x) = \overline{\Gamma_F(x)} = \overline{\mathcal{R}_F(x)^*} \cap x^{-1}\overline{F}$ holds.

6.8.2   Section 6.3

6.4 Let x be an infinite word and let $p(n) = \mathrm{Card}(F(x) \cap A^n)$. Show that x is eventually periodic if and only if $p(n) \leqslant n$ for some n (hint: show that if $p(n) \leqslant n$, then $p(m) = p(m+1)$ for some m with $1 \leqslant m \leqslant n$).

6.5 An element of a factorial set is *conservative* if it is not right-special. Let x be an infinite word and $n \geqslant 1$. Let c be the number of conservative factors of length n in x. Show that if x has a factor of length $n + c$ whose factors of length n are all conservative, then x is eventually periodic.

6.6 Let $A = \{0, 1\}$. The *balance* of two words $u, v \in A^*$ is $\delta(u, v) = \left| |u|_1 - |v|_1 \right|$ where $|u|_1$ is the number of 1 in u. A set of words $X \subseteq A^*$ is *balanced* if for every pair $x, y \in X$ of words of the same length, one has $\delta(x, y) \leqslant 1$. Prove the following statements.

1. A factorial set X is unbalanced if and only if there is a palindrome word w such that $0w0, 1w1 \in X$.
2. If X is balanced, then X has at most $n + 1$ words of length n for all $n \geqslant 1$.
3. If x is Sturmian, then $F(x)$ is balanced.

6.7 Show that the set of factors of a Sturmian word is closed under reversal (hint: use Exercise 6.6).

6.8 Show that the Fibonacci set is Sturmian (hint: show that the left-special words are the prefixes of the $\varphi^n(a)$).

6.9 Show that the complexity of a neutral set on $k + 1$ letters is $kn + 1$.

6.8.3   Section 6.4

6.10 Show that the Tribonacci word is a standard episturmian word.

6.11 Show that the directive word of the Tribonacci word is $\Delta = abcabcabc \cdots$.

6.8.4   Section 6.5

6.12 Show that an Arnoux-Rauzy set is a tree set.

6.9   Solutions

6.9.1   Section 6.2

6.1

1. If $X = \{a\}$ and $v = a^2$, then $\{u \in A^* \mid uv \in X\} = \emptyset$ is distinct from $\{y \in FG(A) \mid y = xv^{-1} \text{ for some } x \in X\} = \{a^{-1}\}$, and so there is ambiguity in this example. The ambiguity does not occur precisely when we have the inclusion

$\{y \in FG(A) \mid y = xv^{-1}$ for some $x \in X\} \subseteq A^*$, that is, precisely when $X \subseteq A^*v$.

2. Let $X = \{a\}$ and $u = v = a^2$. Then $u(Xv^{-1}) = \emptyset$ and $(uX)v^{-1} = \{a\}$.
3. There is no ambiguity because $x\,\mathcal{R}_F(x) \subseteq A^*x$.
4. There is consistency because, for every subset K of A^*, we have the equality $\overline{K} \cap A^* = K$, as the elements of A^* are isolated points of $\widehat{A^*}$.

6.2 With induction on k, one clearly sees that $wX^k \subseteq A^+w$ for all $k \geqslant 1$. Therefore, for each $k \geqslant 1$, the quotient $(wX^k)w^{-1} \subseteq A^+$ coincides with the conjugate wX^kw^{-1} of X in the free group over A, viewing A^+ as a subset of $FG(A)$. In particular, wX^kw^{-1} is nonempty and does not contain the empty word. Moreover, since we know that in $FG(A)$ the equality $(wXw^{-1})^k = wX^kw^{-1}$ holds, we deduce that the equality $(wXw^{-1})^+ = wX^+w^{-1}$ is valid in A^+. Therefore, we have $\overline{(wXw^{-1})^+} = w\overline{X^+}w^{-1}$ (cf. Exercise 4.21).

For the final part, recall that $\mathcal{R}'_F(x) = x\,\mathcal{R}_F(x)x^{-1}$ and $x\,\mathcal{R}_F(x) \subseteq A^*x$. If $\mathcal{R}_F(x) = \emptyset$, then $\mathcal{R}'_F(x)^* = \mathcal{R}_F(x)^* = \{\varepsilon\}$ and we are done. If $\mathcal{R}_F(x) \neq \emptyset$, then we may apply the already proved equality $\overline{(wXw^{-1})^+} = w\overline{X^+}w^{-1}$ by making $X = \mathcal{R}_F(x)$ and $w = x$, getting $\overline{\mathcal{R}'_F(x)^+} = x\overline{\mathcal{R}_F(x)^+}x^{-1}$. Since $\overline{Y^+} = \overline{Y^*} \setminus \{\varepsilon\}$ in $\widehat{A^*}$ for every $Y \subseteq A^+$, we obtain the desired equality.

6.3

Proof of the Inclusion $\overline{\Gamma_F(x)} \subseteq \overline{\mathcal{R}_F(x)^*} \cap x^{-1}\overline{F}$
Recalling that the set $x^{-1}\overline{F}$ is closed (cf. Exercise 4.21), one sees by Proposition 6.2.3 that the topological closure of $\Gamma_F(x)$ is contained in the closed set $\overline{\mathcal{R}_F(x)^*} \cap x^{-1}\overline{F}$.

Proof of the Inclusion $\overline{\mathcal{R}_F(x)^*} \cap x^{-1}\overline{F} \subseteq \Gamma_{\overline{F}}(x)$
As $x\,\mathcal{R}_F(x)^* \subseteq A^*x$, if $y \in \overline{\mathcal{R}_F(x)^*} \cap x^{-1}\overline{F}$ then $xy \in \widehat{A^*}x \cap \overline{F}$, that is $y \in \Gamma_{\overline{F}}(x)$.

Proof of the Inclusion $\Gamma_{\overline{F}}(x) \subseteq \overline{\Gamma_F(x)}$
Let $w \in \Gamma_{\overline{F}}(x)$. Then there is a sequence (v_n) of elements of F converging to xw, and $xw = ux$ for some $u \in \widehat{A^*}$. By Proposition 4.4.17, there are factorizations $v_n = x_nw_n = u_nx'_n$ such that $x = \lim x_n = \lim x'_n$, $w = \lim w_n$ and $u = \lim u_n$. Since $x \in A^*$, for all large enough n we have $x = x_n = x'_n$, whence $w_n \in \Gamma_F(x)$. This yields $w \in \overline{\Gamma_F(x)}$.

6.9.2   Section 6.3

6.4 The condition is clearly necessary since $p(n)$ is bounded for an eventually periodic word (if $x = uvvv \cdots$, then $p(n) \leqslant |uv|$). Conversely, assume that $p(n) \leqslant n$. We have $p(n) \geqslant p(n-1) \geqslant \ldots \geqslant p(1)$. If all inequalities were strict then $p(n) \geqslant n + p(1) > n$ since we may assume $k > 1$. Thus there is some $m \geqslant 1$ such that $p(m) = p(m+1)$.

Consider the Rauzy graph $G_m(F)$ for $F = F(x)$. Since the number of edges is $p(m+1)$, there is one edge going out of each vertex. This implies that the connected components of $G_m(F)$ are simple circuits and thus that x is eventually periodic.

6.5 Let $w = a_1 a_2 \cdots a_{n+c}$ be a factor of length $n + c$ whose factors of length n are all conservative. Set $p_i = a_{i+1} \cdots a_{i+n}$ for $i = 0, \ldots, c$. In the Rauzy graph $G_n(x)$ the path $\pi = (p_0, \ldots, p_c)$ is part of the path of x. Since there are only c conservative vertices, the path π contains a circuit and since each vertex p_i has a unique outgoing edge, the path must stay in this circuit. Thus x is eventually periodic, a contradiction.

6.6

1. The condition is clearly sufficient. Conversely, let $u, v \in X$ be of minimum length n such that $\delta(u, v) \geqslant 2$. We may assume that $u = 0wau'$ and $v = 1wbv'$ with $a, b \in A$ distinct. We have $a = 0$ and $b = 1$ since otherwise we can replace u, v by u', v'. By minimality again, we have $u = 0w0$ and $v = 1w1$.

   Suppose that w is not a palindrome. Then there is a prefix z of w and a letter a such that za is a prefix of w, $\tilde{z}$ is a suffix of w but $a\tilde{z}$ is not a suffix of w. Thus, $b\tilde{z}$ is a suffix of w where $b = 1 - a$. If $a = 0$ and $b = 1$, then $\delta(0z0, 1\tilde{z}1) = 2$, contradicting the minimality of n. But then $u = 0z1u''$ and $v = v''1\tilde{z}0$ with $\delta(u'', v'') = \delta(u, v)$ contradicting again the minimality of n.

2. The case $n = 1$ is clear. For $n = 2$, since X is balanced, it cannot contain 00 and 11. Thus, $\mathrm{Card}(X \cap A^2) \leqslant 3$. Let n be minimum such that $\mathrm{Card}(X \cap A^n) > n+1$. Set $Y = X \cap A^{n-1}$ and $Z = X \cap A^n$. Since $\mathrm{Card}(Z) > \mathrm{Card}(Y)$ there are $y, y' \in Y$ such that $0y, 1y, y'0, y'1 \in Z$. Since X is balanced, we cannot have $y = y'$. Let p be the longest common prefix of y, y'. Then $p0, p1$ are prefixes of y, y' and thus $0p0, 0p1, 1p0, 1p1 \in X$ a contradiction.

3. Assume that x is Sturmian and that $F(x)$ is unbalanced. By 1, there is a palindrome w such that $0w0, 1w1 \in F(x)$. Thus, w is right-special. Set $n = |w| + 1$. We may assume that $0w$ is right-special and $1w$ is not (the other case is symmetric).

   Let v be a word of length $n - 1$ such that $u = 1w1v \in F(x)$. The word u has length $2n$. We prove that all factors of length n of u are conservative. By Exercise 6.5 this implies that x is eventually periodic, a contradiction.

   We need to prove that the unique right-special word of length n, that is $0w$ is not a factor of u. Assume the contrary. Then $w = s0t = t1y$ and $v = yz$. Since w is a palindrome, the first factorization implies $w = \tilde{t}0\tilde{s}$, a contradiction with $w = t1y$.

6.7 Let x be a Sturmian word and let $\tilde{F}(x)$ be the set of reversals of the words of $F(x)$. The set $X = F(x) \cup \tilde{F}(x)$ is balanced. Thus, by Exercise 6.6 $\mathrm{Card}(X \cap A^n) \leqslant n + 1$ for each n and since $\mathrm{Card}(F(x) \cap A^n) = n + 1$, one has $F(x) = \tilde{F}(x)$.

6.8 The words $\varphi^n(a)$ are left special. Indeed, this is true for $n = 0$ and next $\varphi(a\varphi^{n-1}(a)) = ab\varphi^n(a)$ and $\varphi(b\varphi^{n-1}(a)) = a\varphi^n(a)$ shows the property by induction on $n \geqslant 1$. For the converse, we also use induction. If w is left-special, it is a prefix of a word of the form $\varphi(u)$ for some word u which is left-special. By induction hypothesis, u is a prefix of some $\varphi^n(a)$, and thus w is a prefix of $\varphi^{n+1}(a)$.

6.9 Let F be a neutral set and let $p_n = \text{Card}(F \cap A^n)$. The property is true for $n = 0$ and for $n = 1$. Next, assuming the property true up to n, we have

$$p_{n+1} = \sum_{w \in F \cap A^{n-1}} e(w) = \sum_{w \in F \cap A^{n-1}} (r(w) + \ell(w) - 1)$$

$$= p_n + p_n - p_{n-1} = kn + 1 + k = k(n + 1) + 1.$$

6.9.3 Section 6.4

6.10 Since a is a strict left-special word, and since $\varphi(a), \varphi(b), \varphi(c)$ end with distinct letters, every $\varphi^n(a)$ for $n \geqslant 1$ is strict left-special. Assume conversely that w is left-special. Since $bw \in F$, w has to begin with a. Since $uav \in \varphi(A)^*$ implies $u, av \in \varphi(A)^*$, we may assume that w is a prefix of some $\varphi(x)$. This word x is strict left-special and by induction hypothesis is a prefix of some $\varphi^n(a)$.

Finally, let us prove that F is closed under reversal. We use induction again. Consider a word of the form $u = \varphi(v)$. Its reversal is $\tilde{\varphi}(\tilde{v})$ where $\tilde{\varphi}$ is the morphism $a \mapsto ba$, $b \mapsto ca$, $c \mapsto a$. But $\tilde{\varphi} = a^{-1}\varphi a$. Thus, $\tilde{u} = a^{-1}\varphi(\tilde{v})a$, showing by induction that $\tilde{u} \in F$.

6.11 The Tribonacci word is the fixed point x of $\varphi : a \mapsto ab$, $b \mapsto ac$, $c \mapsto a$. One may verify that $\psi_{abc} = \varphi^3$. Thus,

$$\varphi^3(\text{Pal}(\Delta)) = \psi_{abc}(\text{Pal}(\Delta)) = \text{Pal}(abc\Delta)) = \text{Pal}(\Delta)$$

by Justin's Formula (6.4). Thus, we have $x = \text{Pal}(\Delta)$.

6.9.4 Section 6.5

6.12 Let F be an Arnoux-Rauzy set and let $w \in F$. We want to show that the graph $E_F(w)$ is a tree. This is true if w is not bispecial. Assume indeed, that w is not left-special. Then $E_F(w)$ is a tree with root the unique $a \in L(w)$. Otherwise, since there is exactly one word of length $|w| + 1$ which is right-special, there is exactly one $a \in L(w)$ such that a is connected to more than one vertex of $R(w)$ and a is connected to all the vertices of $R(w)$. Symmetrically, since F is closed under reversal, there is a unique $b \in R(w)$ connected to more than one vertex in $L(w)$. Thus $E_F(w)$ has the shape indicated in Fig. 6.9 and is a tree.

Fig. 6.9 The graph $E_F(w)$

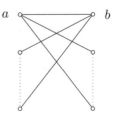

6.10   Notes

The notion of return word has been studied by Durand (1998). Theorem 6.2.4 is Durand (1998, Theorem 4.5). An example of a uniformly recurrent set which is not bounded is given in Durand et al. (2013, Example 3.17).

Sturmian words are one of the best known objects in combinatorics on words. They are characterized in several equivalent ways. See Lothaire (2002, Chapter 2).

The Chacon set (Example 6.3.4) is described in Fogg (2002, Section 5.5.2) where a proof that its complexity is $2n + 1$ can be found.

Theorem 6.3.6 follows from a result in Balková et al. (2008). It is proved in Dolce and Perrin (2017, Corollary 5.3) that every recurrent neutral set is uniformly recurrent.

Episturmian words were extensively studied by Droubay, Justin, and Pirillo (2001) who have introduced the term episturmian and showed that every episturmian word is uniformly recurrent. The particular case of strict episturmian words was introduced before in Arnoux and Rauzy (1991) and these words are often called *Arnoux-Rauzy words*. We have chosen to call Arnoux-Rauzy the set of factors of a strict epistumian word (and not Sturmian as in Berstel et al. (2012)) to be in agreement with a more common use.

Theorem 6.4.2 is from Justin and Vuillon (2000) and Glen and Justin (2009).

Proposition 6.4.5 is from Justin and Vuillon (2000, Theorem 4.4, Corollaries 4.1 and 4.5).

Theorem 6.5.5 is from Berthé et al. (2015, Theorem 4.5) and Theorem 6.5.7 is from Berthé et al. (2015, Theorem 4.7).

For more detailed presentation of Rauzy graphs, see Berthé and Rigo (2010).

Lemma 6.6.2 is Lemma 3.2 in Durand (1998).

Theorem 6.7.3 is from Almeida and Costa (2013, Proposition 5.2).

A word about the connexion of the concepts presented in this chapter with the notion of conjugacy. Sturmian subshifts do not form a class which is closed under conjugacy because conjugacy does not preserve neither the cardinality of the alphabet nor the factor complexity. Moreover, two distinct Sturmian subshifts are never conjugate unless they differ by exchange of the two letters (see Durand and Perrin (2020)). The class of dendric shifts is also not closed under conjugacy, but the wider class of the so called *eventually dendric* shifts is closed under conjugacy (Dolce and Perrin 2020).

Chapter 7
The Schützenberger Group
of a Minimal Set

7.1 Introduction

In this chapter, we study the Schützenberger group of a minimal set. Some of the
results obtained are applied in the next chapter to codes.

Recall that we denote by $G(F)$ the Schützenberger group of $J(F)$, whenever
F is recurrent. The isomorphism classes for $G(F)$ are completely known in the
rational case (i.e., the case where F is recognizable), as we saw in Theorem 5.7.3
and Remark 5.7.4, but the uniformly recurrent case (i.e., the case where F is a
minimal set) is not yet fully understood. In the previous chapter, we saw results
about concrete maximal subgroups of $J(F)$ involving (limit) return sets, when F is
uniformly recurrent. In this chapter, we study abstract properties of $G(F)$, mostly
in the case where F is a minimal set.

We begin by showing in Sect. 7.2 that the Schützenberger group of a transitive
shift is invariant under conjugacy (Theorem 7.2.1). It is however a weak invariant.
Recall, for example, that all nonperiodic transitive shifts have isomorphic Schützenberger groups (Theorem 5.7.3).

In Sect. 7.3, we prove that the tree property is a sufficient condition for $G(F)$
to be a free profinite group (Theorem 7.3.4). The proof uses the Return Theorem
of Chap. 6. This result gives yet another example that $G(F)$ is a weak conjugacy
invariant.

In Sect. 7.4, we explain how to write down a profinite presentation for $G(F)$,
when F is defined by a substitution, and see with some examples how that is useful
in the study of the finite continuous morphic images of $G(F)$. As we shall see, the
group $G(F)$ is not always a free profinite group.

© Springer Nature Switzerland AG 2020
J. Almeida et al., *Profinite Semigroups and Symbolic Dynamics*, Lecture Notes
in Mathematics 2274, https://doi.org/10.1007/978-3-030-55215-2_7

7.2   Invariance of the Schützenberger Group

Consider a shift space X. Suppose that X is transitive, which is the same as to suppose that the language $F(X)$ is recurrent. We use the simplified notation $J(X)$ for $J(F(X))$, and $G(X)$ for $G(F(X))$. We also say that $G(X)$ is the Schützenberger group of X. In the next theorem, we see that the Schützenberger group of a transitive shift space has dynamical meaning: it is a conjugacy invariant.

Theorem 7.2.1 *If X and Y are conjugate shift spaces, then the Schützenberger groups $G(X)$ and $G(Y)$ are isomorphic profinite groups.*

The proof of Theorem 7.2.1 is concluded at the end of this section. First, we need to introduce new material, about a type of continuous maps $\widehat{B^*} \to \widehat{A^*}$ that are a sort of analog of the sliding block codes $B^{\mathbb{Z}} \to A^{\mathbb{Z}}$.

Let $n \in \mathbb{N}$, and take $u \in \widehat{B^*}$. Following Exercise 4.19, we let $i_n(u)$ be the unique prefix of length n of u if u is not a word of length less than n, with $i_n(u) = u$ otherwise. Dually, $t_n(u)$ is the unique suffix of length n of u if u is not a word of length less than n, and $t_n(u) = u$ otherwise. Note that $i_0(u) = \varepsilon = t_0(u)$. One should also have in mind that the maps $i_n : \widehat{B^*} \to \widehat{B^*}$ and $t_n : \widehat{B^*} \to \widehat{B^*}$ thus defined are continuous, as seen in Exercise 4.19.

Let $h : B^{q+1} \to A$ be a block map, where $q \in \mathbb{N}$. The map h can be extended to a map denoted $\hat{h}$ from B^* to A^* as follows. If $|w| \leq q$, we set $\hat{h}(w) = 1$. Otherwise, set

$$\hat{h}(b_1 \cdots b_n) = \hat{h}(b_1 \cdots b_{q+1})\hat{h}(b_2 \cdots b_{q+2}) \cdots \hat{h}(b_{n-q} \cdots b_n),$$

where $b_1, \ldots, b_n \in B$. Note that, for every $u, v \in B^*$, the equalities

$$\hat{h}(uv) = \hat{h}(u) \cdot \hat{h}(t_q(u)v) = \hat{h}(u\, i_q(v)) \cdot \hat{h}(v), \tag{7.1}$$

hold (Exercise 7.1). Observe also that if $q = 0$ then $\hat{h}$ is a morphism of monoids.

Proposition 7.2.2 *Let $q \in \mathbb{N}$. For every block map $h : B^{q+1} \to A$, the map $\hat{h} : B^* \to A^*$ admits an extension to a unique continuous map from $\widehat{B^*}$ to $\widehat{A^*}$.*

Proof We only need to show the existence of the continuous extension, the uniqueness being immediate as B^* is dense in $\widehat{B^*}$.

For the case $q = 0$, apply the universal property of $\widehat{B^*}$ by taking the unique continuous morphism of monoids from $\widehat{B^*}$ to $\widehat{A^*}$ that extends $h : B \to A$.

From hereon, we suppose that $q \neq 0$. By Theorem 4.4.20, it suffices to show that $\hat{h}$ is uniformly continuous with respect to the profinite metric of B^* and the profinite metric of A^*.

Let k be a positive integer. Suppose that $u, v \in B^*$ satisfy $r(\hat{h}(u), \hat{h}(v)) \leq k$. We may then take a morphism of monoids φ from A^* onto a monoid M such that $\mathrm{Card}(M) \leq k$ and $\varphi(\hat{h}(u)) \neq \varphi(\hat{h}(v))$.

Let N be the finite subset of $B^* \times M \times B^*$ whose elements are the triples (x, m, y) such that $x, y \in B^q$, together with the triples $(z, 1, z)$ such that z has length less than q. It is routine to check that N is a monoid for the operation

$$(x, m, y) \cdot (x', m', y') = (i_q(xx'),\ m \cdot \varphi(\hat{h}(yx')) \cdot m',\ t_q(yy')),$$

with the triple $(\varepsilon, 1, \varepsilon)$ as a neutral element (Exercise 7.3). For such monoid structure, let ψ be the unique morphism from B^* to N satisfying $\psi(b) = (b, 1, b)$ whenever $b \in B$ (the morphism ψ is well defined because $q \neq 0$ and so $(b, 1, b)$ indeed belongs to N). With a simple inductive argument on the length of words, one sees that for every $w \in B^*$ the equality $\psi(w) = (i_q(w), \varphi(\hat{h}(w)), t_q(w))$ holds (Exercise 7.4). In particular, we have $\psi(u) \neq \psi(v)$. Let $\beta = \mathrm{Card}(B)$. Note that

$$\mathrm{Card}(N) \leqslant \beta^{2q} \times \mathrm{Card}(M) + (1 + \beta + \beta^2 + \cdots + \beta^{q-1}).$$

Therefore, if $f(k) = \beta^{2q} \times k + (1 + \beta + \beta^2 + \cdots + \beta^{q-1})$, then $\mathrm{Card}(N) \leqslant f(k)$ holds, thus $r(u, v) \leqslant f(k)$. In other words, we have shown that $d(u, v) < 1/f(k)$ implies $d(\hat{h}(u), \hat{h}(v)) < 1/k$, establishing the uniform continuity of $\hat{h}$. $\blacksquare$

For each block map $h : B^{q+1} \to A$, with $q \in \mathbb{N}$, the extension of $\hat{h} : B^* \to A^*$ to a continuous map $\widehat{B^*} \to \widehat{A^*}$ is denoted also by $\hat{h}$ (when $q = 0$, this is consistent with the notation introduced in Chap. 4 for the continuous morphisms $\widehat{B^*} \to \widehat{A^*}$ extending maps $B \to A$). Because i_n and t_n are also continuous on $\widehat{B^*}$, and B^* is dense in $\widehat{B^*}$, the scope of validity of the equalities (7.1) extends to all $u, v \in \widehat{B^*}$. In the special case where q is an even number, we get another equality, highlighted in the following lemma, to be used in the proof of Theorem 7.2.1.

Lemma 7.2.3 *Let $k \in \mathbb{N}$. For every block map $h : B^{2k+1} \to A$, the mapping $\hat{h} : \widehat{B^*} \to \widehat{A^*}$ satisfies the equality*

$$\hat{h}(uv) = \hat{h}(u\, i_k(v)) \cdot \hat{h}(t_k(u)v)$$

for all $u, v \in \widehat{B^*}$.

The proof of Lemma 7.2.3 is left as an exercise (Exercise 7.2).

Theorem 7.2.1 will result from the following more precise statement.

Proposition 7.2.4 *Let X, Y be transitive shift spaces on alphabets A, B, respectively. Suppose that the block map $g : A \to B$ is such that the 1-block code $\gamma = g^{[0,0]}$ is a conjugacy from X onto Y. Consider the unique continuous morphism $\hat{g} : \widehat{A^*} \to \widehat{B^*}$ extending g. The restriction of $\hat{g}$ to a maximal subgroup of $J(X)$ is a continuous isomorphism onto a maximal subgroup of $J(Y)$.*

Proof The inverse $\gamma^{-1} : Y \to X$ has a block map $h : F_{2k+1}(Y) \to A$ with memory and anticipation k, for some natural number k (cf. Exercise 5.6). Consider any extension $B^{2k+1} \to A$ of h, denoting it also by h. We shall need the continuous morphism $\hat{g} : \widehat{A^*} \to \widehat{B^*}$ extending g and the continuous map $\hat{h} : \widehat{B^*} \to \widehat{A^*}$ defined in accordance with Proposition 7.2.2.

Observe that the restriction of the composition $\hat{h}\hat{g}$ to $F_{2k+1}(X)$ is a block map of memory and anticipation k for the identity on X. That is, the implication

$$x \in X \implies \hat{h}\hat{g}(x[i-k, j+k]) = x[i, j] \tag{7.2}$$

holds for all $i, j \in \mathbb{Z}$ such that $i \leqslant j$. Similarly, the restriction of $\hat{g}\hat{h}$ to $F_{2k+1}(Y)$ is a block map of memory and anticipation k for the identity on Y, that is, we have

$$y \in Y \implies \hat{g}\hat{h}(y[i-k, j+k]) = y[i, j] \tag{7.3}$$

for all $i, j \in \mathbb{Z}$ such that $i \leqslant j$.

Let $u \in \overline{F(X)}$. We claim that if u is an infinite pseudoword or a word of length at least $2k + 1$, then

$$u = i_k(u)\hat{h}\hat{g}(u)t_k(u). \tag{7.4}$$

Indeed, set $r = i_k(u)$, $s = t_k(u)$ and $u = rvs$. Let (w_n) be a sequence of elements of $F(X)$ converging to u. We may assume that $w_n = rv_n s$ with v_n converging to v. Then, by (7.2), the equality $\hat{h}\hat{g}(w_n) = v_n$ holds, thus $w_n = r\hat{h}\hat{g}(w_n)s$. Taking the limit, we obtain (7.4), establishing our claim. Note that it follows from Eq. (7.4) that the restriction of $\hat{g}$ to $\overline{F(X)} \setminus A^*$ is injective.

As $g : A \to B$ and $h : F_{2k+1}(Y) \to A$ are respectively block maps of γ and γ^{-1}, the equalities $\hat{g}(F(X)) = F(Y)$ and $\hat{h}(F(Y)) = F(X)$ hold, whence

$$\hat{g}\left(\overline{F(X)}\right) = \overline{F(Y)} \quad \text{and} \quad \hat{h}\left(\overline{F(Y)}\right) = \overline{F(X)}$$

by continuity of $\hat{g}$ and $\hat{h}$.

Let $u \in J(X)$. Take $w \in \overline{F(Y)}$. There is $v \in \overline{F(X)}$ such that $w = \hat{g}(v)$. Since $u \in J(X)$, we have $u = rvs$ for some $r, s \in \widehat{A^*}$, thus $\hat{g}(u) = \hat{g}(r)w\hat{g}(s)$. Hence $\hat{g}(u)$ is an element of $\overline{F(Y)}$ having every element of $\overline{F(Y)}$ as a factor, entailing $\hat{g}(u) \in J(Y)$. We have therefore showed that $\hat{g}(J(X)) \subseteq J(Y)$.

Let H be a maximal subgroup of $J(X)$. As $\hat{g}$ is a continuous morphism of monoids, we already know that $\hat{g}(H)$ is contained in a maximal subgroup K of $J(Y)$, and so it only remains to show that $K = \hat{g}(H)$. Denote the idempotent $\hat{g}(e)$ of K by f. For each $v \in K$, consider the pseudoword

$$v' = \hat{h}\big(t_k(f)v\, i_k(f)\big).$$

Let $r = t_k(f)$ and $s = i_k(f)$. Since rvs is a factor of $v = fvf$, we have $rvs \in \overline{F(Y)}$. Let (w_n) be a sequence of elements of $F(Y)$ converging to rvs. We may as well suppose that there are factorizations $w_n = rv_n s$ such that $v_n \to v$. In view of (7.3), we have $\hat{g}\hat{h}(rv_n s) = v_n$, and therefore $\hat{g}(v') = \hat{g}\hat{h}(rvs) = \lim \hat{g}\hat{h}(rv_n s) = \lim v_n = v$. From the formula

$$\hat{g}(v') = v \tag{7.5}$$

we then get in particular $\hat{g}(f') = f = \hat{g}(e)$. Since $\hat{g}$ is injective on $\overline{F(X)} \setminus A^*$, we deduce that

$$\hat{h}\big(t_k(f) f \, i_k(f)\big) = e.$$

Therefore, for every $v \in K$, from the equality $v = fv$ we obtain the following chain of equalities with the help of Lemma 7.2.3:

$$v' = \hat{h}\big(t_k(f) f \cdot fv \, i_k(f)\big) = \hat{h}\big(t_k(f) f \, i_k(f)\big) \cdot \hat{h}\big(t_k(f) fv \, i_k(f)\big) = e \cdot v'.$$

Hence, we have $v' \leqslant_{\mathcal{R}} e$. Since $e \in J(X)$ and $v' \in \overline{F(X)}$, we also have $e \leqslant_{\mathcal{J}} v'$. Because $\overline{A^*}$ is stable, it follows that $e \,\mathcal{R}\, v'$. Symmetrically, from the equality $v = vf$ we deduce that $e \,\mathcal{L}\, v'$. Therefore, we have $v' \in H$. Looking at (7.5), we see that we established the equality $K = \hat{g}(H)$. ∎

Proof of Theorem 7.2.1 By Corollary 5.4.4, every conjugacy is a composition of a conjugacy which is a 1-block code with a conjugacy which is the inverse of a 1-block code. Thus Theorem 7.2.1 results from Proposition 7.2.4. ∎

7.3   A Sufficient Condition for Freeness

Let F be a uniformly recurrent set on the alphabet A. Consider the natural projection $p_G : \widehat{A^*} \to \widehat{FG(A)}$. Let x be an element of a subgroup of $J(F)$. We look for a condition under which the restriction of p_G to $H(x)$ is an isomorphism. We shall see that the tree property is such a condition. In the process, we investigate when the restriction is surjective. We do that based on results introduced in Chap. 6. This approach is motivated by the observation that, in view of Proposition 6.6.3, the natural projection $p_G(H(x))$ in the free profinite group $\widehat{FG(A)}$ is the intersection of the closures, in the topology of $\widehat{FG(A)}$, of the subgroups $\langle \mathcal{R}_n \rangle$ generated by the return sets $\mathcal{R}_n$ associated to a defining pair $(\ell_n, r_n)_{n \geqslant 1}$ of $H(x)$ (as defined in Sect. 6.6). That is, we have

$$p_G(H(x)) = \bigcap_{n \geqslant 1} \overline{\langle \mathcal{R}_n \rangle}.$$

Therefore, it is not surprising that results like Theorems 6.5.5 and 6.5.7 come into play.

We prove the following result.

Theorem 7.3.1 *Let F be a uniformly recurrent set on the alphabet A, and let ψ : $\widehat{A^*} \to G$ be a continuous morphism from $\widehat{A^*}$ onto a profinite group G. The following conditions are equivalent.*

(i) *The restriction of ψ to each maximal subgroup of $J(F)$ is surjective onto G.*
(ii) *For every $w \in F$, the submonoid $\psi(\mathcal{R}_F(w)^*)$ is dense in G.*

Proof Let $x \in J(F)$ be such that $H(x)$ is a group.

(i) implies (ii). Assume by contradiction that $\psi(\mathcal{R}_F(w)^*)$ is not dense in G. Since w is a factor of x, and since r_1 can be chosen as an arbitrary prefix of x and F is uniformly recurrent, we may assume that r_1 ends with w. Then we have $\mathcal{R}_F(\ell_1 r_1) \subseteq \mathcal{R}_F(w)^*$. This implies that $\psi(\mathcal{R}_F(\ell_1 r_1)^*)$ is not dense in G. Because $\psi(\mathcal{R}_1^*)$ is conjugate to $\psi(\mathcal{R}_F(\ell_1 r_1)^*)$ (cf. Exercise 4.20), the set $\psi(\mathcal{R}_1^*)$ is also not dense in G. Since, according to Proposition 6.6.3, we have $H(x) \subseteq \overline{\mathcal{R}_1^*}$, we conclude that $\psi(H(x))$ is not equal to G.

(ii) implies (i). Let $g \in G$. For each $n \geqslant 1$, the set $\mathcal{R}_n$ is conjugate to $\mathcal{R}_F(\ell_n r_n)$ and thus by (ii) there is $h_n \in \overline{\mathcal{R}_n^*}$ such that $g = \psi(h_n)$. As the image by ψ of the ω-power of an element of $\mathcal{R}_n$ is the neutral element of G, we may as well suppose that $h_n \in \overline{\mathcal{R}_n^*} \setminus A^*$. Let h be an accumulation point of the sequence $(h_n)_n$. Note that $h \notin A^*$, as $\widehat{A^*} \setminus A^*$ is closed. Since the sets $\overline{\mathcal{R}_n^*}$ form a chain (cf. Lemma 6.6.1), we also know that $h \in \bigcap_{n \geqslant 1} \overline{\mathcal{R}_n^*}$. According to Proposition 6.6.3, we have $H(x) = (\bigcap_{n \geqslant 1} \overline{\mathcal{R}_n^*}) \setminus A^*$. As $g = \psi(h)$, it follows that $\psi(H(x)) = G$.   ∎

Corollary 7.3.2 *Let F be a uniformly recurrent set on the alphabet A. The following conditions are equivalent.*

(i) *The restriction to any maximal subgroup of $J(F)$ of the natural projection p_G : $\widehat{A^*} \to \widehat{FG(A)}$ is surjective onto $\widehat{FG(A)}$.*
(ii) *For each $w \in F$ the set $\mathcal{R}_F(w)$ generates the free group $FG(A)$.*

Proof It suffices to apply Theorem 7.3.1, with ψ being the natural projection p_G : $\widehat{A^*} \to \widehat{FG(A)}$, and observe that $\mathcal{R}_F(w)^*$ is dense in $\widehat{FG(A)}$ if and only if $\mathcal{R}_F(w)$ is a generating set of the group $FG(A)$. We explain the latter equivalence. The monoid $\mathcal{R}_F(w)^*$ is dense in $\widehat{FG(A)}$ if and only if $\widehat{FG(A)}$ is the closed subgroup of $\widehat{FG(A)}$ generated by $\mathcal{R}_F(w)$. Because $\mathcal{R}_F(w)$ is finite, it suffices then to apply Corollary 4.6.4 to obtain the equivalence.   ∎

Putting Theorem 6.5.7 and Corollary 7.3.2 together, we get the following.

Corollary 7.3.3 *If F is a uniformly recurrent connected set on the alphabet A, then the restriction to any maximal subgroup of $J(F)$ of the natural projection p_G : $\widehat{A^*} \to \widehat{FG(A)}$ is surjective onto $\widehat{FG(A)}$.*

When restricting to uniformly recurrent tree sets, we obtain freeness, as seen next.

Theorem 7.3.4 *Let F be a uniformly recurrent tree set on the alphabet A. The group $G(F)$ is the free profinite group on A. More precisely, the restriction to any maximal subgroup of $J(F)$ of the natural projection $p_G : \widehat{A^*} \to \widehat{FG(A)}$ is an isomorphism onto $\widehat{FG(A)}$.*

Proof Let G be a maximal subgroup of $J(F)$. By Corollary 7.3.3, the restriction to G of the natural projection $p_G : \widehat{A^*} \to \widehat{FG(A)}$ is surjective. On the other hand, G is the closure of a semigroup generated by $\mathrm{Card}(A)$ elements, by Corollary 6.7.4. Therefore, there is a continuous morphism ψ from $\widehat{FG(A)}$ onto G. Thus $p_G \circ \psi$ is a continuous surjective morphism from $\widehat{FG(A)}$ onto itself. By Proposition 3.12.5, it follows that it is an isomorphism. ∎

Example 7.3.5 Let F be the Fibonacci set. It is a uniformly recurrent tree set on the alphabet $A = \{a, b\}$. Therefore, $G(F)$ is isomorphic to the free profinite group on A. The set $\{\varphi^\omega(a), \varphi^\omega(ba)\}$ is a set of free generators of a maximal subgroup of $J(F)$, where φ denotes the Fibonacci morphism (cf. Example 6.7.5).

7.4   Groups of Substitutive Shifts

Let $\varphi : A^* \to A^*$ be a primitive substitution. Recall from Chap. 5 that $F(\varphi)$ is the uniformly recurrent set of factors of a fixed point of φ, and that $X(\varphi)$ is the associated shift space formed of the biinfinite words having all their factors in $F(\varphi)$. We denote by $J(\varphi)$ the $\mathcal{J}$-class $J\big(F(\varphi)\big)$, and by $G(\varphi)$ the Schützenberger group of $J(\varphi)$.

The unique extension of the primitive substitution $\varphi : A^* \to A^*$ to a continuous endomorphism $\widehat{A^*} \to \widehat{A^*}$ is also denoted by φ in what follows. We begin with the following simple property of $J(\varphi)$.

Proposition 7.4.1 *The inclusions $\varphi^\omega\big(F(\varphi) \setminus \{\varepsilon\}\big) \subseteq J(\varphi)$ and $\varphi\big(J(\varphi)\big) \subseteq J(\varphi)$ hold.*

Proof Note that $\varphi\big(F(\varphi)\big) \subseteq F(\varphi)$ holds directly by the definition of $F(\varphi)$. Since the endomorphism $\varphi : \widehat{A^*} \to \widehat{A^*}$ is continuous, we have in fact the inclusion $\varphi\big(\overline{F(\varphi)}\big) \subseteq \overline{F(\varphi)}$. By Proposition 5.6.14, we also have $J(\varphi) = \overline{F(\varphi)} \setminus A^*$. Therefore, since the nonerasing morphism φ maps infinite pseudowords to infinite pseudowords, we deduce that $\varphi\big(J(\varphi)\big) \subseteq J(\varphi)$. Moreover, from $\varphi\big(\overline{F(\varphi)}\big) \subseteq \overline{F(\varphi)}$ we get $\varphi^k\big(\overline{F(\varphi)}\big) \subseteq \overline{F(\varphi)}$ for every positive integer k, and so, by the definition of the pointwise topology of $\mathrm{End}(\widehat{A^*})$, we get $\varphi^\omega\big(\overline{F(\varphi)}\big) \subseteq \overline{F(\varphi)}$.

Because every pseudoword in $\varphi^\omega(\widehat{A^+})$ is infinite (since φ is expansive), we conclude that $\varphi^\omega\big(\overline{F(\varphi)} \setminus \{\varepsilon\}\big) \subseteq \overline{F(\varphi)} \setminus A^* = J(\varphi)$ ∎

The reverse inclusions do not hold in general (see Exercises 7.9 and 7.10).

Recall from Chap. 5 that a *connection* of φ is a pair (a, b) with $a, b \in A$ such that $ba \in F(\varphi)$, the first letter of $\varphi^\omega(a)$ is a, and the last letter of $\varphi^\omega(b)$ is b. By

Proposition 5.5.10, every primitive substitution has a connection. Remember also that, for a connection (a, b) of φ, there is a *connective power* of φ, which is a finite power $\tilde{\varphi}$ of φ such that the first letter of $\tilde{\varphi}(a)$ is a and the last letter of $\tilde{\varphi}(b)$ is b. One should have in mind the equality $\tilde{\varphi}^{\omega} = \varphi^{\omega}$, which holds as φ^{ω} is the unique idempotent in the closed subsemigroup of $\text{End}(\widehat{A^*})$ generated by φ.

For each connection (a, b) of the primitive substitution φ, we consider the set

$$X(a, b) = a \, \mathcal{R}_{F(\varphi)}(ba) a^{-1},$$

where $\mathcal{R}_{F(\varphi)}(ba)$ is the set of return words of $F(\varphi)$ to ba.

Recall that we denote in a given semigroup (in this case, the free profinite monoid) by $R(x)$ and $L(x)$ respectively the $\mathcal{R}$-class and the $\mathcal{L}$-class of an element x.

Proposition 7.4.2 *Let φ be a primitive substitution and let (a, b) be a connection of φ. Then the elements of $\varphi^{\omega}(X(a, b))$ are contained in the $\mathcal{H}$-class $R(\varphi^{\omega}(a)) \cap L(\varphi^{\omega}(b))$, which is moreover a maximal subgroup of $J(\varphi)$.*

Proof Let $u \in X(a, b)$. Then we have $u \in aA^*b \cap F(\varphi)$. Hence, the infinite pseudoword $\varphi^{\omega}(u)$ belongs to the intersection $\varphi^{\omega}(a) \cdot \widehat{A^*} \cdot \varphi^{\omega}(b) \cap J(\varphi)$ by Proposition 7.4.1. By that same proposition, both pseudowords $\varphi^{\omega}(a)$ and $\varphi^{\omega}(b)$ belong to $J(\varphi)$. Therefore, since $\widehat{A^*}$ is stable, the pseudoword $\varphi^{\omega}(u)$ indeed belongs to the $\mathcal{H}$-class $H = R(\varphi^{\omega}(a)) \cap L(\varphi^{\omega}(b))$.

Since $\varphi^{\omega}(u) \in H$, to show that H is a maximal subgroup, it suffices to show that $\varphi^{\omega}(u)^2 \in H$ (cf. Exercise 3.60). Suppose that w is a finite factor of $\varphi^{\omega}(u)^2$. Then, since $\widehat{A^*}$ is equidivisible, the word w is a factor of $\varphi^{\omega}(u)$, or $w = zt$ for a finite suffix z of $\varphi^{\omega}(u)$ and a finite prefix t of $\varphi^{\omega}(u)$. In the second case, since $\varphi^{\omega}(b)$ is an infinite suffix of $\varphi^{\omega}(u)$ and $\varphi^{\omega}(a)$ is an infinite prefix of $\varphi^{\omega}(u)$, we know that z is a suffix of $\varphi^{\omega}(b)$ and t is a prefix of $\varphi^{\omega}(a)$, and so $w = zt$ is a factor of $\varphi^{\omega}(ba)$. By Proposition 7.4.1, the pseudoword $\varphi^{\omega}(ba)$ also belongs to $J(\varphi)$. Therefore, we are certain that $w \in F(\varphi)$. This shows that $\varphi^{\omega}(u)^2 \in J(\varphi)$ (cf. Remark 5.6.11). ∎

From hereon, we use the notation H_{ba} for the $\mathcal{H}$-class $R(\varphi^{\omega}(a)) \cap L(\varphi^{\omega}(b))$.

The proof of the following result is based on the notion of recognizability of a substitution introduced in Chap. 6.

Theorem 7.4.3 *Let φ be a nonperiodic primitive substitution. Consider a connection (a, b) for φ and a connective power $\tilde{\varphi}$ of φ. Then the equalities $H_{ba} = \tilde{\varphi}(H_{ba}) = \varphi^{\omega}(H_{ba})$ hold.*

Proof Let us first prove that $\tilde{\varphi}(H_{ba}) \subseteq H_{ba}$. Let $u \in X(a, b)$. We have already seen in Proposition 7.4.2 that $\varphi^{\omega}(u) \in H_{ba}$, and that the $\mathcal{H}$-class H_{ba} is a subgroup. Because $\tilde{\varphi}$ is a power of φ and $\varphi(J(\varphi)) \subseteq J(\varphi)$, and since $\tilde{\varphi}$ is a morphism, we know that $\tilde{\varphi}(H_{ba})$ is also a subgroup of $J(\varphi)$. Set $v = \tilde{\varphi}\varphi^{\omega}(u)$. Then we claim that $v \in \tilde{\varphi}(H_{ba}) \cap H_{ba}$. Trivially, from $\varphi^{\omega}(u) \in H_{ba}$ we get $v \in \tilde{\varphi}(H_{ba})$. On the other hand, since $u \in aA^*b$, we have $\tilde{\varphi}(u) \in aA^*b$ by the definition of the connective power $\tilde{\varphi}$. Therefore, taking into account that $\tilde{\varphi}\varphi^{\omega} = \varphi^{\omega}\tilde{\varphi}$ because $\tilde{\varphi}$ is a power of φ, one sees that $v = \varphi^{\omega}\tilde{\varphi}(u) \in \varphi^{\omega}(a) \cdot \widehat{A^*} \cdot \varphi^{\omega}(b)$. As $v \in J(\varphi)$ and $\widehat{A^*}$ is stable, it follows that $v \in R(\varphi^\omega(a)) \cap L(\varphi^\omega(a)) = H_{ba}$. This shows our claim that $v \in \tilde{\varphi}(H_{ba}) \cap H_{ba}$. Hence the subgroup $\tilde{\varphi}(H_{ba})$ intersects the maximal subgroup H_{ba}, forcing $\tilde{\varphi}(H_{ba}) \subseteq H_{ba}$.

To prove the inclusion $H_{ba} \subseteq \tilde{\varphi}(H_{ba})$, we first show that $H_{ba} \subseteq \mathrm{Im}\,\tilde{\varphi}$. For this, consider the biinfinite word $x \in X(\varphi)$ such that $x_{-1} = b$, $x_0 = a$ and $\tilde{\varphi}(x) = x$. By Theorem 5.5.22, $\tilde{\varphi}$ is recognizable on x.

Let $(\ell_n, r_n)_{n \geq 1}$ be a defining pair of H_{ba} and let $\mathcal{R}_n$ be the associated sequence of return sets (see Sect. 6.7). We may assume that all ℓ_n, r_n are in $\mathrm{Im}\,\tilde{\varphi}$ and that the lengths $|\ell_n|, |r_n|$ are larger than the constant m of recognizability of $\tilde{\varphi}$.

By Theorem 6.2.4, the set $F(\varphi)$ is bounded and, by taking a suitable subsequence of $(\ell_n, r_n)_{n \geq 1}$, we may suppose that we have a sequence $(\ell_n, r_n, \mathcal{R}_n)$ such that $\mathcal{R}_n = \{u_{n,1}, \ldots, u_{n,q}\}$ has a fixed number q of elements and that the sequence $(u_{n,1}, \ldots, u_{n_q})$ converges in the product space $\widehat{A^*}^q$ to $\mathcal{R} = \{u_1, \ldots, u_q\}$.

Fix some $n \geq 1$. Since $F(\varphi)$ is uniformly recurrent, and $l_n r_n = x[-|\ell_n|, |r_n|[$, there are i, j such that $0 < i < j$

$$x[-|\ell_n|, |r_n|[= x[i - |\ell_n|, i + |r_n|[= x[j - |\ell_n|, j + |r_n|[\qquad (7.6)$$

and $x[i, j[= u_{n,k}$ for some k (see Fig. 7.1). Because $|\ell_n|, |r_n|$ are larger than the constant m of recognizability of $\tilde{\varphi}$, it follows from (7.6) that $0 \sim_{x,m} i \sim_{x,m} j$. Since $0 \in E_x(\tilde{\varphi})$ by the definition of $E_x(\tilde{\varphi})$, and $\tilde{\varphi}$ is recognizable on x with constant of recognizability m, we deduce that $i, j \in E_x(\tilde{\varphi})$. Therefore, we have $x[0, i[= \varphi(x[0, i'[)$ and $x[0, j[= \varphi(x[0, j'[)$ for some integers i', j' such that $0 < i' < j'$, whence $x[i, j[= \varphi(x[i', j'[)$. We conclude that all $u_{n,k}$ are in the closed subsemigroup $\mathrm{Im}\,\tilde{\varphi}$, and therefore all u_k are also in $\mathrm{Im}\,\tilde{\varphi}$. Thus $H_{ba} \subseteq \mathrm{Im}\,\tilde{\varphi}$ holds by Theorem 6.7.3.

Since $\tilde{\varphi}$ is an arbitrary connective power for the connection (a, b), and since $\tilde{\varphi}^k$ is also a connective power for the same connection, we have in fact $H_{ba} \subseteq \mathrm{Im}\,\tilde{\varphi}^k$, for every positive integer k. Thus, given $g \in H_{ba}$, for every integer $k \geq 1$ there is some $w_k \in \widehat{A^*}$ such that $g = \tilde{\varphi}^{k!}(w_k)$. Let $w \in \widehat{A^*}$ be an accumulation point of the sequence (w_k). Since, by Proposition 3.12.3, the evaluation mapping on continuous endomorphisms of $\widehat{A^*}$ is continuous, and noting that $\tilde{\varphi}^\omega = \varphi^\omega$, we deduce that $g = \tilde{\varphi}^\omega(w) = \varphi^\omega(w)$. Therefore, as φ^ω is idempotent, we obtain $\varphi^\omega(g) = g$. This proves the equality $H_{ba} = \varphi^\omega(H_{ba})$.

Fig. 7.1 The position of $u_{n,k}$

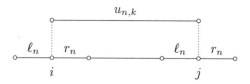

We already proved the inclusion $\tilde{\varphi}(H_{ba}) \subseteq H_{ba}$, from which it immediately follows that $\tilde{\varphi}^k(H_{ba}) \subseteq H_{ba}$ for every positive integer k. Since we have $\tilde{\varphi}^{k!-1} \to \tilde{\varphi}^{\omega-1}$ in $\mathrm{End}(\widehat{A^*})$, and again by the continuity of the evaluation mapping $\mathrm{End}(\widehat{A^*}) \times \widehat{A^*} \to \widehat{A^*}$, we deduce that $\tilde{\varphi}^{\omega-1}(H_{ba}) \subseteq H_{ba}$, in view of the fact that H_{ba} is closed. Therefore, we have

$$H_{ba} = \varphi^{\omega}(H_{ba}) = \tilde{\varphi}^{\omega}(H_{ba}) = \tilde{\varphi}(\tilde{\varphi}^{\omega-1}(H_{ba})) \subseteq \tilde{\varphi}(H_{ba}),$$

thus concluding the proof that $H_{ba} = \tilde{\varphi}(H_{ba})$.   ∎

Note that Theorem 7.4.3 fails if we drop the hypothesis that φ is nonperiodic (Exercise 7.11).

Let X be the code $X(a,b) = a\mathcal{R}_{F(\varphi)}(ba)a^{-1}$, where (a,b) is a connection for the primitive substitution $\varphi : A^* \to A^*$. Let $f : B^* \to X^*$ be a coding morphism for X. By the definition of connective power, one has $\tilde{\varphi}(X) \subseteq X^+$, so that $\tilde{\varphi}$ induces a natural morphism $\tilde{\varphi}_X : B^* \to B^*$. The unique extension of this natural endomorphism of X^* to a continuous endomorphism of $\widehat{FG(B)}$ is denoted by $\tilde{\varphi}_{X,G}$. We then have the commutative diagram (7.7) where $p_G : \widehat{B^*} \to \widehat{FG(B)}$ denotes the natural projection.

$$
\begin{array}{ccccc}
\widehat{A^*} & \xleftarrow{\hat{f}} & \widehat{B^*} & \xrightarrow{p_G} & \widehat{FG(B)} \\
\downarrow{\scriptstyle \tilde{\varphi}} & & \downarrow{\scriptstyle \tilde{\varphi}_X} & & \downarrow{\scriptstyle \tilde{\varphi}_{X,G}} \\
\widehat{A^*} & \xleftarrow{\hat{f}} & \widehat{B^*} & \xrightarrow{p_G} & \widehat{FG(B)}
\end{array}
\qquad (7.7)
$$

Given a finite alphabet A and a closed relation R on $\widehat{FG(A)}$, recall that the profinite group $\langle A \mid R \rangle_G$ is the quotient of $\widehat{FG(A)}$ by the closed congruence generated by R (see Sect. 4.7).

Theorem 7.4.3 is an essential ingredient in the proof of the following result.

Theorem 7.4.4 *Let φ be a nonperiodic primitive substitution over the alphabet A. Let (a,b) be a connection for φ, let $X = X(a,b)$ and let $f : B^* \to A^*$ be a coding morphism for X. Then $G(\varphi)$ admits the presentation*

$$\langle B \mid \tilde{\varphi}_{X,G}^{\omega}(b) = b, \ b \in B \rangle_G.$$

See Proposition 4.7.8 for a characterization of profinite semigroups having a presentation of this form. In preparation for the proof of Theorem 7.4.4, we first prove the following lemma.

Lemma 7.4.5 *Let φ be a nonperiodic primitive substitution over the alphabet A. Let (a,b) be a connection for φ, let $X = X(a,b)$, and let $f : B^* \to A^*$ be a coding morphism for X. Then H_{ba} is a maximal subgroup of $J(\varphi)$ and $H_{ba} = \mathrm{Im}(\varphi^{\omega} \circ \hat{f})$.*

Proof The first assertion is given by Proposition 7.4.2.

Set $K = \text{Im}(\varphi^\omega \circ \hat{f})$. We have already seen in Proposition 7.4.2 that $K \subseteq H_{ba}$.

To prove the reverse inclusion, we first prove that $H_{ba} \subseteq \text{Im}(\hat{f})$. Let $v \in H_{ba}$. Since H_{ba} is a group, we have $v^3 \in H_{ba}$ and thus $bva \in J(\varphi)$. By Proposition 5.6.14, we have $J(\varphi) = \overline{F(\varphi)} \setminus A^*$. Let w_n be a sequence of words in $F(\varphi)$ such that $\lim w_n = bva$. We may assume that every w_n begins and ends with ba. Then $b^{-1} w_n a^{-1} \in X^*$ and thus $v \in \overline{X^*}$. This proves that $H_{ba} \subseteq \text{Im}(\hat{f})$. Since $H_{ba} = \varphi^\omega(H_{ba})$, we obtain $H_{ba} \subseteq \varphi^\omega \circ \hat{f}(\widehat{A^*}) = K$. ∎

Proof of Theorem 7.4.4 Let $H = \text{Im}(\varphi^\omega \circ \hat{f})$. By Lemma 7.4.5, we have $H_{ba} = H$ and H is a group isomorphic to $G(\varphi)$.

Let us show that

$$\text{Ker}(\varphi^\omega \circ \hat{f}) \subseteq \text{Ker}(\tilde{\varphi}_X^\omega). \tag{7.8}$$

Indeed, let $(u, v) \in \text{Ker}(\varphi^\omega \circ \hat{f})$ and let us show that $(u, v) \in \text{Ker}(\tilde{\varphi}_X^\omega)$. Since $\hat{f}$ is injective by Theorem 4.8.3, it suffices to show that they have the same image by $\hat{f} \circ \tilde{\varphi}_X^\omega$. Now, by commutativity of Diagram (7.7), we have $\hat{f} \circ \tilde{\varphi}_X = \tilde{\varphi} \circ \hat{f}$ and thus $\hat{f} \circ \tilde{\varphi}_X^\omega = \tilde{\varphi}^\omega \circ \hat{f}$. Thus $\hat{f} \circ \tilde{\varphi}_X^\omega(u) = \tilde{\varphi}^\omega \circ \hat{f}(u) = \tilde{\varphi}^\omega \circ \hat{f}(v) = \hat{f} \circ \tilde{\varphi}_X^\omega(v)$.

Now Eq. (7.8) implies by Proposition 4.7.6 that H has the profinite semigroup presentation

$$H = \langle B \mid \tilde{\varphi}_X^\omega(b) = b, \ b \in B \rangle_S \tag{7.9}$$

If a profinite group has the semigroup presentation $\langle B \mid R \rangle_S$ then it also has the group presentation $\langle B \mid p_G \times p_G(R) \rangle_G$ where $p_G : \widehat{B^*} \to \widehat{FG(B)}$ is the natural projection. By Diagram (7.7), one has $\tilde{\varphi}_{X,G} \circ p_G = p_G \circ \tilde{\varphi}_X$ and thus H has the profinite group presentation $\langle B \mid \tilde{\varphi}_{X,G}^\omega(b) = b, \ b \in B \rangle_G$. ∎

Example 7.4.6 Let $\tau : a \mapsto ab, \ b \mapsto ba$ be the Thue-Morse morphism. We have seen in Example 5.5.11 that the word aa is a connection for τ and $\tilde{\tau} = \tau^2$ is a connective power of τ. The set $X = a \, \mathcal{R}_{F(\tau)} \, (aa)a^{-1}$ has four elements $x = abba$, $y = ababba, z = abbaba$ and $t = ababbaba$. One has $\tilde{\tau}_X(x) = zxy, \tilde{\tau}_X(y) = ztxy, \tilde{\tau}_X(z) = zxty$ and $\tilde{\tau}_X(t) = ztxty$. By Theorem 7.4.4, the group $G(\tau)$ is the profinite group generated by X with the relations $\tau_{X,G}^\omega(u) = u$ for $u \in X$. Actually, since $\tau^\omega(y)\tau^{\omega-1}(x)\tau^\omega(z) = \tau^\omega(t)$, the relation $xy^{-1}z = t$ is a consequence of the relations above and thus $G(\tau)$ is generated by x, y, z.

In the following lines, we sometimes adopt the following notational flexibility: given a group G and an alphabet A, we may use the same notation h for a mapping $h : A \to G$, for its unique extension to a group morphism $h : FG(A) \to G$, and for its unique extension to a monoid morphism $h : A^* \to G$. Likewise, we may denote by φ an endomorphism of A^* as well the induced endomorphism of $FG(A)$.

Consider a finite group H. Let $\text{Hom}(A^*, H)$ be the finite set of morphisms from A^* to H. Let φ be an endomorphism of A^*. We denote by φ_H the map from $\text{Hom}(A^*, H)$ into itself defined as follows. Consider $h \in \text{Hom}(A^*, H)$. For $w \in A^*$, we define the image of w by $\varphi_H(h)$ to be $\varphi_H(h)(w) = h(\varphi(w))$. This notation extends in an obvious way to the case where φ is an endomorphism of $FG(A)$. In this case, φ_H is a map from $\mathrm{Hom}(FG(A), H)$ into itself.

Notice that if ψ is an endomorphism of $FG(A)$, then

$$(\psi \circ \varphi)_H(h) = h \circ \psi \circ \varphi = \varphi_H(h \circ \psi) = \varphi_H \circ \psi_H(h),$$

whence $(\psi \circ \varphi)_H = \varphi_H \circ \psi_H$. In particular, if φ is invertible as an endomorphism of $FG(A)$, then φ_H is invertible (as a map from $\mathrm{Hom}(FG(A), H)$ into itself) and $(\varphi_H)^{-1} = (\varphi^{-1})_H$.

We say that φ has *finite h-order* if there is an integer $n \geqslant 1$ such that $\varphi_H^n(h) = h$. The least such integer is called the *h-order* of φ. Notice that any substitution φ which is invertible in $FG(A)$ is of finite h-order. Indeed, since $\mathrm{Hom}(A^*, H)$ is finite, there are positive integers n, m with $n < m$ such that $\varphi_H^{n+m} = \varphi_H^n$. Since φ is invertible, so is φ_H, and therefore φ_H^m is the identity.

Example 7.4.7 Let $\varphi : a \mapsto ab,\ b \mapsto a$ be the Fibonacci substitution and let $h : A^* \to \mathbb{Z}/2\mathbb{Z}$ be the parity of the length, that is the morphism into the additive group of integers modulo 2 sending each letter to 1. Then $\varphi^3 : a \mapsto abaab,\ b \mapsto aba$ maps every letter to a word of odd length, so that φ is of h-order 3 (the h-order being a divisor of 3 and not 1).

The following result is a complement to Theorem 7.4.4 (see Exercise 7.13 for a statement relating more precisely both results).

Theorem 7.4.8 *Let φ be a nonperiodic primitive substitution over A and let $h :$ $A^* \to H$ be a morphism onto a finite group. Let (a, b) be a connection of φ and let $X = X(a, b)$. Let $f : B^* \to A^*$ be a coding morphism for X and let $g = h \circ f$. Let ψ be the restriction of $\tilde{\varphi}_X$ to B^*. If*

(i) ψ has finite g-order and
(ii) $g : B^* \to H$ is surjective,

then the restriction of $\hat{h} : \widehat{A^*} \to H$ to each maximal subgroup of $J(\varphi)$ is surjective onto H.

Proof Let K and K' be two maximal subgroups of $J(\varphi)$ and let e and e' be their respective idempotents. Choose $u, v \in \widehat{A^*}$ such that $e' = uev$ and $e \mathrel{\mathcal{R}} ev \mathrel{\mathcal{L}} e'$. By Green's Lemma (cf. Exercise 3.52), we have $K' = uKv$. As $\hat{h}(uv)$ is the idempotent of the group H, we conclude that the images by the morphism $\hat{h}$ of two maximal subgroups of $J(\varphi)$ are conjugate subgroups of H. Thus it is enough to show that $\hat{h}(H_{ba}) = H$.

Let $z \in H$. Since g is surjective, there is some $x \in B^*$ such that $g(x) = z$. Set $y = \hat{f} \circ \tilde{\varphi}_X^\omega(x)$. By (7.7), we have also $y = \tilde{\varphi}^\omega \circ \hat{f}(x)$ and so $y \in H_{ba}$ by Lemma 7.4.5.

We verify that $\hat{h}(y) = z$. Indeed, since ψ has finite g-order, we have $g \circ \psi^n = g$ for some $n \geqslant 1$. This implies that $\hat{g} \circ \tilde{\varphi}_X^\omega = \hat{g}$. Finally, we may compute

$$\hat{h}(y) = \hat{h} \circ \hat{f} \circ \tilde{\varphi}_X^\omega(x)$$
$$= \hat{g} \circ \tilde{\varphi}_X^\omega(x) = \hat{g}(x) = z$$

which concludes the proof.   ∎

Corollary 7.4.9 *Let φ be a nonperiodic primitive substitution over A and let $h : A^* \to H$ be a morphism onto a finite group. If φ has finite h-order, then the restriction of $\hat{h} : \widehat{A^*} \to H$ to each maximal subgroup of $J(\varphi)$ is surjective onto H.*

Proof Let (a, b) be a connection of φ, which we know that exists by Proposition 5.5.10. Let X, f, g, ψ be as in the statement of Theorem 7.4.8. By the hypothesis that φ has finite h-order, we may take $n \geqslant 1$ such that $h \circ \varphi^n = h$. Then, using (7.7), we obtain

$$g \circ \psi^n = h \circ f \circ \psi^n = h \circ f \circ \tilde{\varphi}_X^n$$
$$= h \circ \tilde{\varphi}^n \circ f = h \circ f = g$$

showing that ψ has finite g-order. The conclusion now follows immediately from Theorem 7.4.8.   ∎

We give two examples. In the first one, the morphism is not surjective.

Example 7.4.10 Let φ be as in Example 6.5.3, let $G = \mathbb{Z}/2\mathbb{Z}$ and let $h : A^* \to G$ be the parity of the length. Then $\varphi_G(h) = (0, 0)$ and $\varphi_G(0, 0) = (0, 0)$, where $(0, 0)$ denotes the constant homomorphism $FG(A) \to G$. Thus, φ does not have finite h-order. Actually, any pseudoword in $J(\varphi)$ which is in the image of $\hat{\varphi}$ has even length and so it is mapped by h to 0. Hence, by Theorem 7.4.3, there is a maximal group G in $J(F)$ which contains only pseudowords of even length and therefore $\hat{h}(G) = \{0\}$, showing that the restriction of $\hat{h}$ to G is not surjective.

In the second example, the morphism is surjective.

Example 7.4.11 Let $\varphi : a \mapsto ab$, $b \mapsto a^3b$ be as in the previous example and let $h : A^* \to A_5$ be the morphism from A^* onto the alternating group A_5 defined by $h : a \mapsto (123)$, $b \mapsto (345)$. One may verify that φ has h-order 12. Since A_5 is generated by $\{(123), (345)\}$, by Corollary 7.4.9 it is a quotient of $G(\varphi)$. It is not known whether every finite group is a quotient of $G(\varphi)$.

7.4.1 Proper Substitutions

A substitution φ over A is *proper* if there are letters $a, b \in A$ such that for every $d \in A$, $\varphi(d)$ starts with a and ends with b. Theorem 7.4.4 takes a simpler form for proper substitutions.

Theorem 7.4.12 *Let φ be a nonperiodic proper primitive substitution over a finite alphabet A. Then $G(\varphi)$ admits the presentation*

$$\langle A \mid \varphi_G^\omega(a) = a, a \in A \rangle_G.$$

Proof All the elements of $\widehat{A^*}$ of the form $\varphi^\omega(a)$ with $a \in A$ belong to the same maximal subgroup H of $J(\varphi)$ isomorphic to $G(\varphi)$. By Theorem 7.4.3, $H = \operatorname{Im} \varphi^\omega$.

We apply Proposition 4.7.6 to the following diagram

$$
\begin{array}{ccc}
\widehat{A^*} & \xrightarrow{\varphi^{\omega+1}} & \widehat{A^*} \\
{\scriptstyle \varphi^\omega} \downarrow & & \downarrow {\scriptstyle \varphi^\omega} \\
H & \xrightarrow{\varphi} & H.
\end{array}
$$

Since $\operatorname{Ker} \varphi^\omega \subseteq \operatorname{Ker} \varphi^{\omega+1}$, it follows from Proposition 4.7.6 that the profinite group H admits the presentation $\langle A \mid \varphi^\omega(a) = a, \ a \in A \rangle_S$ and so also the presentation $\langle A \mid \varphi_G^\omega(a) = a, \ a \in A \rangle_G$. ∎

Example 7.4.13 Let $A = \{a, b\}$ and let $\varphi : a \mapsto ab, \ b \mapsto a^3 b$. The morphism φ is proper. Thus, by Theorem 7.4.12, the Schützenberger group of $J(\varphi)$ has the presentation $\langle a, b \mid \varphi_G^\omega(a) = a, \varphi_G^\omega(b) = b \rangle_G$. Since the image of $FG(A)$ by φ is included in the subgroup generated by words of length 2, the relations $\varphi_G^\omega(a) = a$ and $\varphi_G^\omega(b) = b$ are nontrivial and thus $G(F)$ is not a free profinite group of rank two (it is actually not a free profinite group, see Almeida (2005b, Example 7.2)).

7.5   Exercises

7.5.1   *Section 7.2*

7.1 Show that for every natural number q and mapping $h : B^{q+1} \to A$, one has $\hat{h}(uv) = \hat{h}(u) \cdot \hat{h}(t_q(u)v) = \hat{h}(u \, i_q(v)) \cdot \hat{h}(v)$ for every $u, v \in B^*$.

7.2 Prove Lemma 7.2.3.

7.3 Check the following step in the proof of Proposition 7.2.2: N is a monoid.

7.4 Check the following step in the proof of Proposition 7.2.2: for the unique morphism $\psi : B^* \to N$ such that $\psi(b) = (b, 1, b)$ whenever $b \in B$, one has $\psi(w) = (i_q(w), \varphi(\hat{h}(w)), t_q(w))$ for every $w \in B^*$.

7.5.2 Section 7.3

7.5 Show that if F is a uniformly recurrent tree set, and H is a subgroup of finite index n in $FG(A)$, then for every maximal group G in $J(F)$ the intersection $G \cap p_G^{-1}(\overline{H})$ is a subgroup of index n of G.

7.6 Show that if $\text{Card}(A) > 1$ then every finitely generated projective profinite group is contained in some uniformly recurrent $\mathcal{J}$-class of $\widehat{A^*}$. (In particular, this gives a converse to Theorem 4.7.9.)

7.5.3 Section 7.4

7.7 Show that the pair (a, b) is a connection of the Fibonacci morphism φ and that φ^2 is a connective power.

7.8 Show that (a, a) is a connection of the Tribonacci morphism φ and that φ^3 is a connective power.

7.9 Give an example of a primitive nonperiodic substitution $\varphi : A^* \to A^*$ such that $\varphi(J(\varphi)) \subsetneq J(\varphi)$.

7.10 Show that if $u \in \varphi(J(\varphi))$ then $\overrightarrow{u} \in \varphi(A^{\mathbb{N}})$ and $\overleftarrow{u} \in \varphi(A^{\mathbb{Z}-})$.

7.11 Let $\varphi : a \mapsto aba, b \mapsto bab$. Show that (a, b) is a connection but that $H_{ba} \not\subseteq \text{Im}(\varphi^{\omega})$. (Hint: show that $(ab)^{\omega+1}$ does not belong to the closure of the subgroup of H_{ba} generated by $\varphi^{\omega}(ab)$.)

7.12 Give an example of a connection (a, b) of a primitive nonperiodic substitution φ such that $\varphi^{\omega}(ab) \notin H_{ba}$.

7.13 Let φ be a continuous endomorphism of $\widehat{A^+}$. Show that the following conditions are equivalent for a group H.

(i) H is a continuous homomorphic image of the group presented by $\langle A \mid \varphi_G^{\omega}(a) = a, \ a \in A \rangle_G$.
(ii) there is a generating mapping $f : A \to H$ and an integer n with $1 \leqslant n \leqslant \text{Card}(H^A)$ such that $\varphi_H^n(f) = f$.
(iii) there is some generating mapping $f : A \to H$ and an integer $n \geqslant 1$ such that $\varphi_H^n(f) = f$.

7.14 A topological group G is said to be *decidable* if for every finite group H one can decide whether there is a morphism from G onto H. Consider a profinite group with presentation $G = \langle A \mid \varphi_G^{\omega}(a) = a, \ a \in A \rangle_G$ determined by a continuous endomorphism φ of $\widehat{A^+}$ such that, for each $a \in A$, there is an algorithm that computes in each finite group G the operation $\varphi(a)_G$ determined by $\varphi(a)$. Show that G is decidable.

7.6 Solutions

7.6.1 Section 7.2

7.1 By symmetry, it suffices to justify the equality $\hat{h}(uv) = \hat{h}(u) \cdot \hat{h}(t_q(u)v)$. If $|u| \leqslant q$, then $t_q(u) = u$ and $\hat{h}(u) = 1$, and so the equality holds. Suppose that $|u| = n > q$. Let $u = b_1 \cdots b_n$ and $v = b_{n+1} \cdots b_{n+k}$, with $k \geqslant 0$. Then we have:

$$\hat{h}(uv) = h(b_1 \cdots b_{n+k})$$

$$= \hat{h}(b_1 \cdots b_{q+1}) \cdots \hat{h}(b_{n-q} \cdots b_n) \cdot \hat{h}(b_{n-q+1} \cdots b_{n+1}) \cdots \hat{h}(b_{n+k-q} \cdots b_{n+k})$$

$$= \hat{h}(u) \cdot \hat{h}(b_{n-q+1} \cdots b_{n+k})$$

$$= \hat{h}(u) \cdot \hat{h}(\underbrace{b_{n-q+1} \cdots b_n}_{t_q(u)} \cdot b_{n+1} \cdots b_{n+k}) = \hat{h}(u) \cdot \hat{h}(t_q(u)v).$$

7.2 As B^* is dense in $\widehat{B^*}$ and $\hat{h}$, i_k, t_n are continuous in $\widehat{B^*}$, it suffices to check $\hat{h}(uv) = \hat{h}(u\,i_k(v)) \cdot \hat{h}(t_k(u)v)$ for $u, v \in B^*$. If $|v| \leqslant k$, then $i_k(v) = v$ and $|t_k(u)v| < 2k + 1$, thus $\hat{h}(t_k(u)v) = \varepsilon$ and we are done. Suppose that $|v| > k$. Let $w \in B^*$ be such that $v = i_k(v) \cdot w$. From one of the equalities showed in Exercise 7.1, we obtain $\hat{h}(uv) = \hat{h}(u\,i_k(v) \cdot w) = \hat{h}(u\,i_k(v)) \cdot \hat{h}\big(t_{2k}(u\,i_k(v))\,w\big)$. As $|v| > k$, we have $t_{2k}(u\,i_k(v)) = t_k(u)i_k(v)$, thus $t_{2k}(u\,i_k(v))\,w = t_k(u)\,v$, establishing the equality we wished to show.

7.3 We wish to show the equality

$$\big[(x, m, y) \cdot (x', m', y')\big] \cdot (x'', m'', y'') = (x, m, y) \cdot \big[(x', m', y') \cdot (x'', m'', y'')\big]$$

for all $(x, m, y), (x', m', y'), (x'', m'', y'') \in N$. On one hand, we have

$$\big[(x, m, y) \cdot (x', m', y')\big] \cdot (x'', m'', y'') =$$

$$= \big(i_q(xx'),\ m\,\varphi\hat{h}(yx')\,m',\ t_q(yy')\big) \cdot (x'', m'', y'')$$

$$= \big(i_q(i_q(xx')x''),\ m\,\varphi\hat{h}(yx')\,m'\,\varphi\hat{h}(t_q(yy')x'')\,m'',\ t_q(t_q(yy')y'')\big).$$

On the other hand, we also have

$$(x, m, y) \cdot \big[(x', m', y') \cdot (x'', m'', y'')\big] =$$

$$= (x, m, y) \cdot \big(i_q(x'x''),\ m'\varphi\hat{h}(y'x'')\,m'',\ t_q(y'y'')\big)$$

$$= \big(i_q(x\,i_q(x'x'')),\ m\,\varphi\hat{h}(y\,i_q(x'x''))\,m'\,\varphi\hat{h}(y'x'')\,m'',\ t_q(y\,t_q(y'y''))\big).$$

As observed in Exercise 4.19, on the set of words of length at most q the operation $(u, v) \mapsto i_q(uv)$ is associative, thus $i_q(i_q(xx')x'') = i_q(x\, i_q(x'x''))$. Dually, we have $t_q(t_q(yy')y'') = t_q(y\, t_q(y'y''))$. It remains to check the equality

$$m\varphi\hat{h}(yx')m'\varphi\hat{h}(t_q(yy')x'')m'' = m\varphi\hat{h}(yi_q(x'x''))m'\varphi\hat{h}(y'x'')m''. \qquad (7.10)$$

If $(x', m', y') \in B^q \times M \times B^q$ then $i_q(x'x'') = x'$ and $t_q(yy') = y'$, and so (7.10) indeed holds. Finally, suppose that $(x', m', y') \in N \setminus (B^q \times M \times B^q)$. Then we have $(x', m', y') = (x', 1, x')$, whence the left side of (7.10) becomes equal to

$$m\varphi\hat{h}(yx')\varphi\hat{h}(t_q(yx')x'')m'' \qquad (7.11)$$

while the right side of (7.10) becomes equal to

$$m\varphi\hat{h}(yi_q(x'x''))\varphi\hat{h}(x'x'')m''. \qquad (7.12)$$

Applying the equalities shown in Exercise 7.1, we deduce that (7.11) and (7.12) are both equal to $m\varphi\hat{h}(yx'x'')m''$, concluding the proof that the operation on N is associative.

7.4 We use induction on the length of w. The base case $w = \varepsilon$ is trivial. Take $w \in B^*$ for which the formula for ψ holds. Let $b \in B$. Then we have

$$\psi(wb) = (i_q(w), \varphi\hat{h}(w), t_q(w))(b, 1, b)$$

$$= (i_q(i_q(w)b), \varphi\hat{h}(w)\varphi\hat{h}(t_q(w)b), t_q(t_q(w)b)) = (i_q(wb), \varphi\hat{h}(wb), t_q(wb))$$

where in the last step we apply one of the equalities shown in Exercise 7.1.

7.6.2 Section 7.3

7.5 Denote by α the restriction $G \to \widehat{FG(A)}$ of $p_{\mathbf{G}}$ to G. We proved in Theorem 7.3.4 that α is an isomorphism. Therefore, it suffices to observe that $G \cap p_{\mathbf{G}}^{-1}(\overline{H}) = \alpha^{-1}(\overline{H})$.

7.6 Let B be a subset of A such that $\mathrm{Card}(B) = 2$. Consider the Fibonacci set F on the alphabet B. Let G be a maximal subgroup of the uniformly recurrent $\mathcal{J}$-class $J(F)$. Then G is isomorphic to $\widehat{FG(B)}$, as seen in Example 7.3.5. Let P be a projective profinite group generated by n elements, where n is a positive integer. By Proposition 4.3.3, there is a subgroup H of $FG(B)$ of index n. Then H is a free group of rank $n+1$, by Schreier's Index Formula (cf. Exercise 4.12). The topological closure $\overline{H}$ in $\widehat{FG(B)}$ is a free profinite of rank $n+1$ (cf. Theorem 4.6.7). Therefore, there is a continuous onto morphism $g : \overline{H} \to P$. By the definition of projective profinite group, there is a continuous morphism $h : \overline{H} \to P$ such that $g \circ h$ is the identity on $\overline{H}$, whence h is injective. Therefore, the projective group P embeds in $\overline{H}$, and so it embeds in G.

7.6.3 Section 7.4

7.7 One has $ba \in F(\varphi)$ since $\varphi^2(a) = abab$. Next $\varphi(a) = ab$ begins with a and $\varphi^2(b) = ab$ ends with b.

7.8 One has $aa \in F(\varphi)$ since $\varphi^4(a) = abacabaabacaba$ and $\varphi^3(a) = abaca$ ends and begins with a.

7.9 For the Fibonnaci morphism $\varphi : a \mapsto ab$, $b \mapsto a$, we have $\varphi(\widehat{A^*}) \subseteq a\widehat{A^*}$. The pseudoword $\varphi^\omega(a)$ belongs to $J(\varphi)$, whence so does its infinite factor $a^{-1}\varphi^\omega(a)$. Since the first letter of $a^{-1}\varphi^\omega(a)$ is b, one concludes that $a^{-1}\varphi^\omega(a) \in J(\varphi) \setminus \operatorname{Im}\varphi$.

7.10 If $u = \varphi(v)$ for some $v \in J(\varphi)$, then $\overrightarrow{u} = \varphi(\overrightarrow{v})$ and $\overleftarrow{u} = \varphi(\overleftarrow{v})$.

7.11 Note first that $\varphi^n(ab) = (ab)^{3^n}$ for every $n \geqslant 1$, (a, b) is a connection of φ and $\varphi^\omega(ab) \in H_{ba}$. Let K be the closed subgroup of H_{ba} generated by $\varphi^\omega(ab)$. Consider the continuous morphism $\ell_3 : \widehat{A^*} \to \mathbb{Z}/3\mathbb{Z}$ mapping each letter to 1. Clearly, one has $\ell_3(\varphi^n(ab)) = 0$ for every $n \geqslant 1$, thus $\ell_3(K) = 0$. It follows that $(ab)^{\omega+1} \notin K$, since $\ell_3(ab)^{\omega+1} = \ell_3(ab) = 2$. Next, $(ab)^{\omega+1}$ is $\mathcal{H}$-equivalent to $\varphi^\omega(ab)$, that is $(ab)^{\omega+1} \in H_{ba}$. If we had $H_{ba} \subseteq \operatorname{Im}\varphi^\omega$, then we would have $(ab)^{\omega+1} = \varphi^\omega(ab)^{\omega+1} \in K$, a contradiction.

7.12 Take the primitive substitution $\varphi : a \mapsto aca$, $b \mapsto bc$, $c \mapsto acb$. It follows from the property deduced in Exercise 5.15 that if φ is periodic, then $X(\varphi)$ has period at most 9. Since $\varphi^3(a) = acaacbacaacaacbbcacaacbaca$, we conclude that φ is not periodic. Note that (a, b) is a connection of φ, with φ^2 being a connective power. We claim that $ab \notin F(\varphi)$. For that, we show by induction on $n \geqslant 1$ that ab is not a factor of any of the words $\varphi^n(a)$, $\varphi^n(b)$ or $\varphi^n(c)$. The base case $n = 1$ is immediate. Suppose that the property holds for a certain $n \geqslant 1$. Since $\varphi^{n+1}(a) = \varphi^n(a)\varphi^n(c)\varphi^n(a)$ and the first letter of $\varphi^n(c)$ is not b, one concludes that ab is not a factor of $\varphi^{n+1}(a)$. Similarly, ab is not factor of $\varphi^{n+1}(b)$ or $\varphi^{n+1}(c)$. Whence, $ab \notin F(\varphi)$, as claimed. On the other hand, ab is a factor of $\varphi^n(ab)$, since $\varphi(a)$ ends with a and $\varphi(b)$ starts with b. Therefore, we have $\varphi^\omega(ab) \notin J(\varphi)$.

7.13 Let $T = \langle A \mid \varphi_\mathsf{G}^\omega(a) = a, \ a \in A \rangle_\mathsf{G}$ and consider the natural homomorphisms $p : \widehat{A^*} \to \widehat{FG(A)}$ and $\pi : \widehat{FG(A)} \to T$.

(i) $\Rightarrow$ (ii) Suppose that $\theta : T \to H$ is an onto continuous homomorphism. Let $f = \theta \circ \pi \circ p|_A$, which is a generating mapping for H. Then, for every $k \geqslant 1$, the following equalities hold:

$$\varphi_H^k(f) = f \circ \varphi^k = \theta \circ \pi \circ p \circ \varphi^k$$

$$= \theta \circ \pi \circ \varphi_\mathsf{G}^k \circ p|_{A^*}.$$

Hence, for every $a \in A$, we have

$$\varphi_H^\omega(f)(a) = \theta \circ \pi \circ \varphi_G^\omega(a)$$
$$= \theta \circ \pi \circ \varphi_G^0(a) = \varphi_G^0(f)(a)$$

thereby showing that $\varphi_H^\omega(f) = f$. Since the set $\mathrm{Hom}(A^*, H)$ of homomorphisms from A^* into H has $\mathrm{Card}(H^A)$ elements, condition (ii) is true.

(ii) $\Rightarrow$ (iii) is trivial. To prove that (iii) $\Rightarrow$ (i), we prove that there is a continuous morphism $\theta : T \to H$ such that the following diagram commutes, where $\eta : FG(A) \to H$ is the natural morphism such that $\hat{f} = \eta \circ p$:

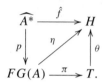

For this, it is enough to prove that $\eta(\varphi_G^\omega(a)) = \eta(a)$ for each $a \in A$. But, since there is $n \geqslant 1$ such that $\varphi_H^n(f) = f$ and since $\varphi_G \circ p = p \circ \varphi$, we have $\eta(\varphi_G^\omega(p(a))) = \eta(p(a))$ and finally $\eta(\varphi_G^\omega(a)) = \eta(a)$ as desired.

7.14 This follows directly from Exercise 7.13.

7.7 Notes

Theorem 7.2.1 is from Costa (2006). The statement had been previously announced in Almeida (2005c) but complete proofs based on the methods presented there were published only later: in Almeida and Costa (2016), as a byproduct of other results, and only for the minimal case; and in the thesis of Costa (2007), with a more complicated proof holding for all transitive shifts. The proof presented in this chapter follows the ideas in the proof given in Costa (2006). The methods employed in Almeida and Costa (2016) are very different and technically more sophisticated. Proposition 7.2.2 is from Almeida and Azevedo (1993). Theorem 7.2.1 provides an invariant of recurrent subshifts under conjugacy. A few words are in order to place the question of conjugacy of subshifts in a proper context. First of all, conjugacy is not known to be decidable, even for the simple class of shifts called *shifts of finite type*, that is, subshifts defined by a finite set of forbidden blocks. Next, many invariants are known which allow one to easily distinguish subshifts which are not conjugate. One such invariant is the *entropy*, which is $\lim(\log p_n)/n$ where p_n is the factor complexity. Few invariants are known for minimal subshifts with the notable exception of the dimension group, an ordered Abelian group which can be used to characterize the orbit equivalence of subshifts (see Putnam (2018)).

Theorem 7.3.4 is from Almeida and Costa (2016, Theorem 6.5).

The results of Sect. 7.4 appear in Almeida and Costa (2013).

More precisely, Theorem 7.4.3 is from Almeida and Costa (2013, Theorem 5.6). Theorem 7.4.4 is the main result in Almeida and Costa (2013, Theorem 6.2).

In Almeida and Costa (2013, Section 7.3) one finds a more detailed presentation of Example 7.4.6. An application of the presentation given for $G(\tau)$ is the deduction, also made in Almeida and Costa (2013, Section 7.3), that $G(\tau)$ is not a free profinite group, and not even a relatively free profinite group.

Theorem 7.4.8 is essentially in Almeida and Costa (2013, Proposition 3.2) (using also Theorem 7.4.4). Example 7.4.11 is from Almeida and Costa (2013, Section 7.2). Theorem 7.4.12 is from Almeida and Costa (2013, Theorem 6.4)

Finally, Exercise 7.13 is taken from Almeida and Costa (2013, Proposition 3.2).

Chapter 8
Groups of Bifix Codes

8.1 Introduction

In this chapter, we show how one may associate to some finite automata permutation groups. This applies in particular to automata recognizing the submonoid generated by a bifix code.

There is a rich interplay between three kinds of properties, namely

1. properties of finite monoids expressing the syntactic characteristics of recognizable sets,
2. the relativization of recognizable sets to a given factorial set (replacing a recognizable set L by $L \cap F$ where F is a factorial, possibly non recognizable set),
3. properties of the free profinite monoid and, in particular, of the $\mathcal{J}$-class $J(F)$ corresponding to a uniformly recurrent set F.

The result of this interplay is additional information on the syntactic properties of recognizable sets, in particular the submonoids generated by bifix codes.

In the first section (Sect. 8.2), we treat in more detail the notions of codes and prefix codes already met in Sect. 4.8.

In Sect. 8.3, we focus on bifix codes and study in particular bifix codes included in a factorial set F. We define the parses of a word and the F-degree of a bifix code as the maximal number of parses of a word of F.

Looking at the special case of tree sets, we prove in Sect. 8.4 an important result relating, when F is a tree set, F-complete bifix codes to bases of subgroups of finite index of the free group (Theorem 8.4.2, called the Finite Index Basis Theorem).

In Sect. 8.5, for when X is an F-complete bifix and F is a recurrent set, we define the F-minimum $\mathcal{J}$-class $J_X(F)$ of the syntactic monoid of X^*, and consider its Schützenbeger group, called the F-group $G_X(F)$. In the case where $F = A^*$, the F-group $G_X(F)$ is called the group of X, and is denoted by G_X. We see the F-group as a permutation group of degree equal to the F-degree of the code.

© Springer Nature Switzerland AG 2020
J. Almeida et al., *Profinite Semigroups and Symbolic Dynamics*, Lecture Notes
in Mathematics 2274, https://doi.org/10.1007/978-3-030-55215-2_8

As an example of what we may expect from the relativization process described by the mapping $Z \mapsto Z \cap F$, we establish in Sect. 8.6 necessary and sufficient conditions for the permutation groups G_Z and $G_{Z \cap F}(F)$ to be equivalent. This is done in what we call the Charged Code Theorem (Theorem 8.6.3), which holds when Z is a group code and F is a tree set, but also in many other cases. In the statement of this theorem we use the notion of charged code, which has a concise formulation in terms of the profinite group $G(F)$. This formulation guides us to a new (although less elementary) algebraic-topological proof of the Finite Index Basis theorem.

In Sect. 8.7, we make a study of the bifix codes in the free profinite monoid. The insight there acquired helps us in Sect. 8.8, dedicated to the proof of the Charged Code Theorem. As a byproduct of the proof, we deduce at the end of Sect. 8.8 a companion theorem (Theorem 8.8.4) concerning an equivalence between $G_X(F)$ and G_Z described by sets of return words.

8.2 Prefix Codes in Factorial Sets

In this section, we study prefix codes in a factorial set F. We define F-maximal prefix codes and prove that, when F is factorial, they coincide with right F-complete prefix codes (Proposition 8.2.7).

To facilitate the reading of this chapter, we recall some notions already met in Chap. 4. Given a subsemigroup S, a generating set X of S of A^+ is a *minimal generating set* (sometimes also called a *basis*) if X does not contain properly a generating set of S. It turns out that $X = S \setminus S^2$ is the unique minimal generating set of S (cf. Exercise 4.36).

If M is a submonoid of A^*, then we also say that X is a minimal generating set (or a basis) of M if X is a minimal generating set of the subsemigroup $M \setminus \{\varepsilon\}$ of A^+.

A nonempty set X contained in A^+ is a *code* when the subsemigroup X^+ of A^+ satisfies the following property: if $x_1 \cdots x_n = y_1 \cdots y_m$, with $x_i, y_j \in X$ for all i, j, then $n = m$ and $x_i = y_i$ for every $i \in \{1, \ldots, n\}$. In other words, $X \subseteq A^+$ is a code when the subsemigroup X^+ of A^+ is the free semigroup generated by X. In particular, the code $X \subseteq A^+$ is the minimal generating set of $X^* \subseteq A^*$.

Example 8.2.1 For the alphabet $A = \{a, b\}$, let $X = \{ab, ba, b\}$. Then X is a minimal generating set of the submonoid X^* of A^*. On the other hand, X is not a code, since $bab = b \cdot ab = ba \cdot b$.

The examples in which we are more interested concern recognizable codes. One should bear in mind the following fact (whose proof is part of Exercise 4.36).

Proposition 8.2.2 *A code* $X \subseteq A^+$ *is recognizable if and only if the submonoid* X^* *is recognizable.*

8.2.1   Prefix Codes

The *prefix order* is the name given to the $\leqslant_{\mathcal{R}}$-order in A^*, as $u \leqslant_{\mathcal{R}} v$ is synonymous of v being a prefix of u, whenever $u, v \in A^*$.

Two words u, v are *prefix-comparable* if one is a prefix of the other. Thus u and v are prefix-comparable if and only if there are words x, y such that $ux = vy$ or, equivalently, if and only if $uA^* \cap vA^* \neq \emptyset$. The *suffix order*, and the notion of suffix-comparable words, are defined symmetrically.

A set $X \subseteq A^+$ of nonempty words is a *prefix code* if any two distinct elements of X are incomparable for the prefix order. A prefix code is a code.

The dual notion of a *suffix* code is defined symmetrically with respect to the suffix order.

Example 8.2.3 The set $X = \{a, ba\}$ is a prefix code, but not a suffix code.

The submonoid M generated by a prefix code satisfies the following property: if $u, uv \in M$ then $v \in M$. Such a submonoid of A^* is said to be *right unitary*. One can show that conversely, any nontrivial right unitary submonoid of A^* is generated by a prefix code (Exercise 8.2). The symmetric notion of a *left unitary* submonoid is defined by the condition $v, uv \in M$ implies $u \in M$.

8.2.2   Maximal Prefix Codes

Let F be a subset of A^*. A set $X \subseteq A^*$ is *right dense* in $F \subseteq A^*$, or *right F-dense*, if every $u \in F$ is a prefix of some word in X.

A set $X \subseteq F$ is *right complete* in F, or *right F-complete*, if X^* is right dense in F, that is, if every word in F is a prefix of X^*.

A prefix code $X \subseteq F$ is *maximal* in F, or *F-maximal*, if it is not properly contained in any other prefix code $Y \subseteq F$. The notion symmetric to that of F-maximal prefix code is the notion of F-maximal suffix code.

Example 8.2.4 Let $F = \{ab, aab, aabb\}$. The prefix code $X = \{ab, aab\}$ is F-maximal but not F-complete.

When $F = A^*$, one talks of a *maximal* or *complete* prefix code instead of an F-maximal or F-complete prefix code.

Example 8.2.5 The prefix code $X = \{aa, ab, ba, bba, bbb\}$ is maximal and complete.

We proceed with the following simple observation.

Proposition 8.2.6 *Let F be a subset of A^*. For every prefix code $X \subseteq F$, the following conditions are equivalent:*

  (i) every element of F is prefix-comparable with some element of X,
  (ii) X is an F-maximal prefix code.

Proof (i) ⇒ (ii) By assumption, every word $u \in F$ is prefix-comparable with some word of X. This implies that if $u \notin X$, then $X \cup u$ is no longer a prefix code. Thus X is an F-maximal prefix code.

(ii) ⇒ (i) Assume that $u \in F$ is not prefix-comparable to any word in X. Then $X \cup u$ is a prefix code, and X is not an F-maximal prefix code.   ■

In Proposition 8.2.6, the set F could be any subset of A^*. In the following proposition, F is required to be factorial.

Proposition 8.2.7 *Let F be a factorial subset of A^*. For every prefix code X contained in F, the following conditions are equivalent:*

  (i) every element of F is prefix-comparable with some element of X,
 (ii) XA^* is right F-dense,
(iii) X is right F-complete,
(iv) X is an F-maximal prefix code.

Proof (i) ⇒ (ii) Let $u \in F$. Let $x \in X$ be prefix-comparable with u. Then there exist v, w such that $uv = xw$. Thus XA^* is right F-dense.

(ii) ⇒ (iii) Consider a word $u \in F$. Let us show that u is a prefix of X^*. Since XA^* is right F-dense, one has $uw = xw'$ for some word $x \in X$ and $w, w' \in A^*$. If u is a prefix of X, then we are done. Otherwise, x is a prefix of u. Thus, $u = xu'$ for some $u' \in A^*$. Since u is in F and since F is factorial, we have $u' \in F$. Since $x \neq 1$, we have $|u'| < |u|$. Arguing by induction, the word u' is a prefix of X^*. Hence, u is a prefix of X^*.

(iii) ⇒ (i) Let $u \in F$. Then, u is a prefix of X^* and, consequently, u is prefix-comparable with a word in X.

(i) ⇔ (iv) This is just a special case of Proposition 8.2.6.   ■

Remark 8.2.8 Note that if the prefix code X is known and finite, then Proposition 8.2.7 gives an algorithm to decide if X is right F-complete, provided we can compute the list of words of the factorial set F with length less or equal than the maximum length of an element of X.

Example 8.2.9 The set $X = \{a, ba\}$ is a maximal prefix code in the Fibonacci set F since XA^* is right F-dense.

Example 8.2.10 Let F be the Fibonacci set. The set $X = \{aaba, ab, baa, baba\}$ is a right F-complete prefix code. Indeed, every word of F with length at most 4 is prefix-comparable with some element of X. Using the dual of Proposition 8.2.7, one sees that X is also a left F-complete suffix code.

8.2.3   Minimal Automata of Prefix Codes

The following observation is helpful.

Proposition 8.2.11 *If X is a prefix code, then the initial state of the minimal automaton $\mathcal{A}(X^*)$ of X^* is the unique terminal state of $\mathcal{A}(X^*)$.*

Proof We follow the model for $\mathcal{A}(X^*)$, introduced in Chap. 4, that uses left quotients of X^* as states. The initial state of $\mathcal{A}(X^*)$ is X^*. Let L be a terminal state. Then $L = u^{-1}X^*$ for some word u such that $u \in X^*$. Since X^* is right unitary (Exercise 8.2), we have $L = X^*$.   ∎

A deterministic automaton $\mathcal{A} = (Q, i, T)$ on the alphabet A is called *complete* if $q \cdot a$ is defined for every $q \in Q$ and every $a \in A$.

Proposition 8.2.12 *Let $X \subseteq A^+$ be a prefix code, and let $\mathcal{A} = (Q, i, i)$ be the minimal automaton of X^*, on the alphabet A. Then X is a right complete prefix code if and only if the automaton $\mathcal{A}$ is complete.*

Proof Suppose first that $\mathcal{A}$ is complete. Let $u \in A^*$. Since $\mathcal{A}$ is complete, the state $q = i \cdot u$ is defined. Because $\mathcal{A}$ is trim, there is $v \in A^*$ such that $i = q \cdot v = i \cdot uv$, thus $uv \in X^*$. Therefore, X is right complete, by definition.

Conversely, suppose that X is right complete. Let $q \in Q$. Let $a \in A$. Since $\mathcal{A}$ is trim, there is $u \in A^*$ such that $i \cdot u = q$. Because X is right complete, there is $v \in A^*$ such that $uav \in X^*$. Therefore, $i = i \cdot uav = q \cdot av$. In particular, the state $q \cdot a$ is defined, thus showing that $\mathcal{A}$ is complete.   ∎

Proposition 8.2.12 is helpful in situations like the one in the following example, in which we are dealing with an infinite recognizable code whose minimal automaton is known.

Example 8.2.13 Let $Z = \{aa, ab, ba\} \cup b^2(a^+b)^*b = \{aa, ab, ba\} \cup b(ba^+)^*b^2$. The code Z is a prefix code. The minimal automaton of Z^* is depicted in Fig. 8.1. It is a complete automaton, thus Z is a right complete bifix code. The code Z is also a suffix code. In fact, Z is closed under reversal. In particular, Z is a left complete suffix code.

Fig. 8.1 Minimal automaton of Z^* for the code $Z = \{aa, ab, ba\} \cup b^2(a^+b)^*b$

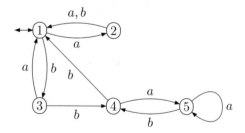

More generally, we have the following property. In its statement, we use the notation $q \cdot z = \emptyset$, where q is a state and z is a word, for meaning that $q \cdot z$ is not defined, that is, $q \cdot z \neq \emptyset$ means that q belongs to the domain of z in the transition monoid of the automaton.

Proposition 8.2.14 *Let $X \subseteq A^+$ be a prefix code contained in a factorial set $F \subseteq A^*$. Let $\mathcal{A} = (Q, i, i)$ be the minimal automaton of X^*, on the alphabet A. Then X is right F-complete if and only if whenever u labels a path $i \to q$ in $\mathcal{A}$ such that i appears only at the beginning of the path, one has $q \cdot a \neq \emptyset$ for every $a \in A$ such that $ua \in F$.*

The proof of Proposition 8.2.14 is left as an exercise (Exercise 8.3). When X is finite, we have an algorithm similar to the one described in Remark 8.2.10, which may be useful when the minimal automaton of X^* is known.

Example 8.2.15 Let F be the Thue-Morse set and consider the prefix code

$$X = \{ab, ba, a^2b^2, b^2a^2, a^2bab^2, b^2aba^2\}$$

The minimal automaton of X^* is represented in Fig. 8.2. It has 11 states, labeled by the integers $1, \ldots, 11$, with the initial state being labeled by 1. The prefix code X is right F-complete. To see that, we may use Proposition 8.2.14. The only states q where $q \cdot z = \emptyset$ for some $z \in A$ are $4, 5, 8, 10, 11$. One has $4 \cdot a = \emptyset$. The only path $1 \to 4$ where 1 appears only at the beginning is labeled by aa, and $aaa \notin F$. Looking similarly at the other states $5, 8, 10, 11$, we conclude that X is right F-complete by Proposition 8.2.14. Note that X and F are closed under reversal, and so X is also a left F-complete suffix code.

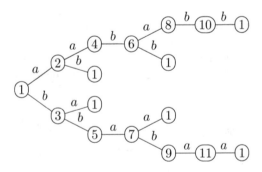

Fig. 8.2 Minimal automaton of $\{ab, ba, a^2b^2, b^2a^2, a^2bab^2, b^2aba^2\}^*$

8.3   Bifix Codes in Recurrent Sets

In this section, we study bifix codes contained in a recurrent set. A set X of nonempty words is a *bifix code* if any two distinct elements of X are incomparable for the prefix order and for the suffix order. In other words, the bifix codes are the codes which are simultaneously prefix codes and suffix codes.

8.3.1   Group Codes

We next describe an important class of bifix codes. For every subgroup H of the free group $FG(A)$, the submonoid $M = H \cap A^*$ of the free monoid A^* is generated by a bifix code, provided $M \neq \{\varepsilon\}$. Indeed, M is left and right unitary and thus the minimal generating set X of M is a bifix code. In this case, the intersection with A^* of the subgroup of $FG(A)$ generated by the code X is equal to X^*.

The case where the intersection group $H \cap A^*$ is recognizable deserves special attention. Recall from Chap. 4 that a *group automaton* (also called *permutation automaton*) is a finite, trim, deterministic automaton $\mathcal{A} = (Q, i, i)$ whose transition monoid is a group. The group automata on A are the Stallings automata of subgroups of $FG(A)$ of finite index, and the transition monoid of the Stallings automaton of a subgroup H of finite index is the representation of H on its right cosets (cf. Proposition 4.3.3). Note that every group automaton $\mathcal{A}$ is a reduced automaton, and so it is the minimal automaton of the language recognized by $\mathcal{A}$.

Proposition 8.3.1 *The following conditions are equivalent for a submonoid M of A^*.*

  (i) *M is recognized by a group automaton on the alphabet A;*
 (ii) *M is recognized by a finite group generated by A;*
(iii) *the syntactic monoid of M is a finite group generated by A;*
(iv) *$M = H \cap A^*$ for some subgroup of finite index of $FG(A)$.*

Proof (i) $\Rightarrow$ (ii) The transition monoid of a group automaton is a finite group.

(ii) $\Leftrightarrow$ (iii) The equivalence holds because the syntactic monoid of M recognizes M, and it is a morphic image of every monoid recognizing M (Proposition 4.2.14).

(ii) $\Rightarrow$ (iv) Suppose that M is recognized by a morphism $\varphi : A^* \to G$ onto a finite group G. Since G is finite, its submonoid $\varphi(M)$ is a group. Consider the group morphism $\psi : FG(A) \to G$ extending φ. Then $H = \psi^{-1}(\varphi(M))$ is a subgroup of $FG(A)$ with the same index in $FG(A)$ as that of $\varphi(M)$ in G. Finally, one has $H \cap A^* = \varphi^{-1}(\varphi(M)) = M$, where the last equality holds because φ recognizes M.

(iv) $\Rightarrow$ (i) If $\mathcal{A}$ is the Stallings automaton of a subgroup H of $FG(A)$ of finite index, then $H \cap A^*$ is the language recognized by $\mathcal{A}$.   ∎

A code $Z \subseteq A^+$ such that $Z^* \subseteq A^*$ satisfies the equivalent conditions (i)–(iv) is a *group code*. Hence, group codes are recognizable bifix codes.

For each group code $Z \subseteq A^+$, there is a unique subgroup H of $FG(A)$ of finite index such that $Z^* = H \cap A^*$, as H is the topological closure of Z^* in the profinite topology of $FG(A)$ (cf. Exercise 4.27 and Corollary 4.6.4). We shall prove later in this chapter, in Theorem 8.4.2, that H has a basis contained in Z.

Example 8.3.2 For every positive integer n, the bifix code A^n is a group code on the alphabet A. Indeed, one has $A^n = H \cap A^*$ for the kernel H of the morphism $\psi : FG(A) \to \mathbb{Z}/n\mathbb{Z}$ such that $\psi(a) = 1$ for every $a \in A$. The set $\{aa, ba, bb\}$ is a basis of H.

There are recognizable bifix codes which are not group codes.

Example 8.3.3 The set $X = \{aa, ab, ba\}$ is clearly a bifix code. The subgroup generated by X contains bb since $bb = ba(aa)^{-1}ab$. Since $bb \notin X^*$, we conclude that there is no subgroup H of $FG(A)$ such that $X^* = H \cap A^*$.

   Similarly, the bifix code $Z = \{aa, ab, ba\} \cup b^2(a^+b)^*b$ from Example 8.2.13 is also not a group code.

If $\mathcal{A} = (Q, i, i)$ is a group automaton on the alphabet A, it follows from the definition of minimal generating set that the group code Z generating the submonoid of A^* recognized by $\mathcal{A}$ consists of the words labeling nonempty cycles from i to i such that i appears in the cycle only at the beginning and at the end.

Example 8.3.4 The code $Z = a \cup ba^*b$ is the group code such that Z^* is the language recognized by the group automaton in Fig. 8.3.

8.3.2 Parses

The notion of parse is of great importance for studying bifix codes. A *parse* of a word w with respect to a set X is a triple (v, x, u) such that $w = vxu$ with $v \in A^* \setminus A^*X$, $x \in X^*$ and $u \in A^* \setminus XA^*$.

Proposition 8.3.5 *Let X be a bifix code. For any factorization $w = uv$ of w, there is a unique parse (s, yz, p) of w with $y, z \in X^*$, $sy = u$ and $v = zp$.*

Proof There are words $z \in X^*$ and $p \in A^* \setminus XA^*$ such that $v = zp$. Symmetrically, there exist $y \in X^*$ and $s \in A^* \setminus A^*X$ such that $u = sy$. Then (s, yz, p) is a parse of w which satisfies the conditions of the statement. Suppose that the parse $(s', y'z', p')$ also satisfies the conditions. Since $sy = s'y'$, we may suppose without

Fig. 8.3 A group automaton

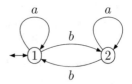

loss of generality that $s' = st$ and $y = ty'$ for some $t \in A^*$. Because X^* is left unitary, we obtain $t \in X^*$. As $s \in A^* \setminus A^*X$, we conclude that $t = \varepsilon$, $s = s'$ and $y = y'$. Symmetrically, we have $p = p'$ and $z = z'$. ∎

The number of parses of a word w with respect to X is denoted by $\delta_X(w)$. The function $\delta_X : A^* \to \mathbb{N}$ is the *parse enumerator* with respect to X.

Example 8.3.6 Let $X = \emptyset$. Then $\delta_X(w) = |w| + 1$.

Proposition 8.3.7 *Let X be a prefix code. For every word w, the number $\delta_X(w)$ is equal to the number of prefixes of w which have no suffix in X.*

Proof For every prefix v of w which is in $A^* \setminus A^*X$, there is a unique parse of w of the form (v, x, u). Since any parse is obtained in this way, the statement is proved. ∎

Proposition 8.3.7 has a dual statement for suffix codes.

Note that, as a consequence of Proposition 8.3.7, we have for two prefix codes X, Y, and for all words w,

$$X \subseteq Y \implies \delta_Y(w) \leqslant \delta_X(w). \tag{8.1}$$

Indeed, a word without suffixes in Y is also a word without suffixes in X.

Proposition 8.3.8 *Let X be a prefix code. For every $u \in A^*$ and $a \in A$, one has*

$$\delta_X(ua) = \begin{cases} \delta_X(u) & \text{if } ua \in A^*X \\ \delta_X(u) + 1 & \text{otherwise} \end{cases} \tag{8.2}$$

Proof This follows directly from Proposition 8.3.7. ∎

Proposition 8.3.8 has a dual for suffix codes expressing $\delta_X(au)$ in terms of $\delta_X(u)$.

Lemma 8.3.9 *For a bifix code X and for all $u, v, w \in A^*$ one has*

$$\delta_X(v) \leqslant \delta_X(uvw). \tag{8.3}$$

Moreover, if $uvw \in X$ and $u, w \in A^+$ then the inequality is strict, that is,

$$\delta_X(v) < \delta_X(uvw). \tag{8.4}$$

Proof Each parse (s, x, p) of v can be extended to a parse (s', yxz, p') of uvw such that $s'y = us$, $pw = zp'$ and $y, z \in X^*$, and this extended parse is unique by Proposition 8.3.5. Suppose that $uvw \in X$. Since $u, w \neq \varepsilon$, if $s' = \varepsilon$ then $y = us \in X^+$ is a proper prefix of uvw, contradicting that X is a prefix code. Therefore, uvw has at least one parse which is not obtained by extending a parse of v, namely $(\varepsilon, uvw, \varepsilon)$. ∎

In Exercise 8.8 one finds a characterization of the functions which are a parse enumerator.

8.3.3   Complete Bifix Codes

Let F be a set of words. A set $X \subseteq F$ is said to be F-*thin*, if there exists a word of F which is not a factor of any word in X. When $F = A^*$, we say simply thin instead of A^*-thin.

Every finite set is thin and provided F is infinite, every finite set $X \subseteq F$ is F-thin. We note the following important link with recognizability.

Proposition 8.3.10   *Every recognizable code is thin.*

Proof Let X be a recognizable code. Let $\varphi : A^* \to M$ be a morphism onto a finite monoid recognizing X. Let J be the minimum ideal of M. We show that $\varphi(X) \cap J = \emptyset$, which implies that X is thin. Indeed, if $x \in X$ is such that $\varphi(x) \in J$, there is an $n \geqslant 2$ such that $\varphi(x^n) = \varphi(x)$. But then $x^n \in X$, a contradiction with the hypothesis that X is a code. ∎

Next, we note the following elementary observation.

Proposition 8.3.11   *Let F be a uniformly recurrent set. Then any F-thin set is finite.*

Proof Let $w \in F$ be a word which is not a factor of any word in the F-thin set X. Since F is uniformly recurrent, every long enough word of F has w as a factor. Thus X is finite. ∎

An *internal factor* of a word x is a word v such that $x = uvw$ with u, w nonempty. For a set $X \subseteq A^*$, denote by

$$I(X) = \{w \in A^* \mid A^+ w A^+ \cap X \neq \emptyset\}$$

the set of internal factors of words in X. For abbreviation purposes, an element of $I(X)$ may be said to be an *internal factor of X*.

When F is biextendable, a set $X \subseteq F$ is F-thin if and only if $F \setminus I(X) \neq \emptyset$. Indeed, the condition is trivially necessary. Conversely, if w is in $F \setminus I(X)$, let $a, b \in A$ be such that $awb \in F$. Since awb cannot be a factor of a word in X, it follows that X is F-thin.

The notion of internal factor appears in the proof of the following result, which improves Proposition 8.3.10 in the case of prefix codes.

Proposition 8.3.12   *Let F be a biextendable set. Every recognizable prefix code contained in F is F-thin.*

Proof Let $X \subseteq A^+$ be a recognizable prefix code contained in $F \subseteq A^*$. Suppose that X is not F-thin, i.e., every word of F is a factor of some element of X. Since $X \subseteq F$ and F is biextendable, it follows that every element of X is an internal factor of some element of X. Thus, we can find a sequence (x_n) of words of X such that each x_n is an internal factor of x_{n+1}. We can also choose for each $n \geqslant 1$ a factorisation $x_n = u_n v_n$ such that u_n is a proper suffix of u_{n+1} and v_n a proper prefix of v_{n+1} (this is easily proved by induction on n). Let $i \xrightarrow{u_n} p_n \xrightarrow{v_n} t_n$ be the path starting at the initial state with label $x_n = u_n v_n$ in the minimal automaton $\mathcal{A}$ of X. Since there is a finite number of states, we can find $n < m$ such that $p_n = p_m$. Then $i \xrightarrow{u_m} p_m \xrightarrow{v_n} t_n$ is a path in $\mathcal{A}$ and thus $u_m v_n$ is in X although it is a proper prefix of x_m, a contradiction.   ∎

We say that a bifix code $X \subseteq F$ is F-*maximal* if it is not properly contained in any other bifix code $Y \subseteq F$.

Remark 8.3.13 Trivially, a bifix code X which is an F-maximal prefix code is also an F-maximal bifix code. The converse is not true in general (see Exercise 8.10) but it is true under the additional hypotheses that F is recurrent and that X is F-thin (see the Notes Section for references).

We say that a bifix code $X \subseteq F$ is F-*complete* if X is both left F-complete and right F-complete.

When $F = A^*$, one says simply maximal or complete instead of F-maximal or F-complete.

Proposition 8.3.14 *Let F be a factorial set. An F-complete bifix code is F-maximal.*

Proof If X is F-complete, then it is an F-maximal prefix code by Proposition 8.2.6. Hence, it is also an F-maximal bifix code.   ∎

In view of Proposition 8.2.6 and Remark 8.3.13, the converse of Proposition 8.3.14 does not hold in general, but it does when F is recurrent.

Example 8.3.15 Every group code $Z \subseteq A^+$ is a complete bifix code, and therefore a maximal bifix code. Indeed, we know that there is a subgroup H of $FG(A)$ of finite index such that $Z^* = H \cap A^*$, and if n is the order of the finite group which is the representation of $F(A)$ on the cosets of H, then $u \in A^*$ implies $u^n \in H \cap A^* = Z^*$.

Let $F \subseteq A^*$ be a factorial set. The F-*degree*, denoted $d_X(F)$, of a set $X \subseteq A^*$ is the maximal number of parses of words of F with respect to X, that is

$$d_X(F) = \max_{w \in F} \delta_X(w).$$

The F-degree of a set X is finite or infinite.

The A^*-degree of $X \subseteq A^*$ is called the *degree* of X, and is denoted d_X. Note that $d_X(F) = d_{X \cap F}(F)$, and that $d_X(F) \leqslant d_X$.

Observe also that if X is an F-complete bifix code then, for every $w \in F$, the number $\delta_X(w)$ is equal to the number of prefixes of w that are proper suffixes of X. Indeed, this follows from Proposition 8.3.7 since a word p that has no suffix in X, is a suffix of X.

Theorem 8.3.16 *Let F be a recurrent set and let $X \subseteq F$ be a bifix code. Then X is an F-thin and F-complete bifix code if and only if its F-degree $d_X(F)$ is finite. In this case, the set of internal factors of X is*

$$I(X) = \{w \in F \mid \delta_X(w) < d_X(F)\}. \tag{8.5}$$

Proof Assume first that X is an F-thin and F-complete bifix code. Since X is F-thin, $F \setminus I(X)$ is not empty. Let $u \in F \setminus I(X)$ and $w \in F$. Because F is recurrent, there is a word $v \in F$ such that $uvw \in F$. Since $u \in I(X)$, the prefixes of uvw that are proper suffixes of X are prefixes of u. It follows that $\delta_X(uvw) = \delta_X(u)$.

On the other hand, since by Eq. (8.3) we have $\delta_X(w) \leqslant \delta_X(uvw)$, we get $\delta_X(w) \leqslant \delta_X(u)$. This shows that δ_X is bounded, and thus that the F-degree of X is finite. Moreover, this shows that $F \setminus I(X)$ is contained in the set of words of F with maximal value of δ_X. Conversely, consider $w \in I(X)$. Then, there exist $w' \in X$ and $p, s \in A^+$ such that $w' = pws$. Equation (8.4) yields $\delta_X(w') > \delta_X(w)$ and, thus, $\delta_X(w)$ is not maximal. This proves Eq. (8.5).

Conversely, let $w \in F$ be a word with $\delta_X(w) = d_X(F)$. For any nonempty word $u \in F$ such that $uw \in F$ we have $uw \in XA^*$. Indeed, set $u = au'$ with $a \in A$ and $u' \in F$. Then $\delta_X(au'w) \geqslant \delta_X(u'w) \geqslant \delta_X(w)$ by Eq. (8.3). This implies $\delta_X(au'w) = \delta_X(u'w) = \delta_X(w)$. By the dual of Eq. (8.2) we obtain that $uw \in XA^*$.

This implies first that X is F-thin and next that XA^* is right F-dense. Indeed, suppose that w is an internal factor of a word in X. Let $p, s \in F \setminus \{\varepsilon\}$ be such that $pws \in X$. Since $pw \in F$, the previous argument shows that $pw \in XA^*$, a contradiction. It follows that $w \in F \setminus I(X)$. This shows that X is F-thin.

Next, and since F is recurrent, for every $v \in F$, there is a word $u \in F$ such that $vuw \in F$. We deduce that $vuw \in XA^*$ by using again the above argument. Hence, XA^* is right F-dense and X is right F-complete. The proof that it is left F-complete is symmetric. ∎

Example 8.3.17 Let F be the Fibonacci set. The set $X = \{aaba, ab, baa, baba\}$ is a bifix code which is F-complete, as was seen in Example 8.2.10. By Theorem 8.3.16, it has F-degree 3. Indeed, the word $aaba$ has three parses $(\varepsilon, aaba, \varepsilon)$, (a, ab, a) and (aa, ε, ba) and it is in $F \setminus I(X)$.

We use Theorem 8.3.16 in the proof of the following proposition.

Proposition 8.3.18 *Let H be a subgroup of finite index d of $FG(A)$. Then the group code $Z \subseteq A^+$ such that $Z^* = H \cap A^*$ is a recognizable complete bifix code of degree d.*

Proof We know that Z is a recognizable bifix code by the definition of group code, and we saw in Example 8.3.15 that group codes are complete. Since the code Z is recognizable, it is thin, by Proposition 8.3.10. It remains to show that the degree of Z is d.

By the definition of thin code, we may take a word $w \in A^*$ which is not an internal factor of Z. By Theorem 8.3.16, it suffices to show that w has d parses. Since H has finite index d, we may choose a set R of d representatives of the right cosets of H such that $R \subseteq A^*$ (cf. Exercise 4.10). As Z is complete, for each $r \in R$ there is $v \in A^*$ for which $rwv \in Z^*$. Because w is not an internal factor of Z, it follows from $rwv \in Z^*$ that there is some parse (s_r, x_r, p_r) of w such that $rs_r \in Z^* \subseteq H$. If $r, r' \in R$ satisfy $s_r = s_{r'}$, then $r, r' \in Hs_r^{-1}$, which implies $r = r'$, by the definition of R. This shows that $\delta_Z(w) \geqslant d$.

On the other hand, suppose that (s, x, p) and (s', x', p') are parses of w such that $Hp = Hp'$. Without loss of generality, we may suppose that $p' = tp$ for some $t \in A^*$. Canceling p in $Hp = Htp$, we get $t \in H \cap A^* = Z^*$. By the definition of parse, p' has no prefix in Z, thus $t = \varepsilon$ and $p = p'$. This shows that $\delta_Z(w) \leqslant d$, by the dual of Proposition 8.3.7. Combining with the conclusion of the previous paragraph, we obtain $\delta_Z(w) = d$. ∎

In other words, Proposition 8.3.18 says that if $Z \subseteq A^+$ is a group code such that Z^* is recognized by a group automaton with d states, then Z is a complete bifix code of degree d (cf. Exercise 8.11).

Example 8.3.19 The group code $Z = a \cup ba^*b$ has degree 2, as it is the minimal generating set of a group automaton with two vertices (cf. Fig. 8.3).

The following result establishes the link between complete bifix codes and F-complete ones.

Theorem 8.3.20 *Let $F \subseteq A^*$ be a recurrent set. For every thin complete bifix code $Z \subseteq A^+$ of degree d, the set $X = Z \cap F$ is an F-thin and F-complete bifix code. One has $d_X(F) \leqslant d$, with equality when Z is finite.*

Proof Recall that $d_X(F) = d_{Z \cap F}(F) = d_Z(F) \leqslant d$. Thus $d_X(F)$ is finite and, by Theorem 8.3.16, X is an F-thin and F-complete bifix code. If Z is finite, then there is in F a word u which is not an internal factor of Z, and such word u has d parses by Theorem 8.3.16. It follows that $d_Z(F) = d$, whence $d_X(F) = d$. ∎

Since we shall focus primarily on recognizable codes and uniformly recurrent sets, it is convenient to highlight the following consequence of Theorem 8.3.20.

Corollary 8.3.21 *Let $F \subseteq A^*$ be a recurrent set. For every recognizable complete bifix code $Z \subseteq A^+$ of degree d, the set $X = Z \cap F$ is an F-complete bifix code of F-degree at most d. Moreover, X is finite if F is uniformly recurrent.*

Proof Every recognizable code is thin (Proposition 8.3.10). Therefore, by Theorem 8.3.20, the set X is an F-thin and F-complete bifix code of degree at most d. If F is uniformly recurrent, then X is finite by Proposition 8.3.11. ∎

As a special case, we have the following useful statement.

Corollary 8.3.22 *Let $F \subseteq A^*$ be a recurrent set, and let H be a subgroup of finite index d of $FG(A)$. Consider the group code $Z \subseteq A^+$ such that $Z^* = H \cap A^*$. Then the set $X = Z \cap F$ is an F-complete bifix code of F-degree at most d. Moreover, X is finite if F is uniformly recurrent.*

Proof This follows directly from Proposition 8.3.18 and Corollary 8.3.21. ∎

Let us see some examples concerning Theorem 8.3.20 and their corollaries (Corollaries 8.3.21 and 8.3.22).

Example 8.3.23 The group code $Z = a \cup ba^*b$ was seen in Example 8.3.19 to be of degree 2. Let F be the Fibonacci set. By Corollary 8.3.22, the set $Z \cap F = \{a, bab, baab\}$ is an F-complete bifix code of F-degree at most 2. On the other hand, bab has two parses, namely $(\varepsilon, bab, \varepsilon)$ and (b, a, b). Therefore, the F-degree of X is 2.

The next example is similar to Example 8.3.23.

Example 8.3.24 Let F be the Thue–Morse set. Consider again the group code $Z = a \cup ba^*b$. According to Corollary 8.3.22, the set $Z \cap F = \{a, bab, baab, bb\}$ is an F-complete bifix code of F-degree at most 2. In fact, $d_X(F) = 2$, as bab has two parses (the same as in Example 8.3.23).

The following examples show that a strict inequality can hold in Theorem 8.3.20. The second example shows that this may happen even if all letters occur in the words of F.

Example 8.3.25 Consider again the complete bifix code $Z = a \cup ba^*b$, of degree 2. Let $F = a^*$. Then F is a recurrent set. We have $X = Z \cap F = \{a\}$. The F-degree of X is 1.

Example 8.3.26 Let $A = \{a, b\}$ and let $Z = \{aa, ab, ba\} \cup b^2(a^+b)^*b$. Then Z is a complete bifix code (cf. Example 8.2.13). Note also that Z is recognizable, thus thin by Proposition 8.3.10. Applying Theorem 8.3.16, one sees that Z is of degree 3, because the word bba is not an internal factor and has 3 parses. Let $K = \{aa, ab, ba\}$. Let F be the Fibonacci set. Then K is an F-maximal bifix code and $d_K(F) = 2$. This may easily be seen directly, or we may observe that $K = F \cap A^2$ and apply Theorem 8.3.20 to the finite group code A^2 of degree 2. Since K is F-maximal and is clearly contained in $Z \cap F$, we conclude that $K = Z \cap F$. This shows that $d_{Z \cap F}(F) < d_Z$. Note that the code Z and the uniformly recurrent set F satisfy the conditions of Theorem 8.3.20 (and Corollary 8.3.21).

The following result will be used later.

Theorem 8.3.27 *Let F be a recurrent set and let $X \subseteq F$ be an F-thin and F-complete bifix code of F-degree d. The set S of proper nonempty suffixes of X is a disjoint union of $d - 1$ F-maximal prefix codes.*

Proof For $1 \leqslant i \leqslant d$, let $S_i = \{s \in S \mid \delta_X(s) = i\}$. Then $S = \bigcup_{i=1}^{d} S_i$ is a partition of the set of proper nonempty suffixes of X. First note that $S_1 = \emptyset$: indeed, if s is a proper nonempty suffix of X and $s = xt$ with $x \in X^*$ and $t \notin XA^*$, then $(s, \varepsilon, \varepsilon)$ and (ε, x, t) are distinct parses of s. Let us show that for $2 \leqslant i \leqslant d$, the set S_i is an F-maximal prefix code.

Suppose that $s, su \in S_i$ with u nonempty. Let $v \in A^*$ and $a \in A$ be such that $u = va$. Because X is a suffix code, from $su \in S$ we get $su \notin A^*X$. It follows from Proposition 8.3.8 that $\delta_X(su) = \delta_X(sv) + 1$. Applying Lemma 8.3.9, we then obtain

Fig. 8.4 The partition of the set of proper suffixes

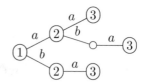

$\delta_X(s) \leqslant \delta_X(sv) < \delta_X(su)$, which is a contradiction with $\delta_X(s) = \delta_X(su) = i$. Thus, S_i is a prefix code. Since X is F-thin and F is recurrent, every word in F is a prefix of a word $w = a_1 \cdots a_n$ which is not an internal factor of X, and thus such that $\delta_X(w) = d$ by Theorem 8.3.16. Since the successive values of $\delta_X(u)$ on the prefixes $a_1 \cdots a_k$ of w increase at most by one at each step by Proposition 8.3.8, there is a prefix of w in S_i. Thus $S_i A^*$ is right F-dense, and so S_i is F-maximal by Proposition 8.2.7. ∎

A dual statement holds for the set of proper prefixes of X.

Example 8.3.28 Let F be the Fibonacci set and let $X = \{aaba, ab, baa, baba\}$ be the F-maximal bifix code of F-degree 3 of Example 8.3.17. The set S of proper suffixes of X is the union of $S_1 = \{\varepsilon\}$, $S_2 = \{a, b\}$ and $S_3 = \{aa, aba, ba\}$ where S_2, S_3 are F-maximal prefix codes (see Fig. 8.4).

8.4 Bifix Codes in Tree Sets

In this section, our attention goes to F-complete bifix codes where F is a uniformly recurrent tree set. The first result holds for all uniformly recurrent neutral sets.

8.4.1 Cardinality Theorem

The following statement shows in particular that in a uniformly recurrent neutral set F, all finite F-complete bifix codes of a given F-degree have the same cardinality.

Theorem 8.4.1 *Let F be a uniformly recurrent neutral set, with set of letters A, and let X be a finite F-complete bifix code of F-degree d. Then*

$$\text{Card}(X) - 1 = d(\text{Card}(A) - 1). \tag{8.6}$$

Proof Recall that $r(w)$ denotes the number of letters $a \in A$ such that $wa \in F$, when $w \in F$. Let $\rho(w) = r(w) - 1$ and consider the set P of proper prefixes of X. Then, by Euler's formula for planar graphs, we have the equality $r(P) = \text{Card}(X) + \text{Card}(P) - 1$, so that $\rho(P) = \text{Card}(X) - 1$.

Next, by the dual of Theorem 8.3.27, the set P of proper prefixes of X is the union of $\{\varepsilon\}$ and of a disjoint set of $d - 1$ F-maximal suffix codes P_i. By Proposition 6.3.5, we have $\rho(P_i) = \rho(\varepsilon)$ for $2 \leqslant i \leqslant d$. Since $\rho(\varepsilon) = \mathrm{Card}(A) - 1$, the chain of equalities

$$\mathrm{Card}(X) - 1 = \rho(P) = \rho(\varepsilon) + \sum_{i=2}^{d} \rho(P_i)$$

$$= d(\mathrm{Card}(A) - 1)$$

holds.   ∎

Formula (8.6) is reminiscent of the famous *Schreier's Index Formula*, which is the equality $\mathrm{Card}(X) - 1 = d(\mathrm{Card}(A) - 1)$ for X a basis of a subgroup H of finite index d of the free group $FG(A)$ (Exercise 4.12).

8.4.2 The Finite Index Basis Theorem

We will prove the following important result.

Theorem 8.4.2 *Let F be a uniformly recurrent tree set. Let H be a subgroup of finite index d of the free group $FG(A)$. Let Z be the bifix code such that $H \cap A^* = Z^*$. Then $X = Z \cap F$ is an F-complete bifix code of F-degree d which is a basis of H.*

Theorem 8.4.2 is actually a particular case of a stronger statement asserting that, in a uniformly recurrent tree set F, a finite bifix code is F-complete of F-degree d if and only if it is a basis of a subgroup of index d of the free group $FG(A)$ (see the Notes Section for a reference). We will not need the stronger form since all F-complete bifix codes that we will meet are already of the form given in Theorem 8.4.2.

Example 8.4.3 Let F be the Fibonacci set. The bifix code $X = \{a, bab, baab\}$ is an F-complete bifix code of F-degree 2. This may be seen as a consequence of Theorem 8.4.2, because if H is the kernel of the morphism $FG(A) \to \mathbb{Z}/2\mathbb{Z}$ sending a to 0 and b to 1, then X is the intersection of F with the minimal generating set of $H \cap A^*$. As expected by Theorem 8.4.2, the code X is a basis of H.

Next is an example showing that the conclusion of Theorem 8.4.2 may fail if F is not a tree set.

Example 8.4.4 Let F be the Thue–Morse set. The intersection of F with the group code $Z = a \cup ba^*b$ is $Y = \{a, baab, bab, bb\}$, an F-complete bifix code of F-degree 2, already met in Example 8.3.24. The code Y is not a basis of the subgroup H of $FG(A)$ generated by Z. Indeed, H has rank three, as already seen (cf. Example 8.4.3).

Proof The set X is a finite F-complete bifix code of degree at most d by Corollary 8.3.22. We denote by N the subgroup of $FG(A)$ generated by X.

Let $u \in F$ be such that $\delta_X(u) = d_X(F)$. Let Q be the set of suffixes of u which are proper prefixes of X. Note that $\mathrm{Card}(Q) = d_X(F)$ by the dual of Proposition 8.3.7, because the elements of F with no prefix in X are precisely the proper prefixes of F, as X is F-complete.

Let us first show that the map $q \in Q \mapsto Nq$ is injective. Indeed, assume that $p, q \in Q$ are such that $Np = Nq$. Note that the elements of Q are comparable for the suffix order. Assuming that q is longer than p, we have $q = tp$ for some $t \in F$. Then $Np = Nq$ implies $Nt = N$ and thus $t \in N \cap F$. Since $N \cap F \subseteq H \cap F = Z^* \cap F = X^*$, we conclude that $t \in X^*$, and thus $t = \varepsilon$ since q is a proper prefix of X. It follows that $p = q$.

We consider the set

$$V = \{v \in FG(A) \mid Qv \subseteq NQ\}$$

For every $v \in V$, the map $p \mapsto q$ if $pv \in Nq$ is a permutation of Q. Indeed, assume that for $p, q, r \in Q$, we have $pv, qv \in Nr$. Then $rv^{-1} \in Np \cap Nq$ implies $Np \cap Nq \neq \emptyset$ which implies $p = q$ as we have seen. Since Q is finite, the map in question must be a permutation.

Next V is a subgroup of $FG(A)$. Indeed, if $v, w \in V$, then $Qvw \subseteq NQw \subseteq NQ$. On the other hand, for $v \in V$, given $q \in Q$, by the preceding paragraph there is $p \in Q$ such that $pv \in Nq$, so that $pv = hq$ for some $h \in N$. Hence, $qv^{-1} = h^{-1}q \in NQ$, which shows that $Qv^{-1} \subseteq NQ$ and so $v^{-1} \in V$.

Let us show now that the set $\mathcal{R}_F(u)$ of return words to u is contained in V. Indeed, let $y \in \mathcal{R}_F(u)$ and $q \in Q$. Since q is a suffix of u, qy is a suffix of uy, and since uy is in F (by definition of $\mathcal{R}_F(u)$), also qy is in F. Since X is an F-complete bifix code, qy is a prefix of a word in X^* and thus there is a proper prefix r of X such that $qy \in X^*r$. We verify that the word r is a suffix of u. Since $y \in \mathcal{R}_F(u)$, there is a word y' such that $uy = y'u$. Consequently, r is a suffix of $y'u$, and in fact the word r is a suffix of u. Indeed, one has $|r| \leqslant |u|$ since otherwise u is in the set $I(X)$ of internal factors of X, and this is not the case by the choice of u and Theorem 8.3.16. Hence, we have $r \in Q$ (see Fig. 8.5). Since $X^* \subseteq N$ and $r \in Q$, we have $qy \in NQ$, so that $y \in V$.

By Theorem 6.5.5, the group generated by $\mathcal{R}_F(u)$ is the free group on A. Since $\mathcal{R}_F(u) \subseteq V$, and since V is a subgroup of $FG(A)$, we have $V = FG(A)$. Thus, the inclusion $Qw \subseteq NQ$ holds for every $w \in FG(A)$. Since $\varepsilon \in Q$, we have in particular $w \in NQ$. It follows that $FG(A) = NQ$. Hence, as $\mathrm{Card}(Q) = d_X(F)$

Fig. 8.5 A word $y \in \mathcal{R}_F(u)$

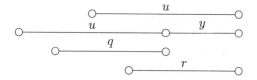

and we showed that the mapping $q \in Q \mapsto Nq$ is injective, the index of N is $d_X(F)$. And since H has index d and $N \subseteq H$, we deduce that $d_X(F) \geqslant d$, that is, $d_X(F) = d$, whence $N = H$. Finally, since, by Theorem 8.4.1, we have $\text{Card}(X) - 1 = d(\text{Card}(A) - 1)$, we conclude by Schreier's Index Formula that X is a basis of H.   ∎

Remark 8.4.5 Note that we have shown in particular the well known fact that every subgroup of $FG(A)$ of finite index has a basis contained in A^+. For $A = \{a, b\}$, an example of a finitely subgroup of $FG(A)$ which has no basis contained in A^+ is the subgroup H generated by ab^{-1}, for which one has $H \cap A^+ = \emptyset$.

8.5 The Syntactic Monoid of a Recognizable Bifix Code

In this section, we look at some properties of the syntactic monoid $M(X^*)$ of a recognizable bifix code $X \subseteq F$, with $F \subseteq A^*$. An emphasis is given to the image of F in $M(X^*)$. We begin with a more general setting, holding for arbitrary finite deterministic automata.

8.5.1 The F-Minimum 𝒥-Class

Consider a finite deterministic automaton $\mathcal{A} = (Q, i, T)$ on the alphabet A. In some parts of this section, it is not necessary to prescribe the initial state and the terminal states, in which case we may view $\mathcal{A}$ just as a labeled multigraph.

Let $M(\mathcal{A})$ be the transition monoid of $\mathcal{A}$, and let $\varphi_\mathcal{A} : A^* \to M(\mathcal{A})$ be the corresponding transition morphism. Let $u \in A^*$. The image of $\varphi_\mathcal{A}(u)$ is denoted $\text{Im}_\mathcal{A}(u)$. The *rank of u in $\mathcal{A}$*, denoted $\text{rank}_\mathcal{A}(u)$, is the rank of the partial right transformation $\varphi_\mathcal{A}(u)$, that is, the cardinality of the set $\text{Im}_\mathcal{A}(u)$ (cf. Exercise 3.48). Let F be a recurrent set. We denote by $\text{rank}_\mathcal{A}(F)$ the *F-minimal rank*, defined as the minimum of the ranks in $\mathcal{A}$ of elements of F, that is,

$$\text{rank}_\mathcal{A}(F) = \min\{\text{rank}_\mathcal{A}(u) \mid u \in F\}.$$

The following result gives a method to compute the F-minimal rank.

Theorem 8.5.1 *Consider a recurrent set F and a finite deterministic automaton $\mathcal{A}$. Let w be in F. Then w has rank equal to $\text{rank}_\mathcal{A}(F)$ if and only if $\text{rank}_\mathcal{A}(wz) = \text{rank}_\mathcal{A}(w)$ for every $z \in \mathcal{R}_F(w)$.*

Proof Assume first that $\text{rank}_\mathcal{A}(w) = \text{rank}_\mathcal{A}(F)$. If z is in $\mathcal{R}_F(w)$, then wz is in F. Since $\text{rank}_\mathcal{A}(wz) \leqslant \text{rank}_\mathcal{A}(w)$ and $\text{rank}_\mathcal{A}(w)$ is minimal, this forces $\text{rank}_\mathcal{A}(wz) = \text{rank}_\mathcal{A}(w)$.

Conversely, take $w \in F$ such that $\mathrm{rank}_A(wz) = \mathrm{rank}_A(w)$ whenever $z \in \mathcal{R}_F(w)$. Let $I = \mathrm{Im}_A(w)$. For each $r \in \mathcal{R}_F(w)$, we have $I \cdot r = \mathrm{Im}_A(wr) \subseteq \mathrm{Im}_A(w) = I$. Because $\mathrm{rank}_A(wr) = \mathrm{rank}_A(w)$, this implies $I \cdot r = I$. Therefore, since the set $\Gamma_F(w) = \{z \in F \mid wz \in A^*w \cap F\}$ is contained in $\mathcal{R}_F(w)^*$ by Proposition 6.2.3, we conclude that

$$\Gamma_F(w) \subseteq \{z \in F \mid I \cdot z = I\}. \tag{8.7}$$

Let u be a word of F of minimal rank. Since F is recurrent, there exist words v, v' such that $wvuv'w \in F$. Then $vuv'w$ is in $\Gamma_F(w)$ and thus $I \cdot vuv'w = I$ by (8.7). This implies that

$$\mathrm{rank}_A(u) \geqslant \mathrm{rank}_A(wvuv'w) = \mathrm{Card}(I \cdot vuv'w) = \mathrm{Card}(I) = \mathrm{rank}_A(w).$$

Thus w has minimal rank in F.   ∎

Theorem 8.5.1 can be used to compute the F-minimal rank of a finite deterministic automaton in an effective way, for a uniformly recurrent set F, provided one can compute effectively the finite sets $\mathcal{R}_F(w)$ for $w \in F$.

Example 8.5.2 Let F be the Fibonacci set and let A be the automaton given by its transitions in the table of Fig. 8.6. One has $\mathrm{Im}_A(a^2) = \{1, 2, 4\}$. The action on the 3-element sets of states of the automaton is also shown in Fig. 8.6. Note that $\mathrm{Im}_A(a^2)$ is fixed by each element of $\mathcal{R}_F(a^2) = \{baba^2, ba^2\}$. By Theorem 8.5.1, we obtain $\mathrm{rank}_A(F) = 3$.

Proposition 8.5.3 *Consider a recurrent set F and a finite deterministic automaton A. Let $d = \mathrm{rank}_A(F)$. The set of elements of $\varphi_A(F)$ of rank d is included in a regular $\mathcal{J}$-class. Moreover, if $u, v \in F$ are such that $\mathrm{rank}_A(u) = d$ then $\varphi_A(v)$ is a factor of $\varphi_A(u)$.*

	1	2	3	4	5	6	7	8	9
a	2	4	1	–	7	8	4	1	3
b	3	5	–	6	–	–	9	1	–

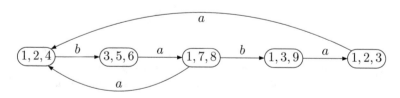

Fig. 8.6 An automaton of F-degree 3

Proof Take $u, v \in F$ such that $\text{rank}_A(u) = d$. Set $\varphi = \varphi_A$. Since F is recurrent, there are w, t such that $uwvtu \in F$. Set $z = wvtu$. Because $uz \in F$ and $\text{Im}_A(uz) \subseteq \text{Im}_A(z) \subseteq \text{Im}_A(u)$, it follows from the minimality of $\text{rank}_A(u)$ that $\text{Im}_A(uz) = \text{Im}_A(u)$. Therefore, $\varphi(z)$ restricts to a permutation on the finite set $\text{Im}_A(u)$. Hence, the idempotent power $\varphi(z)^\omega$, which is well defined as $M(A)$ is finite, restricts to the identity on $\text{Im}_A(u)$, thus $\varphi(u) = \varphi(u)\varphi(z)^\omega$. Since on the other hand u is a factor of z, we deduce that $\varphi(z)^\omega \mathcal{J} \varphi(u)$, and so the $\mathcal{J}$-class of $\varphi(u)$ is regular. Moreover, v is a factor of z, thus $\varphi(v)$ is a factor of $\varphi(u)$. In particular, $\text{rank}_A(v) = d$ implies $\varphi(v) \mathcal{J} \varphi(u)$.   ∎

In the setting of Proposition 8.5.3, we denote by $J_A(F)$ the $\mathcal{J}$-class of $M(A)$ whose elements have rank d. It is said to be the *F-minimum $\mathcal{J}$-class* of the transition monoid $M(A)$.

An alternative characterization of $J_A(F)$ is that it is the $\mathcal{J}$-class of $M(A)$ containing the image of the regular $\mathcal{J}$-class $J(F)$ of $\widehat{A^*}$ by the unique continuous morphism $\hat{\varphi}_A : \widehat{A^*} \to M(A)$ extending φ_A (Exercise 8.12).

The Schützenberger group of the regular $\mathcal{J}$-class $J_A(F)$ is denoted by $G_A(F)$. It is said to be the *F-group* of A.

8.5.2   The F-Group as a Permutation Group

Formally, a *permutation group* is a pair (I, G) consisting of a set I and a subgroup G of the group of permutations of I, acting on the right of I. The permutation group (I, G) is said to be *transitive* if, for every $p, p' \in I$, there is $g \in G$ such that $p \cdot g = p'$. The cardinality of the set I is the *degree* of the permutation group (I, G).

Example 8.5.4 The transition monoid G of a group automaton (Q, i, i) is a transitive permutation group (Q, G).

Example 8.5.5 If G is a group, then (G, G) is a transitive permutation group, for the action $h \cdot g = hg$.

Two permutation groups (I, G) and (I', G') are said to be *equivalent* if there is a pair (β, ζ) formed by a bijection $\beta : I \to I'$ and an isomorphism $\zeta : G \to G'$ such that

$$\beta(p \cdot g) = \beta(p) \cdot \zeta(g)$$

for every $p \in I$ and $g \in G$. Such a pair (β, ζ) is an *equivalence* between the permutation groups. This is the isomorphism notion for permutation groups.

In Exercise 8.13, the reader is invited to fill the details justifying the validity of the following example.

Example 8.5.6 Let T be a subsemigroup of the monoid of partial right transformations on a set Q, and let H be a maximal subgroup of T. Denote by $\text{Im}(H)$ the image of an element of H, a set depending on H only. Given $t \in H$, the restriction $q \in \text{Im}(H) \mapsto q \cdot t \in \text{Im}(H)$, here denoted $r(t)$, is a well-defined permutation on $\text{Im}(H)$. The mapping $r : t \mapsto r(t)$ thus defined is an injective morphism from H into the group of permutations on $\text{Im}(H)$. Identifying each element t of H with $r(t)$, the pair $(\text{Im}(H), H)$ becomes a transitive permutation group. If H' is another maximal subgroup of T in the same $\mathcal{D}$-class of H, then the permutation groups $(\text{Im}(H), H)$ and $(\text{Im}(H'), H')$ are equivalent.

If D is a regular $\mathcal{D}$-class of a monoid of right transformations, then we may view the Schützenberger group $G(D)$ as a transitive permutation group, by identifying it with the isomorphism class of the permutation groups of the form $(\text{Im}(H), H)$, with H a maximal subgroup contained in D (cf. Example 8.5.6).

We look again at the automaton $\mathcal{A}$ which we have been considering along this section. Let $u \in A^*$, and let $I = \text{Im}_\mathcal{A}(u)$. The *stabilizer* of I is the submonoid $\text{Stab}(I)$ of A^* consisting of the words $v \in A^*$ such that $I \cdot v = I$.

Since I is finite, if $v \in \text{Stab}(I)$ then the restriction of $\varphi_\mathcal{A}(v)$ to I is a permutation on I, denoted $\pi_I(v)$. This defines a morphism π_I from $\text{Stab}(I)$ into the group of permutations of I. The following is a basic property of the maximal subgroups of $M(\mathcal{A})$, whose proof is left as an exercise (Exercise 8.14).

Lemma 8.5.7 *Suppose that e is an idempotent of $M(\mathcal{A})$ that is $\mathcal{L}$-equivalent to $\varphi_\mathcal{A}(u)$. Let V be the maximal subgroup of $M(\mathcal{A})$ containing e. The mapping $\rho_I :$ $\pi_I(\text{Stab}(I)) \rightarrow V$, sending $\pi_I(v)$ to $e\varphi_\mathcal{A}(v)$, for $v \in \text{Stab}(I)$, is a well-defined isomorphism of groups. Moreover, the pair (id_I, ρ_I), where id_I is the identity on I, is an equivalence of permutation groups.*

Let us look at the special case where $\varphi_\mathcal{A}(u) \in J_\mathcal{A}(F)$. We view $G_\mathcal{A}(F)$ as a transitive permutation group of degree $\text{rank}_\mathcal{A}(F)$, equivalent to the permutation group $(I, \pi_I(\text{Stab}(I)))$. In this context, the following lemma is helpful.

Lemma 8.5.8 *Let $u \in F$ be such that $\varphi_\mathcal{A}(u) \in J_\mathcal{A}(F)$. The set $\mathcal{R}_F(u)$ is contained in $\text{Stab}(I)$.*

Proof Let $v \in \mathcal{R}_F(u)$. Because $uv \in F$ and $\varphi_\mathcal{A}(u) \in J_\mathcal{A}(F)$, one knows that $\varphi_\mathcal{A}(uv) \in J_\mathcal{A}(F)$, by the $\mathcal{J}$-minimality of $J_\mathcal{A}(F)$. Therefore the set $Quv = Iv$ has the same cardinality as $Qu = I$. On the other hand, since $uv \in A^*u$, we have $Iv = Quv \subseteq Qu = I$. Since these are finite sets, we obtain $Iv = I$. ∎

The following result provides additional motivation for looking at free profinite monoids in the study of recognizable codes.

Proposition 8.5.9 *Let $\mathcal{A} = (Q, i, T)$ be a finite deterministic automaton with transition morphism $\varphi : A^* \rightarrow M(\mathcal{A})$, and consider the unique continuous morphism $\hat{\varphi} : \widehat{A^*} \rightarrow M(\mathcal{A})$ extending φ. Suppose that the set $F \subseteq A^*$ is uniformly recurrent. The following conditions are equivalent.*

(i) *For every $u \in F$ such that $\varphi(u) \in J_\mathcal{A}(F)$, and for the set $I = \text{Im}_\mathcal{A}(u)$, the group $\pi_I(\text{Stab}(I))$ is generated by $\pi_I(\mathcal{R}_F(u))$.*

(ii) For some maximal subgroup K of $J(F)$, the group $\hat{\varphi}(K)$ is a maximal subgroup of $J_A(F)$.

(iii) For every maximal subgroup K of $J(F)$, the group $\hat{\varphi}(K)$ is a maximal subgroup of $J_A(F)$.

Proof (ii) $\Leftrightarrow$ (iii) Immediate by Exercise 3.66.

(iii) $\Rightarrow$ (i) Set $\pi = \pi_I$. As $u \in F$, there is an idempotent f in $J(F)$ having u as a suffix (cf. Exercise 5.28). Let K be the maximal subgroup of $J(F)$ containing f, and let V be the maximal subgroup $\hat{\varphi}(K)$ of $M(\mathcal{A})$. Take $e = \hat{\varphi}(f)$. Note that, since $e \in J_A(F)$, the elements e and $\varphi(u)$ are $\mathcal{L}$-equivalent, as u is a suffix of f and $M(\mathcal{A})$ is a stable semigroup (cf. Proposition 3.6.5). We may then consider the morphism $\rho = \rho_I$ as in Lemma 8.5.7. By the definition of ρ, we have

$$\rho \circ \pi(\mathcal{R}_F(u)^*) = \hat{\varphi}(f \cdot \mathcal{R}_F(u)^*) = \hat{\varphi}(f \cdot \overline{\mathcal{R}_F(u)^*}), \tag{8.8}$$

the last equality holding by continuity of $\hat{\varphi}$. Let $z \in K$. Note that u is a suffix of $z = zf$. Consider a factorization $f = wu$. We have $z = wuz$, whence uz belongs to the factorial set $\overline{F}$. From $uz \in \overline{F} \cap \widehat{A^*}u$ we obtain $z \in \overline{\mathcal{R}_F(u)^*}$ (cf. Exercise 6.3). We have therefore shown the inclusion $K \subseteq \overline{\mathcal{R}_F(u)^*}$. Observing that $fK = K$, we deduce from (8.8) that $\rho \circ \pi(\mathcal{R}_F(u)^*) \supseteq \hat{\varphi}(K) = V$, and so $\pi(\mathcal{R}_F(u))$ generates $\rho^{-1}(V) = \pi(\mathrm{Stab}(I))$.

(i) $\Rightarrow$ (iii) Let K be a maximal subgroup of $J(F)$, and let f be its neutral element. Consider a defining pair $(\ell_n, r_n)_{n \geqslant 1}$ of K, and the associated sequence of central return sets $(\mathcal{R}_n)_n$, as introduced in Sect. 6.6. Recall also that the inclusion $\mathcal{R}_{n+1}^+ \subseteq \mathcal{R}_n^+$ holds for all $n \geqslant 1$ (Lemma 6.6.1). Since $M(\mathcal{A})$ is finite, by taking a subsequence of $(\ell_n, r_n)_{n \geqslant 1}$ we may in fact suppose that the sets $\hat{\varphi}(\overline{\mathcal{R}_n^+} \setminus A^*)$ are the same set, which we denote by T. Given $s \in T$, choose for each $k \geqslant 1$ a pseudoword $w_k \in \overline{\mathcal{R}_k^+} \setminus A^*$ such that $s = \hat{\varphi}(w_k)$. Let w be an accumulation point of (w_k). Since $w_n \in \overline{\mathcal{R}_k^+} \setminus A^*$ for every $n \geqslant k$, we conclude that $w \in \bigcap_{k \geqslant 1} \overline{\mathcal{R}_k^+} \setminus A^*$. By Proposition 6.6.3, this means that $w \in K$. Note that $s = \hat{\varphi}(w)$, by continuity of $\hat{\varphi}$. We have therefore shown that

$$T = \hat{\varphi}(K) = \hat{\varphi}(\overline{\mathcal{R}_n^+} \setminus A^*) \tag{8.9}$$

for every $n \geqslant 1$.

By taking a subsequence, one may also suppose that $(\ell_n, r_n)_{n \geqslant 1}$ converges in $\widehat{A^*} \times \widehat{A^*}$ to (ℓ, r). Since ℓ_n is a suffix of f and r_n is a prefix of f, we know that the pseudowords ℓ and r are respectively an infinite suffix and an infinite prefix of f, whence

$$\ell \,\mathcal{R}\, f \,\mathcal{L}\, r \tag{8.10}$$

by Proposition 5.6.14 and because $\widehat{A^*}$ is a stable semigroup. Using the continuity of $\hat{\varphi}$, we take once more a subsequence of $(\ell_n, r_n)_{n \geqslant 1}$ in order to become reduced to the case where $\hat{\varphi}(\ell_n) = \hat{\varphi}(\ell)$ and $\hat{\varphi}(r_n) = \hat{\varphi}(r)$ for all $n \geqslant 1$. Take $p = \ell_1$, $q = r_1$, and let $z \in \widehat{A^*}$ be such that $f = qzp$. Consider also the word $u = pq$. Note that u is a factor of $f = (qzp)^2$, thus $u \in F$. From (8.10) we obtain

$$\hat{\varphi}(p) \, \mathcal{R} \, \hat{\varphi}(f) \, \mathcal{L} \, \hat{\varphi}(q).$$

As we also have $\hat{\varphi}(f) \in J_A(F)$ (by Exercise 8.12), and $f \leqslant_{\mathcal{J}} u \leqslant_{\mathcal{J}} p$, we see that

$$\{\hat{\varphi}(p), \hat{\varphi}(q), \hat{\varphi}(u), \hat{\varphi}(f)\} \subseteq J_A(F).$$

The equality $\mathcal{R}_1 = q \, \mathcal{R}_F(u)q^{-1}$ entails $\overline{\mathcal{R}_1^+} = q\overline{\mathcal{R}_F(u)^+}q^{-1}$ (cf. Exercise 6.2). Letting $X = \overline{\mathcal{R}_F(u)^+} \setminus A^*$, we then have

$$\overline{\mathcal{R}_1^+} \setminus A^* = qXq^{-1}. \tag{8.11}$$

Therefore, the following chain of equalities holds:

$$\begin{aligned}
\hat{\varphi}(K) &= \hat{\varphi}(fKf) \\
&= \hat{\varphi}(fqXq^{-1}f) \qquad \text{(by (8.9) and (8.11))} \\
&= \hat{\varphi}(fqXq^{-1}qzp) \\
&= \hat{\varphi}(qzpqXzp) \\
&= \hat{\varphi}(q \cdot zuX \cdot zp) \\
&= \hat{\varphi}(q) \cdot \hat{\varphi}(zuX) \cdot \hat{\varphi}(zp). \tag{8.12}
\end{aligned}$$

Note that $zu = zpq = q^{-1}(qzp)q = q^{-1}fq$ is an idempotent which is $\mathcal{J}$-equivalent to f. As $zu \in J(F)$, we deduce that the idempotent $e = \hat{\varphi}(zu)$ is $\mathcal{L}$-equivalent to $\varphi(u)$ (cf. Exercise 8.12). Denote by V the maximal subgroup of $M(\mathcal{A})$ containing e. Let $I = \mathrm{Im}_A(u)$. Set $\pi = \pi_I$ and consider the isomorphism $\rho = \rho_I$ from $\pi(\mathrm{Stab}(I))$ to V, as in Lemma 8.5.7. By Lemma 8.5.8, the inclusion $\mathcal{R}_F(u)^* \subseteq \mathrm{Stab}(I)$ holds, and by hypothesis, one has

$$\pi(\mathrm{Stab}(I)) = \pi(\mathcal{R}_F(u)^+),$$

thus

$$V = \rho(\pi(\mathcal{R}_F(u)^+)) = e\varphi(\mathcal{R}_F(u)^+) = \hat{\varphi}(zu\overline{\mathcal{R}_F(u)^+}) \tag{8.13}$$

where the last equality holds by continuity of $\hat{\varphi}$. Since $uzu \in \overline{F}$, we have $zu \in X$ (cf. Exercise 6.3), and so

$$zu\overline{\mathcal{R}_F(u)^+} = zu(zu\overline{\mathcal{R}_F(u)^+}) \subseteq zu(\overline{\mathcal{R}_F(u)^+} \setminus A^*) \subseteq zu\overline{\mathcal{R}_F(u)^+},$$

which yields $zu\overline{\mathcal{R}_F(u)^+} = zuX$. Therefore, from (8.13) we get $V = \hat{\varphi}(zuX)$, and (8.12) gives

$$\hat{\varphi}(K) = \hat{\varphi}(q) \cdot V \cdot \hat{\varphi}(zp). \tag{8.14}$$

Let $x = \hat{\varphi}(q)$ and $y = \hat{\varphi}(zp)$. Since xy is the idempotent of the maximal subgroup of $J_A(F)$ containing $\hat{\varphi}(K)$, and yx is the idempotent of the maximal subgroup V, it follows from (8.14) and from the arguments used in the proof of Proposition 3.6.11 (alternatively, from Green's Lemma, cf. Exercise 3.52) that $\hat{\varphi}(K)$ is a maximal subgroup. ∎

8.5.3   The Minimal Automaton of a Recognizable Bifix Code

We now focus on the case of a recognizable bifix code $X \subseteq A^+$ and its minimal automaton $\mathcal{A}(X^*)$. We denote by h_X the syntactic morphism of X^*, that is, the transition morphism of $\mathcal{A}(X^*)$, mapping A^* onto the syntactic monoid $M(X^*)$.

Let $F \subseteq A^*$ be a recurrent set. The F-minimum $\mathcal{J}$-class of $M(X^*)$, established by Proposition 8.5.3, is denoted simply by $J_X(F)$. The Schützenberger group of $J_X(F)$ is denoted by $G_X(F)$, and is called the F-group of X. When $F = A^*$ we use the simplified notation J_X and G_X for $J_X(F)$ and $G_X(F)$, respectively, and say that G_X is the group of X. Note that J_X is just the minimum ideal of the syntactic monoid $M(X^*)$.

Also, we may use the notation rank_X and Im_X for $\text{rank}_{\mathcal{A}(X^*)}$ and $\text{Im}_{\mathcal{A}(X^*)}$, respectively.

In the next proposition we use the fact that in the minimal automaton $\mathcal{A}(X^*)$ of a prefix code, the initial state is the unique terminal state (Proposition 8.2.11).

Proposition 8.5.10 *Let $F \subseteq A^*$ be a recurrent set, and let $X \subseteq A^+$ be a recognizable F-complete bifix code with minimal automaton $\mathcal{A} = (Q, i, i)$. Then the equality*

$$d_X(F) = \text{rank}_X(F)$$

holds. Moreover, for each $w \in F$, if P_w is the set of suffixes of w with no prefix in X, then the mapping

$$p \in P_w \mapsto i \cdot p \in Q \tag{8.15}$$

is injective. Finally, if the word $w \in F$ satisfies $\delta_X(w) = d_X(F)$, then

$$\text{Im}_X(w) = \{i \cdot p \mid p \in P_w\} \tag{8.16}$$

and $h_X(w) \in J_X(F)$.

Proof For each $w \in F \subseteq A^*$, set $I_w = \{i \cdot p \mid p \in P_w\}$.

Let $p, p' \in P_w$. Then p, p' are suffix-comparable. Assume that $p' = tp$ for some $t \in A^*$. Suppose that $i \cdot p = i \cdot p'$. Since X is right F-complete, we may take $z \in A^*$ such that $pz \in X^*$. Then, since the initial state i is also the unique terminal state of $\mathcal{A}$, one has $i = i \cdot pz = i \cdot p'z$, entailing $tpz = p'z \in X^*$. Since X^* is left unitary, it follows that $t \in X^*$. As $p' = tp \notin XA^*$, we deduce that $t = \varepsilon$ and $p = p'$. Therefore, the mapping (8.15) is injective, and so

$$\mathrm{Card}(I_w) = \mathrm{Card}(P_w) = \delta_X(w), \tag{8.17}$$

where the last equality holds by the dual of Proposition 8.3.7.

Let $p \in P_w$. Then w has a parse of the form (s, x, p). As X is left F-complete, there is $z \in A^*$ with $zs \in X^*$. Therefore, $i \cdot p = (i \cdot zsx) \cdot p = (i \cdot z) \cdot w$, yielding $I_w \subseteq \mathrm{Im}_X(w)$, whence $\delta_X(w) = \mathrm{Card}(I_w) \leqslant \mathrm{rank}_X(w)$.

Suppose now that $w \in F$ satisfies $\delta_X(w) = d_X(F)$. In order to show (8.16), it remains to prove $I_w \supseteq \mathrm{Im}_X(w)$. Let $q \in \mathrm{Im}_X(w)$, and let q' be a state such that $q = q' \cdot w$. Because $\mathcal{A}$ is trim, there are $u, v \in A^*$ with $i \cdot u = q'$ and $q \cdot v = i$. Therefore, we have $i \cdot uwv = i$, thus $uwv \in X^*$. Because w is not an internal factor of X by Theorem 8.3.16, it follows that there is a parse $w = (s, x, p)$ of w such that $us, pv \in X^*$. From $usx \in X^*$, we deduce that $i \cdot p = (i \cdot usx) \cdot p = (i \cdot u) \cdot w = q' \cdot w = q$. Since $p \in P_w$, we conclude that $q \in I_w$, thus establishing (8.16). And since the mapping (8.15) is injective, we deduce that $\mathrm{rank}_X(w) = \mathrm{Card}(P_w) = d_X(F)$.

Together with a word $w \in F$ such that $\delta_X(w) = d_X(F)$, consider also $w' \in F$ such that $\mathrm{rank}_X(w') = \mathrm{rank}_X(F)$. Because F is recurrent, there is $t \in A^*$ with $wtw' \in F$. Then $\delta_X(w) \leqslant \delta_X(wtw')$ holds by Lemma 8.3.9, so that $\delta_X(wtw') = d_X(F)$ by the maximality of $d_X(F)$. Accordingly to what was proved in the previous paragraph, it follows that $\mathrm{rank}_X(wtw') = d_X(F)$. On the other hand, $\mathrm{rank}_X(wtw') \leqslant \mathrm{rank}_X(w') = \mathrm{rank}_X(F)$, and so we have $\mathrm{rank}_X(wtw') = \mathrm{rank}_X(F)$ by the minimality of $\mathrm{rank}_X(F)$. Putting all together, we obtain $\mathrm{rank}_X(F) = d_X(F)$. This shows that $h_X(w) \in J_X(F)$. ∎

It is convenient to highlight the following immediate consequence of the equality $\mathrm{rank}_X(F) = d_X(F)$ in Proposition 8.5.10 and of the characterization of $J_X(F)$, stemming from Proposition 8.5.10, as being the $\mathcal{J}$-class of $M(X^*)$ containing the elements of F-minimum rank.

Corollary 8.5.11 *Let $F \subseteq A^*$ be a recurrent set, and let $X \subseteq A^+$ be a recognizable F-complete bifix code. Then the degree of the permutation group $G_X(F)$ is $d_X(F)$. In particular, if $Z \subseteq A^+$ is a recognizable complete bifix code, then G_Z has degree d_Z.*

We have seen in Proposition 8.5.10 that if $w \in F$ satisfies $\delta_X(w) = d_X(F)$, then $h_X(w) \in J_X(F)$, whenever X is a recognizable F-complete bifix code, where F is a recurrent set. The converse does not hold.

Example 8.5.12 Consider the group code $Z = a^* \cup ba^*b$, having as syntactic monoid the group $G_Z = \mathbb{Z}/2\mathbb{Z}$. Then we have $\delta_Z(a) = 1 < \mathrm{rank}_A(a) = 2$.

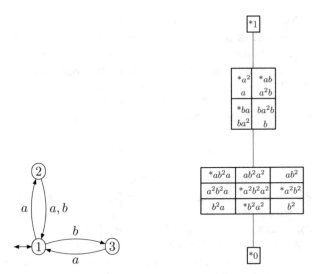

Fig. 8.7 Minimal automaton of $\{aa, ab, ba\}^*$ and eggbox diagram of its transition monoid

Next we see an example where $J_X(F)$ is not the minimum ideal of $M(X^*)$.

Example 8.5.13 Let F be the Fibonacci set, and consider the F-complete bifix code $X = \{aa, ab, ba\} = F \cap A^2$. The minimal automaton and the eggbox diagram of the syntactic monoid of X^* are shown in Fig. 8.7. Note that $\delta_X(a) = 2 = d_X(F)$, and so the F-minimum $\mathcal{J}$-class of $M(X^*)$ is the $\mathcal{J}$-class of a. Therefore, the F-group of X is the cyclic group $\mathbb{Z}/2\mathbb{Z}$.

As a manifestation of the fact that $G_X(F)$ has degree two, observe, for example, that the maximal subgroup of ab in $M(X^*)$, consisting of the transformations $ab = [1, 3, 3]$ and $a^2b = [3, 1, 1]$, acts by permutations on $\mathrm{Im}_X(ab) = \{1, 3\}$.

8.6 The Charged Code Theorem

In this section we introduce the notion of charged code, and use it to formulate in a concise manner a result that gives necessary and sufficient conditions for the groups G_Z and $G_{Z \cap F}(F)$ to be equivalent permutations groups, when $F \subseteq A^*$ is uniformly recurrent and $Z \subseteq A^+$ is a recognizable bifix code. This result, which we call the Charged Code Theorem, applies to tree sets and group codes, but also to many other cases, as we shall see in examples.

8.6.1 Charged Codes

Let Z be a recognizable bifix code $Z \subseteq A^+$. Consider the unique continuous morphism $\hat{h}_Z$ extending the syntactical morphism $h_Z : A^* \to M(Z^*)$ to $\widehat{A^*}$. Let $F \subseteq A^*$ be a recurrent set. We say that Z is F-*charged* if for some maximal subgroup K of $J(F)$, the image of K by the syntactic morphism $\hat{h}_Z$ is a maximal subgroup of the minimum ideal J_Z. Equivalently, Z is F-charged if for every maximal subgroup K of $J(F)$, the image $\hat{h}_Z(K)$ is a maximal subgroup of J_Z, cf. Exercise 3.66.

We wish to use the notion of charged code to obtain results that hold beyond tree sets. Therefore, we begin to observe the following fact.

Proposition 8.6.1 *If $F \subseteq A^*$ is a uniformly recurrent connected set, then every group code $Z \subseteq A^+$ is F-charged.*

Proof Take any maximal subgroup K of $J(F)$. By the universal property of $\widehat{FG(A)}$, there is a unique continuous morphism $\bar{h}_Z$ from $\widehat{FG(A)}$ onto the finite group G_Z such that $\bar{h}_Z(a) = \hat{h}_Z(a)$ for every $a \in A$. This is illustrated by the commutative Diagram (8.18), where p_G is the canonical projection from $\widehat{A^*}$ onto $\widehat{FG(A)}$.

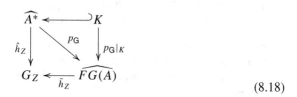

$$(8.18)$$

Because F is a uniformly recurrent connected set, we have $p_G(K) = \widehat{FG(A)}$, by Corollary 7.3.3. As $\hat{h}_Z = \bar{h}_Z \circ p_G$, it follows that $\hat{h}_Z(K) = \bar{h}_Z(\widehat{FG(A)}) = G_Z$. ∎

The proof we give for the following proposition is inspired by the proof of Proposition 8.6.1.

Proposition 8.6.2 *Let F be a uniformly recurrent connected set. Let H be a subgroup of finite index d of the free group $FG(A)$. Let Z be the group code such that $H \cap A^* = Z^*$. Then $X = Z \cap F$ is a finite F-complete bifix code that generates H.*

Proof We already know that X is a finite F-complete bifix code by Corollary 8.3.22.

Let N be the subgroup of $FG(A)$ generated by X. Let $Hp_1, \ldots, Hp_d$ be the right cosets of H. Take a maximal subgroup K of $J(F)$. Consider the commutative Diagram (8.18) from the proof of Proposition 8.6.1, the notation $\bar{h}_Z$ having here the same meaning as there. Let $u \in FG(A)$. Let $p \in \{p_1, \ldots, p_d\}$ be such that $u \in Hp$. By Corollary 7.3.3, there are $v, q \in K$ such that $p_G(v) = u$ and $p_G(q) = p$. Then, the pseudoword $w = vq^{\omega-1}$ is an element of the group K such that $p_G(w) = up^{-1} \in H$. Since H has finite index in $FG(A)$, the sets H and $H \cap A^*$ have the same

242   8 Groups of Bifix Codes

topological closure in $\widehat{FG(A)}$ (cf. Exercise 4.27), and so we have $p_G(w) \in p_G(\overline{Z^*})$. Applying $\bar{h}_Z$, we deduce that $\hat{h}_Z(w) \in \hat{h}_Z(\overline{Z^*})$. Since h_Z recognizes Z^*, it follows that $w \in \overline{Z^*}$. As $w \in \overline{F}$ and $\overline{Z^*} \cap \overline{F} = \overline{X^*}$ (cf. Exercise 5.18), we deduce that $w \in \overline{X^*}$, thus $up^{-1} = p_G(w)$ is in the topological closure of N in $\widehat{FG(A)}$. Since N is finitely generated, we have $up^{-1} \in N$ by Corollary 4.6.4, that is $u \in Np$. This shows that $FG(A) = Np_1 \cup \ldots \cup Np_d$, and so N has index at most d. Since $N \subseteq H$, we conclude that N has index d and that $N = H$. ∎

With Proposition 8.6.2 at our disposal, we are only a very small step away of completing a new proof of the Finite Index Basis Theorem (Theorem 8.4.2). In Exercise 8.18 we propose the completion of this new proof.[1]

8.6.2 The Charged Code Theorem: Statement and Examples

It is convenient to consider the following variation on the notion of F-charged code. For a recurrent set $F \subseteq A^*$, let us say that a recognizable code $X \subseteq A^+$ is *weakly F-charged* when for some (equivalently, for every) maximal subgroup K of $J(F)$, the image of K by the syntactic morphism $\hat{h}_X$ is a maximal subgroup of the F-minimum ideal $J_X(F)$. Note that an F-charged code is weakly F-charged.

We now state what we call the Charged Code Theorem.

Theorem 8.6.3 (Charged Code Theorem) *Let $F \subseteq A^*$ be a uniformly recurrent set and let $Z \subseteq A^+$ be a recognizable complete bifix code. Then the intersection $X = Z \cap F$ is a finite F-complete bifix code, and the following conditions are equivalent:*

(i) *Z is F-charged;*
(ii) *the permutation groups $G_X(F)$ and G_Z are equivalent, and X is weakly F-charged;*
(iii) *$d_X(F) = d_Z$, $\mathrm{Card}(G_X(F)) = \mathrm{Card}(G_Z)$, and X is weakly F-charged.*

Due to the abundance of technicalities that it involves, the proof of Theorem 8.6.3 is deferred to Sect. 8.8.

Remark 8.6.4 Under the hypothesis of Theorem 8.6.3, we already know that X is a finite F-complete code such that $d_F(X) \leqslant d$, by Corollary 8.3.21. The core of Theorem 8.6.3 is that $G_X(F)$ and G_Z are equivalent permutation groups when Z is F-charged. In view of Corollary 8.5.11, the equivalence between of $G_X(F)$ and G_Z implies the equality $d_F(X) = d_Z$.

---

[1]This new proof is less elementary than that in Sect. 8.4. The Return Theorem (Theorem 6.5.5) is explicitly used in the proof deduced in Sect. 8.4. Proposition 8.6.1 depends on Corollary 7.3.3, which depends on the generalization of the Return Theorem to connected sets (Theorem 6.5.7).

Remark 8.6.5 We are mostly interested in uniformly recurrent sets, but in fact Theorem 8.6.3 has a generalization where F may be recurrent. The proof of this generalization is more complicated. See the Notes Section.

For the sake of clarity and for an easy reference, we underline the following direct consequence of the Charged Code Theorem.

Corollary 8.6.6 *Let* $F \subseteq A^*$ *be a uniformly recurrent connected set and let* $Z \subseteq A^+$ *be a group code. Set* $X = Z \cap F$. *Then* $d_X(F) = d_Z$, *and* $G_X(F)$ *is equivalent to* G_Z.

Proof Immediate in view of Proposition 8.6.1 and Theorem 8.6.3. ∎

Example 8.6.7 Let F be the Fibonacci set, and consider the group code $Z = A^2$. We saw in Example 8.5.13 that the F-group of $X = \{aa, ab, ba\} = Z \cap F$ is the cyclic group $\mathbb{Z}/2\mathbb{Z}$. This can be seen directly from Corollary 8.6.6, as one clearly has $G_Z = \mathbb{Z}/2\mathbb{Z}$.

The next statement is similar to Corollary 8.6.6, but covers other types of examples.

Corollary 8.6.8 *Let* φ *be a primitive nonperiodic substitution over the alphabet* A. *Let* $Z \subseteq A^+$ *be a group code. Set* $F = F(\varphi)$ *and* $X = Z \cap F$. *Consider the syntactic morphism* $h : A^* \to G_Z$. *If* φ *has finite* h-*order, then* $d_X(F) = d_Z$ *and* $G_X(F)$ *is equivalent to* G_Z.

Proof By Corollary 7.4.9, the code Z is F-charged, and so we are done by Theorem 8.6.3. ∎

Let us see a concrete example fitting Corollary 8.6.8, concerning a uniformly recurrent set which is neither connected nor neutral.

Example 8.6.9 Let $A = \{a, b\}$ and consider the nonperiodic primitive morphism $\varphi : a \mapsto ab$, $b \mapsto a^3 b$. Let $F = F(\varphi)$. As seen in Examples 6.5.3 and 6.3.3, the set F is neither connected nor neutral. Consider the morphism $h : A^* \to A_5$ from A^* onto the alternating group of degree 5 defined by $h : a \mapsto (123)$, $b \mapsto (345)$. We view h as the transition morphism of an automaton $\mathcal{A}$ on the states $\{1, 2, 3, 4, 5\}$. The automaton $\mathcal{A}$ is a group automaton, in which we prescribe 1 to be the initial state. Let Z be the group code generating the submonoid of A^* consisting of the words stabilizing 1 in $\mathcal{A}$. Remember that $\mathcal{A}$ is the minimal automaton of Z^*, as $\mathcal{A}$ is a group automaton, and so h is the syntactic morphism of Z^*. We have seen in Example 7.4.11 that φ has h-order 12 and thus, by Corollary 7.4.9 the code Z is F-charged.

The intersection $X = Z \cap F$ is an F-maximal bifix code (cf. Corollary 8.3.21) which has eight elements. It is represented in Fig. 8.8 with the states of the minimal automaton indicated on its nodes. One way of verifying that the code in Fig. 8.8 is indeed $Z \cap F$ consists in checking that it is contained in $Z \cap F$ and that it is an F-maximal prefix code, which may be easily done (cf. Propositions 8.2.7 and 8.2.14).

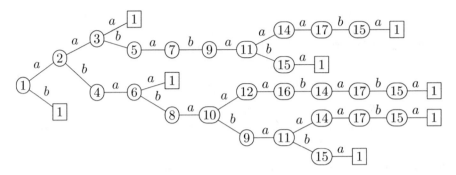

Fig. 8.8 The bifix code X

In agreement with the Charged Code Theorem (Theorem 8.6.3), the F-degree of X is 5. The word a^3 has rank 5 in $\mathcal{A}$, and so the syntactic image of a^3 belongs to $J_X(F)$, by Proposition 8.5.10. The set of return words of a^3 is $\mathcal{R}_F(a^3) = \{babaaa, babababaaa\}$. The corresponding permutations defined on the image $\{1, 2, 3, 16, 17\}$ of a^3 are respectively

$$(1, 2, 16, 3, 17) \quad (1, 17, 16, 2, 3)$$

which generate A_5, as expected by the Charged Code Theorem (ensuring that $G_X(F)$ is weakly charged) and by Proposition 8.5.9.

In the next example, we see a weakly charged code for which the Charged Code Theorem is not satisfied.

Example 8.6.10 Let F be the Thue-Morse set and consider the bifix code

$$X = \{ab, ba, a^2b^2, b^2a^2, a^2bab^2, b^2aba^2\}$$

whose minimal automaton $\mathcal{A}$ is represented on the left of Fig. 8.9, with the initial state being the node labeled 1. As seen in Example 8.2.15, X is an F-complete bifix code.

In $\mathcal{A}$, the word a^2 has rank 3 and image $I = \{1, 2, 4\}$. The action on the images accessible from I is given in Fig. 8.10.

All words with image $\{1, 2, 4\}$ end with a^2. The paths returning for the first time to $\{1, 2, 4\}$ are labeled by the set $\mathcal{R}_F(a^2) = \{b^2a^2, bab^2aba^2, bab^2a^2, b^2aba^2\}$. Then $\mathrm{rank}_A(F) = 3$ by Theorem 8.5.1, which means that $d_X(F) = 3$ by Proposition 8.5.10. Moreover, each of the words of $\mathcal{R}_F(a^2)$ defines the trivial permutation on the set $\{1, 2, 4\}$. Since the stabilizer of $\{1, 2, 4\}$ is generated by $\mathcal{R}_F(a^2)$, we conclude that $G_X(F)$ is trivial.

Consider the group automaton $\mathcal{B}$ represented on the right of Fig. 8.9, corresponding to the map sending each word to the difference modulo 3 of the number of occurrences of a and b. Take one of its states as initial state. Let Z be the group

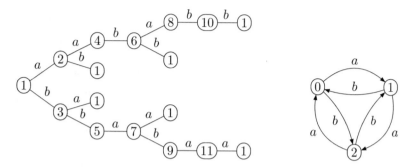

Fig. 8.9 An F-complete bifix code of F-degree 3 and trivial F-group

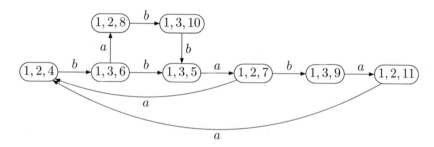

Fig. 8.10 The action on the minimal images

code generating the submonoid of A^* recognized by $\mathcal{B}$. Note that $X \subseteq Z \cap F$. Since X is F-maximal by Proposition 8.3.14, it follows that $X = Z \cap F$. One has $d_Z = 3 = d_X(F)$, but $G_Z = \mathbb{Z}/3\mathbb{Z}$ is not isomorphic with $G_X(F)$.

We close this section by presenting an example not involving a group code.

Example 8.6.11 Consider the complete bifix code $Z = \{aa, ab, ba\} \cup b^2(a^+b)^*b$. In Fig. 8.11 we find the minimal automaton of Z^* together with the eggbox diagram of its transition monoid. Observe that the rank of the minimum ideal J_Z of $M(Z^*)$ is 3, whence $d_Z = 3$ by Proposition 8.5.10.

Take any uniformly recurrent connected set F whose set of letters is $A = \{a, b\}$. Let K be a maximal subgroup of $J(F)$. Let ψ be the unique continuous endomorphism of $\widehat{A^*}$ such that $\psi(a) = a$ and $\psi(b) = ab^2a^2$. Then the set $F' = \psi(F)$ is clearly a uniformly recurrent subset of A^*. We claim that Z is F'-charged.

To show the claim, first observe that $\psi(K) \subseteq \overline{F} \setminus A^*$, thus $\psi(K) \subseteq K'$ for some maximal subgroup K' of $J(F')$, by Proposition 5.6.14. Consider the maximal subgroup H of J_Z containing $h_Z(a)$. It is a group of permutations on the image of a in $\mathcal{A}(Z^*)$, the set $\{1, 2, 5\}$. As $h_Z(a) = [2, 1, 1, 5, 5]$ has order 2 and $h_Z(ab^2a^2) = [2, 5, 5, 1, 1]$ has order 3, the group H is generated by $\{h_Z(a), h_Z(ab^2a^2)\}$, and it is isomorphic to S_3. Therefore, for $\lambda = h_Z \circ \psi$, one has $\lambda(\widehat{A^*}) = H$. Let $\bar{\lambda}$ be

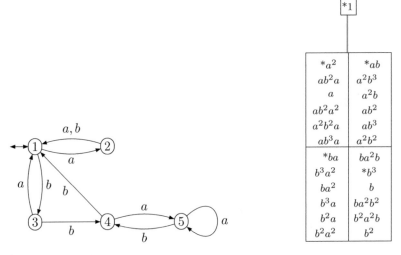

Fig. 8.11 Minimal automaton $\mathcal{A}(Z^*)$ and eggbox diagram of $M(Z^*)$, for $Z = \{aa, ab, ba\} \cup b^2(a^+b)^*b$

the unique continuous morphism from $\widehat{FG(A)}$ to H such that $\bar{\lambda} \circ p_G = \lambda$, where p_G is the natural projection $\widehat{A^*} \to \widehat{FG(A)}$. Since, by Corollary 7.3.3, the equality $p_G(K) = p_G(\widehat{A^*})$ holds, we have

$$\hat{h}_Z(K') \supseteq \hat{h}_Z \circ \psi(K) = \lambda(K) = \bar{\lambda} \circ p_G(K) = \bar{\lambda} \circ p_G(\widehat{A^*}) = \lambda(\widehat{A^*}) = H,$$

thus $\hat{h}_Z(K') = H$, establishing the claim.

Let $X = Z \cap F'$. By the Charged Code Theorem (Theorem 8.6.3), we know that $d_{F'}(X) = 3$ and that $G_{F'}(X)$ is isomorphic to S_3. As a more concrete example, if F is the Fibonacci set, then $X = \{aa, ab, ba, b^2a^4b^2, b^2a^5b^2\}$. For example, $b^2a^4b^2 \in X$ has three parses with respect to X, namely $(1, b^2a^4b^2, 1)$, (b^2, a^4, b^2) and (b, ba^4b, b).

8.7 Bifix Codes in the Free Profinite Monoid

A submonoid N of a monoid M is called *stable in M* if for any $u, v, w \in M$, whenever $u, vw, uv, w \in N$ then $v \in N$ (this should not be confused with the notion of a stable semigroup introduced in Chap. 3). Recall that N is called *right unitary* if for every $u, v \in M$, $u, uv \in N$ implies $v \in N$.

A submonoid of A^* is stable if and only if it is generated by a code (as seen in Exercise 5.12) and it is right unitary if and only if it is generated by a prefix code (as seen in Exercise 8.2).

The following statement extends these notions to pseudowords.

Proposition 8.7.1 *Let N be a recognizable submonoid of A^*. If N is stable in A^* (resp. right unitary), then the submonoid $\overline{N}$ of $\widehat{A^*}$ is stable in $\widehat{A^*}$ (resp. right unitary).*

Proof The set $\overline{N}$ is clearly a submonoid of $\widehat{A^*}$ (cf. Exercise 3.35). Assume that N is stable in A^*. Let $u, v, w \in \widehat{A^*}$ be such that $uv, w, u, vw \in \overline{N}$. Let (u_n), (v_n) and (w_n) be sequences of words converging to u, v and w respectively with $u_n, w_n \in N$. By Theorem 4.4.9, the closure $\overline{N}$ of the submonoid N is open in $\widehat{A^*}$ and $N = \overline{N} \cap A^*$. Therefore, for large enough n we have $u_n v_n, v_n w_n \in N$. Since N is stable, this implies $v_n \in N$ for large enough n, which implies $v \in \overline{N}$. Thus $\overline{N}$ is stable in $\widehat{A^*}$. The proof when N is right unitary is similar. ∎

When X is a prefix code, every word w may be written in a unique way $w = xp$ with $x \in X^*$ and $p \in A^* \setminus XA^*$, a word without any prefix in X. When X is a maximal prefix code, a word which has no prefix in X is a proper prefix of a word in X and thus every word w may be written in a unique way $w = xp$ with $x \in X^*$ and p a proper prefix of X. This extends to pseudowords as follows.

Proposition 8.7.2 *Consider a subset F of A^*. Let X be a finite right F-complete prefix code, and let P be the set of proper prefixes of X. Every pseudoword w of $\overline{F}$ has a unique factorization $w = xp$ with $x \in \overline{X^*}$ and $p \in P$.*

Proof Let (w_n) be a sequence of words of F converging to w. Since X is right F-complete, for each n we have some factorization $w_n = x_n p_n$ with $x_n \in X^*$ and $p_n \in P$. Taking a subsequence of (x_n, p_n), we may assume that the sequences (x_n) and (p_n) converge to $x \in \overline{X^*}$ and $p \in \overline{P}$. Since X is finite, P is finite and thus $\overline{P} = P$. This proves the existence of a factorization of the form described in the statement. To establish uniqueness, consider $xp = x'p'$ with $x, x' \in \overline{X^*}$ and $p, p' \in P$. Then, as $\widehat{A^*}$ is equidivisible (Proposition 4.4.18), and assuming that $|p| \geqslant |p'|$, we have $p = tp'$ and $xt = x'$ for some $t \in A^*$. Since $\overline{X^*}$ is right unitary (cf. Proposition 8.7.1), we deduce that $t \in X^*$ and thus $t = \varepsilon$. ∎

Corollary 8.7.3 *For a subset F of A^*, let X be a finite right F-complete prefix code. If a pseudoword of $\overline{F}$ has no prefix in X, then it is a proper prefix of X.*

Proof Since X is finite, we have $\overline{X^+} = X \cdot \overline{X^*}$ (cf. Exercise 3.35). Therefore, if $x \in \overline{X^*}$ has no prefix in X then $x = \varepsilon$. It suffices to combine this fact with Proposition 8.7.2. ∎

A *parse* of a pseudoword $w \in \widehat{A^*}$ with respect to X is a triple (v, x, u) with $w = vxu$ such that v has no suffix in X, $x \in \overline{X^*}$ and u has no prefix in X. We use the notation $\delta_X(w)$ for the number of parses of w with respect to X.

The following provides a generalization of Proposition 8.3.7 to pseudowords, in some situations.

Proposition 8.7.4 *Consider a factorial subset* F *of* A^*. *Let* X *be a finite* F-*complete bifix code. For every pseudoword* $w \in \overline{F}$, *the mappings*

$$(s, x, p) \mapsto s \qquad and \qquad (s, x, p) \mapsto p$$

are bijections from the set of parses of w, *with respect to* X, *to the set of prefixes of* w *which have no suffix in* X, *and to the set of suffixes of* w *which have no prefix in* X, *respectively.*

Proof By symmetry, it is enough to show that the first mapping is bijective. Let s be a prefix of w with no suffix in X. Consider a factorization $w = sv$. Since F is factorial, we have $v \in \overline{F}$, and so, by Proposition 8.7.2, there is a factorization $v = xp$ with $x \in \overline{X^*}$ and p a proper prefix of X. Therefore, the triple (s, x, p) is a parse of w, with respect to X. By the dual of Corollary 8.7.3, the fact that s has no suffix in X yields $s \in A^*$. Hence, if (s, x', p') is also a parse of w with respect to X, we may cancel the word s in the equality $sxp = sx'p'$ (Exercise 4.20), thus obtaining $x = x'$ and $p = p'$ by Proposition 8.7.2. ∎

Observe that, for w and X as in Proposition 8.7.4, every parse (s, x, p) of w with respect to X is such that $s, p \in A^*$, by Corollary 8.7.3.

The next result is not surprising, in view of Theorem 8.3.16, which is used in its proof.

Proposition 8.7.5 *Let* F *be a recurrent set and let* X *be a finite* F-*maximal bifix code. The number of parses of any element of* $\overline{F} \setminus A^*$ *is equal to* $d_X(F)$.

Proof Let $w \in \overline{F} \setminus A^*$. Note that $\delta_X(w)$ is finite by Corollary 8.7.3 and Proposition 8.7.4. Moreover, again by Proposition 8.7.4, the set Q of suffixes of w with no prefix in X has $\delta_X(w)$ elements. Let q be the longest word in Q. As each element of Q is a suffix of q, it follows, once more from Proposition 8.7.4, that $\delta_X(q) \geqslant \mathrm{Card}(Q) = \delta_X(w)$. Since $q \in F$, we deduce that $d_X(F) \geqslant \delta_X(w)$.

Let (w_n) be a sequence of elements of F converging to w. Since, by Theorem 8.3.16, each long enough word of F has $d_X(F)$ parses, and $w \notin A^*$, we may assume that all w_n have $d_X(F)$ of parses. We may then enumerate the parses of w_n as $(s_{n,i}, x_{n,i}, p_{n,i})$, with $i \in \{1, \ldots, d_X(F)\}$. By compactness, we may suppose that $(s_{n,i}, x_{n,i}, p_{n,i}) \to (s_i, x_i, p_i)$, for each i. Note that $x_i \in \overline{X^*}$. On the other hand, since $s_{n,i}$ is a proper suffix of X and $p_{n,i}$ is a proper prefix of X (cf. Corollary 8.7.3), and X is finite, the sequences $(s_{n,i})$ and $(p_{n,i})$ are ultimately constant. This shows that $\delta_X(w) \geqslant d_X(F)$, thus $\delta_X(w) = d_X(F)$ by the preceding paragraph. ∎

8.8   Proof of the Charged Code Theorem

We are now ready to prove the Charged Code Theorem (Theorem 8.6.3), whose statement is repeated below, within a longer, more informative, statement. Inside the proof, we state and prove a couple of intermediate lemmas needed to establish the theorem. As a byproduct of this proof, we deduce at the end of the section a companion result avoiding the use of pseudowords that are not words.

Theorem 8.8.1 *Let $F \subseteq A^*$ be a uniformly recurrent set and let $Z \subseteq A^+$ be a recognizable complete bifix code. Then the intersection $X = Z \cap F$ is a finite F-complete bifix code, and the following conditions are equivalent:*

  (i) Z *is F-charged;*
  (ii) the permutation groups $G_X(F)$ and G_Z are equivalent, and X is weakly F-charged;
 (iii) $d_X(F) = d_Z$, $\mathrm{Card}(G_X(F)) = \mathrm{Card}(G_Z)$, *and X is weakly F-charged.*

If these equivalent conditions hold, and K is a maximal subgroup of F with idempotent e, and we let $K_X = \hat{h}_X(K)$ and $K_Z = \hat{h}_Z(K)$ be the images of K under the continuous syntactic morphisms $\hat{h}_X : A^* \to M(X^*)$ and $\hat{h}_Z : A^* \to M(Z^*)$, then there is an isomorphism $\zeta : K_X \to K_Z$ such that Diagram (8.19) commutes.

$$(8.19)$$

Moreover, if P_e is the set of finite suffixes of e with no prefix in X, and for each $Y \in \{X, Z\}$ the set $\mathrm{Im}_Y(e)$ is the image of $\hat{h}_Y(e)$ in the set of states Q_Y of the minimal automaton $\mathcal{A}(Y^*) = (Q_Y, i_Y, i_Y)$ of Y^*, then

$$\mathrm{Im}_Y(e) = \{i_Y \cdot p \mid p \in P_e\}$$

and we have a bijection $\beta : \mathrm{Im}_X(e) \to \mathrm{Im}_Z(e)$, defined by $\beta(i_X \cdot p) = i_Z \cdot p$, with $p \in P_e$, such that the pair (β, ζ) is an equivalence of permutation groups.

Proof We already know by Corollary 8.3.21 that X is a finite F-complete bifix code of F-degree at most d_Z. Because Z is recognizable, the equality

$$\overline{Z^*} \cap \overline{F} = \overline{X^*}$$

holds (cf. Exercise 5.18). We shall use this equality along the proof.

For $Y \in \{X, Z\}$ and $u \in \widehat{A^*}$, we use the notation $q \cdot u$ for $q \cdot \hat{h}_Y(u)$, and $\mathrm{Im}_Y(u)$ denotes the set $\{q \cdot u \mid q \in Q_Y\}$. Also, one should bear in mind, along the proof, the equivalence

$$i_Y \cdot u = i_Y \Leftrightarrow u \in \overline{Y^*},$$

which holds because $\overline{Y^*} = \hat{h}_Y^{-1}(\hat{h}_Y(Y^*))$, as Y^* is recognized by h_Y.

For each $w \in \widehat{A^*}$, let P_w be the set of suffixes of w which have no prefix in Z. Since $\overline{F}$ is factorial and $X = Z \cap F$, it is clear that $w \in \overline{F}$ implies that P_w is the set of suffixes of w which have no prefix in X. In particular, P_w is finite whenever $w \in \overline{F}$, thanks to Propositions 8.7.4 and 8.7.5.

Take a maximal subgroup K of $J(F)$, and let e be its idempotent. Denoting by $\|V\|$ the maximum length of a word in a finite subset V of A^*, we let $\ell = \|P_e \cup X\|$. For each positive integer n, denote by e_n the suffix of length n of e. Let $(e_{n_k})_{k \geqslant 1}$ be a subsequence of $(e_n)_{n \geqslant 1}$ converging in $\widehat{A^*}$ to a pseudoword e'. Note that $e' \in J(F)$ by Proposition 5.6.14. By continuity of $\hat{h}_X$ and $\hat{h}_Z$, there is an integer $k_0 \geqslant 1$ such that $k \geqslant k_0$ implies $n_k > \ell$ and also the equalities $\hat{h}_X(e') = \hat{h}_X(e_{n_k})$ and $\hat{h}_Z(e') = \hat{h}_Z(e_{n_k})$. Set $w = e_{n_{k_0}}$. We have $e \mathcal{L} e'$ by Proposition 5.6.15, from which it follows that

$$\hat{h}_X(e) \mathcal{L} \hat{h}_X(w) \quad \text{and} \quad \hat{h}_Z(e) \mathcal{L} \hat{h}_Z(w), \tag{8.20}$$

thus

$$\mathrm{Im}_X(e) = \mathrm{Im}_X(w) \quad \text{and} \quad \mathrm{Im}_Z(e) = \mathrm{Im}_Z(w).$$

Moreover, because $P_w \subseteq P_e$ and $|w| > \ell \geqslant \|P_e\|$, we have

$$P_e = P_w.$$

Take $y \in A^*$ such that $\delta_Z(y) = d_Z$. Because the word $z = yw$ satisfies $\delta_Z(z) \geqslant \delta_Z(y)$ by Proposition 8.3.8, we have in fact $\delta_Z(z) = d_Z$ as the number of parses of y with respect to Z is maximum. Moreover, one has $P_e = P_w \subseteq P_z$. Since $|w| > \ell \geqslant \|X\|$, the word w is not an internal factor of X, and so $\delta_X(w) = d_X(F)$ by Theorem 8.3.16. In particular, we have $\hat{h}_X(w) \in J_X(F)$ by Proposition 8.5.10, thus $\hat{h}_X(J(F)) \subseteq J_X(F)$ (an inclusion also seen in Exercise 8.12).

Due to the equalities $P_e = P_w$, $\mathrm{Im}_X(e) = \mathrm{Im}_X(w)$, $\delta_X(w) = d_X(F)$ and $\delta_Z(z) = d_Z$, we have

$$\mathrm{Card}(P_e) = d_X(F) \leqslant d_Z = \mathrm{Card}(P_z), \tag{8.21}$$

by the dual of Proposition 8.3.7 and moreover, by Proposition 8.5.10 applied respectively to the recurrent sets F and A^*, the equalities

$$\mathrm{Im}_X(e) = \{i_X \cdot p \in P_e\}, \qquad \mathrm{Im}_Z(z) = \{i_Z \cdot p \mid p \in P_z\}, \tag{8.22}$$

hold, and the mappings

$$p \in P_e \mapsto i_X \cdot p \in \mathrm{Im}_X(e) \quad \text{and} \quad p \in P_z \mapsto i_Z \cdot p \in \mathrm{Im}_Z(z) \tag{8.23}$$

are bijective. Therefore, thanks to the inclusion $P_e \subseteq P_z$, we know that the mapping $\beta : i_X \cdot p \mapsto i_Z \cdot p$, with $p \in P_e$, is a well-defined injective function from $\mathrm{Im}_X(e)$ into $\mathrm{Im}_Z(z)$.

For later reference in the proof of the theorem, we isolate the following fact.

Lemma 8.8.2 *If $d_X(F) = d_Z$, then $P_e = P_z$ and the images of e and z by $\hat{h}_Z$ are $\mathcal{L}$-equivalent elements belonging to J_Z.*

Proof Since $\mathrm{Card}(P_w) = d_X(F)$, $\mathrm{Card}(P_z) = d_Z$ and $P_w \subseteq P_z$, it is immediate that $d_X(F) = d_Z$ implies $P_w = P_z$, thus $\delta_Z(w) = d_Z = \delta_Z(z)$ (we are using the dual of Proposition 8.3.7). Then, by Proposition 8.5.10, both $\hat{h}_Z(w)$ and $\hat{h}_Z(z)$ belong to $J_Z = J_Z(A^*)$. Because w is a suffix of z and $M(Z^*)$ is a stable semigroup (Proposition 3.6.5), we conclude that $\hat{h}_Z(w)\mathcal{L}\hat{h}_Z(z)$. Since $P_w = P_e$ and $\hat{h}_Z(w)\mathcal{L}\hat{h}_Z(e)$ by the choice of w, we are done. ∎

We need another auxiliary lemma.

Lemma 8.8.3 *Consider the maximal subgroup $K \subseteq J(F)$ and the idempotent $e \in K$. Take the maximal subgroup K_X of $J_X(F)$ containing $\hat{h}_X(K)$ and the maximal subgroup K_Z of $M(Z^*)$ containing $\hat{h}_Z(K)$. Let $W = \hat{h}_X^{-1}(K_X)$. If $d_X(F) = d_Z$, then $K_Z \subseteq J_Z$ and there is an injective morphism $\zeta : K_X \to K_Z$ such that Diagram (8.24) commutes.*

$$
\begin{array}{ccc}
eWe & \xrightarrow{\ \hat{h}_Z\ } & K_Z \\[2pt]
{\scriptstyle \hat{h}_X}\Big\downarrow & \nearrow{\scriptstyle \zeta} & \\[2pt]
K_X & &
\end{array}
\tag{8.24}
$$

Proof Since $d_X(F) = d_Z$, we know that $\hat{h}_Z(e) \in J_Z$ by Lemma 8.8.2, and so K_Z is indeed a maximal subgroup of J_Z.

Note that $\hat{h}_X(eWe) = K_X$ and $K \subseteq eWe$. Let $u, v \in eWe$ be such that $\hat{h}_X(u) = \hat{h}_X(v)$. Let $p \in P_e$. Since $\hat{h}_X(u) \in K_X$, one has $\hat{h}_X(u^\omega) = \hat{h}_X(e)$. On the other hand, as seen in (8.22), the sate $i_X \cdot p$ is defined and belongs to $\mathrm{Im}_X(e)$. Hence, we have $i_X \cdot p = i_X \cdot pu^\omega = (i_X \cdot pu) \cdot u^{\omega-1} \neq \emptyset$, which implies that $i_X \cdot pu \neq \emptyset$. Similarly, we also have $i_X \cdot pv \neq \emptyset$. Because $\hat{h}_X(u) = \hat{h}_X(v)$, it follows that $i_X \cdot pu = i_X \cdot pv \neq \emptyset$. Therefore, as $\mathcal{A}(X^*)$ is trim, there is $p' \in A^*$ such that $i_X = i_X \cdot pup' = i_X \cdot pvp'$, thus $pup', pvp' \in \overline{X^*} \subseteq \overline{Z^*}$, which in turn implies that

$$
i_Z = i_Z \cdot pup' = i_Z \cdot pvp'. \tag{8.25}
$$

Since $\hat{h}_Z(e)$ belongs to the minimum ideal J_Z, one also has $\hat{h}_Z(ep') \in J_Z$, and $\hat{h}_Z(e) = \hat{h}_Z(ep'p'')$ for some $p'' \in A^*$, as $M(Z^*)$ is a stable semigroup. Hence, and since $u = ue$ and $v = ve$, from (8.25), we obtain

$$
i_Z \cdot pu = i_Z \cdot pv \tag{8.26}
$$

for every $p \in P_e$. By Lemma 8.8.2, the hypothesis $d_X(F) = d_Z$ implies that $\mathrm{Im}_Z(e) = \mathrm{Im}_Z(z) = i_Z \cdot P_z = i_Z \cdot P_e$. Since $u = eu$ and $v = ev$, it then follows from (8.26) holding for every $p \in P_e$ that $\hat{h}_Z(u) = \hat{h}_Z(v)$. We showed that $\hat{h}_X(u) = \hat{h}_X(v)$ implies $\hat{h}_Z(u) = \hat{h}_Z(v)$ when $u, v \in eWe$. This means that there is a map $\zeta : K_X \to K_Z$ that makes Diagram (8.24) commutative. Because eWe is a subsemigroup of $\widehat{A^*}$, the map ζ is a group morphism.

We proceed to show that ζ is injective. Let $u \in eWe$. Let p be an arbitrary element of P_e. Recall again from (8.22) that $\mathrm{Im}_X(e) = i_X \cdot P_e$, and so we have $i_X \cdot p = i_X \cdot pe$. As $u = eu$, to prove that ζ is injective we only need to show that $\hat{h}_Z(u) = \hat{h}_Z(e)$ implies $i_X \cdot pu = i_X \cdot p$. As seen in the preceding paragraph, from $p \in P_e$ and $u \in eWe$ we get $i_X \cdot pu \neq \emptyset$. Since the equality $u = ue$ implies $\mathrm{Im}_X(pu) \subseteq \mathrm{Im}_X(e) = i_X \cdot P_e$, there is $s \in P_e$ with

$$i_X \cdot pu = i_X \cdot s.$$

Because se is a suffix of the idempotent e, one has $se \in \overline{F}$. Hence, there is a factorization $se = rr'$ with $r \in \overline{X^*}$ and $r' \in A^*$, by Proposition 8.7.2. Because se is an infinite pseudoword, we must have $r \neq \varepsilon$, thus $r = r_0 r_1$ for some $r_0 \in X$ and $r_1 \in \widehat{A^*}$. Since $s, r_0 \in A^*$ are prefix-comparable (as $\widehat{A^*}$ is equidivisible) and $s \notin XA^*$, it follows that $r_0 = st$ for some $t \in A^*$. Then, $i_X = i_X \cdot st = i_X \cdot put$, which entails that $put \in \overline{X^*} \subseteq \overline{Z^*}$. Since $\hat{h}_Z(u) = \hat{h}_Z(e)$, we obtain $pet \in \overline{Z^*}$. Let s' be such that $e = s's$. Since $pet = pes'st \in \overline{Z^*}$ and $st \in X$, we have $pes' \in \overline{Z^*}$, as $\overline{Z^*}$ is left unitary in $\widehat{A^*}$ (cf. Proposition 8.7.1). As pes' is a factor of e, it follows that $pes' \in \overline{Z^*} \cap \overline{F} \subseteq \overline{X^*}$, thus $i_X \cdot s = (i_X \cdot pes') \cdot s = i_X \cdot pe = i_X \cdot p$, which yields the equality $i_X \cdot pu = i_X \cdot p$, and thereby concludes the proof that ζ is injective. ∎

Retaining the notation introduced in Lemma 8.8.3, we proceed with the proof of the theorem. Suppose that the conditions (iii) in the statement of the theorem hold. Then, from $d_X(F) = d_Z$ and Lemma 8.8.3 we obtain an injective morphism ζ making commutative Diagram (8.19). Since Lemma 8.8.3 also yields $K_Z \subseteq J_Z$, from $\mathrm{Card}(G_X(F)) = \mathrm{Card}(G_Z)$, we get that ζ is an isomorphism. It then follows that $\hat{h}_Z(K) = K_Z$, as $\hat{h}_X(K) = K_X$ by the definition of weakly charged. This shows the implication (iii) $\Rightarrow$ (i).

The implication (ii) $\Rightarrow$ (iii) follows from Corollary 8.5.11.

It remains to show the implication (i) $\Rightarrow$ (ii).

Suppose that $\hat{h}_Z(K) = K_Z$. Remember that e is the idempotent in K, and recall also the meaning of the words w, y, z introduced in the first paragraphs of the proof of Theorem 8.8.1. We begin by proving that $d_X(F) = d_Z$. Suppose that P_e is strictly contained in P_Z. Let $p \in P_z \setminus P_e$. Then, since $z = yw$ and $P_e = P_w$, the word w must be a proper suffix of p. As $\hat{h}_Z(e) \mathcal{L} \hat{h}_Z(w)$, it follows that $\hat{h}_Z(p) \leqslant_\mathcal{L} \hat{h}_Z(e)$, thus $\hat{h}_Z(p) = \hat{h}_Z(pe)$. On the other hand, by Lemma 8.8.2, $\hat{h}_Z(e)$ belongs to the minimum ideal J_Z. Since $M(Z^*)$ is stable, we deduce that $\hat{h}_Z(ep) = \hat{h}_Z(epe)$ belongs to the $\mathcal{H}$-class K_Z. By the hypothesis $\hat{h}_Z(K) = K_Z$, we may take $u \in K$ such that $\hat{h}_Z(ep) = \hat{h}_Z(u)$. By Proposition 8.7.2, there are $x \in \overline{X^*}$ and $r \in P_u$ such that $u = xr$. Note that $u \mathcal{L} e$ yields $P_u = P_e \subseteq P_z$. Because $\hat{h}_Z(ep) = \hat{h}_Z(u)$, and $e \in \overline{Z^*}$ as e is idempotent (cf. Exercise 8.16), we have $i_Z \cdot p = i_Z \cdot ep = i_Z \cdot u = (i_Z \cdot x) \cdot r = i_Z \cdot r$. As $p, r \in P_z$, applying Proposition 8.7.2, we deduce from $i_Z \cdot p = i_Z \cdot r$ that $p = r \in P_e$ by (8.23). This is a contradiction with our assumption

$p \notin P_e$. To avoid the contradiction, we then must have $P_e = P_z$, showing that $d_X(F) = d_Z$ in view of (8.21). Then, since we have the bijections (8.23) and the equality $\mathrm{Im}_Z(e) = \mathrm{Im}_Z(z)$ holds by Lemma 8.8.2, we conclude that $\beta : \mathrm{Im}_X(e) \rightarrow \mathrm{Im}_Z(e)$ is a bijection.

By Lemma 8.8.3, the equality $d_X(F) = d_Z$ entails the existence of an injective morphism $\zeta : K_X \rightarrow K_Z$ such that Diagram (8.19) commutes. Moreover, the hypothesis $\hat{h}_Z(K) = K_Z$ forces ζ to be an isomorphism, whence $\hat{h}_X(K) = K_X$.

It remains to show that

$$\beta(q \cdot g) = \beta(q) \cdot \zeta(g),$$

for every $g \in K_X$ and $q \in \mathrm{Im}_X(e)$. For any such g, q we may take $u \in K$ and $p \in P_e$ such that $g = \hat{h}_X(u)$ and $q = i_X \cdot p$. As $eu = u$, the fact that p is a suffix of e implies that pu is in the $\mathcal{L}$-class containing K, thus $P_{pu} = P_e$ and $pu \in \overline{F}$. Therefore, there are $x \in X^*$ and $s \in P_e$ such that $pu = xs$, by Proposition 8.7.2. It follows that

$$\beta(q \cdot g) = \beta(i_X \cdot pu) = \beta(i_X \cdot xs) = \beta(i_X \cdot s) = i_Z \cdot s = i_Z \cdot xs = (i_Z \cdot p) \cdot u = \beta(q) \cdot \zeta(g),$$

where the last equality holds because Diagram (8.19) commutes.

We have therefore established the implication (i) $\Rightarrow$ (ii), and in the process we also proved the remaining parts of the theorem. ∎

In Exercise 8.20, one sees an improvement of the implication (iii) $\Rightarrow$ (i) in Theorem 8.8.1 for the case where Z is a group code. For most of the proof of Theorem 8.8.1, there is not much difference between assuming that Z is bifix or just a group code.

8.8.1 A Byproduct of the Proof of the Charged Code Theorem

The following statement concerns the framework of Lemma 8.5.7 (recall also Lemma 8.5.8).

Theorem 8.8.4 *Let $F \subseteq A^*$ be a uniformly recurrent set and let $Z \subseteq A^+$ be a recognizable complete bifix code that is F-charged. Set $X = Z \cap F$. Let $u \in F$ be such that $\delta_Z(u) = d_Z$. For each $Y \in \{X, Z\}$, let $I_Y = \mathrm{Im}_Y(u) \subseteq Q_Y$ be the image of u in the minimal automaton $\mathcal{A}(Y^*) = (Q_Y, i_Y, i_Y)$ of Y^*, and let $\pi_Y = \pi_{I_Y}$ be the natural morphism from $\mathrm{Stab}(I_Y)$ into the group of permutations on I_Y, as introduced before Lemma 8.5.7. Then, we have*

$$\pi_X(\mathcal{R}_F(u)^*) = \pi_X(I_X) = G_X(F) \tag{8.27}$$

and

$$\pi_Z(\mathcal{R}_F(u)^*) = \pi_Z(I_Z) = G_Z, \tag{8.28}$$

and there is an isomorphism ξ such that the diagram

$$
\begin{array}{ccc}
\mathcal{R}_F(u)^* & \xrightarrow{\ \pi_Z\ } & \pi_Z(I_Z) \\
{\scriptstyle \pi_X} \downarrow & \nearrow {\scriptstyle \xi} & \\
\pi_X(I_X) & &
\end{array}
\tag{8.29}
$$

commutes. Moreover, if P_u is the set of suffixes of u with no prefixes in X, then the mapping $\beta : \mathrm{Im}_X(u) \to \mathrm{Im}_Z(u)$ defined by $\beta(i_X \cdot p) = i_Z \cdot p$, with $p \in P_u$, is a well-defined bijection for which the pair (β, ξ) is an equivalence of permutation groups.

Proof By Theorem 8.8.1, we have $d_X(F) = d_Z$, thus $\delta_X(u) = d_X(F)$. It follows that $h_X(u) \in J_X(F)$ and $h_Z(u) \in J_Z$, by Proposition 8.5.10, where in the second case the recurrent set under consideration is A^*. We may take an idempotent $e \in J(F)$ such that u is a suffix of e (cf. Exercise 5.28). Let K be the maximal subgroup of $J(F)$ containing e. As $M(X^*)$ and $M(Z^*)$ are finite semigroups, we know that $\hat{h}_X(e)\mathcal{L}\hat{h}_X(u)$ and $\hat{h}_Z(e)\mathcal{L}\hat{h}_Z(u)$ (cf. Proposition 3.6.5). For each $Y \in \{X, Z\}$, let K_Y be the maximal subgroup of $M(Y^*)$ containing $\hat{h}_Y(e)$. For the remainder of the proof, we consider the associated isomorphism $\rho_Y : \pi_Y(\mathrm{Stab}(I_Y)) \to K_Y$, as in Lemma 8.5.7.

Since X is weakly F-charged (again by Theorem 8.8.1), we may apply Proposition 8.5.9 to $\mathcal{A}(X^*)$ and deduce that the equalities (8.27) hold.

Let ζ be as in Lemma 8.8.3, which is part of the proof of Theorem 8.6.3, and whose conditions apply here. By the last part of Theorem 8.8.1, the pair (β, ζ) is a well-defined equivalence of permutation groups. Consider the morphism $\xi = \rho_Z^{-1}\zeta\rho_X$ (cf. Diagram (8.30)).

$$
\begin{array}{ccc}
K_X & \xrightarrow{\ \zeta\ } & K_Z \\
{\scriptstyle \rho_X} \uparrow & & \uparrow {\scriptstyle \rho_Z} \\
\pi_X(\mathrm{Stab}(I_X)) & \xrightarrow{\ \xi\ } & \pi_Z(\mathrm{Stab}(I_Z))
\end{array}
\tag{8.30}
$$

Composing the already established equivalences, we see that the pair (β, ξ) is also an equivalence of permutation groups.

Let $v \in \mathcal{R}_F(u)^*$. Note that $v \in \mathrm{Stab}(I_X)$ by Lemma 8.5.8. Also, noting that $\mathcal{R}_F(u)^* \subseteq \mathcal{R}_{A^*}(u)^*$ and replacing in the application of Lemma 8.5.8 the set F by

A^*, we see that $v \in \text{Stab}(I_Z)$. For each $Y \in \{X, Z\}$, and by the definition of ρ_Y (recall Lemma 8.5.7), we have

$$\rho_Y(\pi_Y(v)) = \hat{h}_Y(ev) = \hat{h}_Y(eve). \tag{8.31}$$

It follows that

$$\xi\pi_X(v) = \rho_Z^{-1}\varsigma\rho_X\pi_X(v) = \rho_Z^{-1}\varsigma\hat{h}_X(eve).$$

The pseudoword eve belongs to the set eWe considered in Lemma 8.8.3. In view of that lemma, the equality $\varsigma\hat{h}_X(eve) = \hat{h}_Z(eve)$ holds. Therefore, applying (8.31) in the case $Y = Z$, we get $\pi_Z(v) = \xi\pi_X(v)$. This shows that Diagram (8.29) commutes, which together with (8.27) also shows that the equalities in (8.28) hold. ∎

8.9 Exercises

8.9.1 Section 8.2

8.1 A subset of A^* is *prefix-closed* if it contains the prefixes of its elements. Let $X \subseteq A^+$ be a prefix code. Show that $R = A^* \setminus XA^*$ is prefix-closed and that, conversely, if $R \subseteq A^*$ is prefix-closed, then $X = RA \setminus R$ is a prefix code and $R = A^* \setminus XA^*$.

8.2 Show that a nontrivial submonoid of A^* is generated by a prefix code if and only if it is right unitary.

8.3 Prove Proposition 8.2.14.

8.9.2 Section 8.3

8.4 A *formal series* with coefficients in a field K is a function $S : A^* \to K$. We denote by (S, w) the image by S of the word w. The sum and the product of two series S, T are defined by

$$(S + T, w) = (S, w) + (T, w), \quad (ST, w) = \sum_{uv=w} (S, u)(T, v).$$

For $k \in K$, we denote simply by k the series with value k on the empty word and 0 on nonempty words. If $(S, \varepsilon) = 0$, we define the series S^* as the infinite sum

$$S^* = 1 + S + S^2 + \dots$$

which is well defined since each sum (S^*, w) is finite. Show that the set of series is a ring with neutral element 1 and that S^* is the inverse of $1 - S$.

8.5 We denote by $\underline{X}$ the *characteristic series* of a set $X \subseteq A^*$. It is defined to be the series with rational coefficients given by the formula

$$(\underline{X}, u) = \begin{cases} 1 & \text{if } u \in X, \\ 0 & \text{otherwise.} \end{cases}$$

Let $X \subseteq A^+$ and let $M = X^*$. Show that X is a code if and only if $(\underline{X})^* = \underline{M}$.

8.6 Show that if X is a prefix code and $U = A^* \setminus XA^*$, then

$$\underline{A^*} = \underline{X^*U} \quad \text{and} \quad \underline{X} - 1 = \underline{U}(\underline{A} - 1). \tag{8.32}$$

8.7 The *indicator* of a set $X \subseteq A^*$ is the series L_X defined for $w \in A^*$ by $(L_X, w) = \delta_X(w)$. Show that if X is a prefix code and $V = A^* \setminus A^*X$, then

$$\underline{V} = L_X(1 - \underline{A}). \tag{8.33}$$

If X is bifix, one has

$$1 - \underline{X} = (1 - \underline{A})L_X(1 - \underline{A}). \tag{8.34}$$

8.8 Show that a function $\delta : A^* \to \mathbb{N}$ is the parse enumerator of some bifix code if and only if it satisfies the following conditions.

(i) For every $a \in A$ and $w \in A^*$

$$0 \leqslant \delta(aw) - \delta(w) \leqslant 1. \tag{8.35}$$

(ii) For every $w \in A^*$ and $a \in A$

$$0 \leqslant \delta(wa) - \delta(w) \leqslant 1. \tag{8.36}$$

(iii) For every $a, b \in A$ and $w \in A^*$

$$\delta(aw) + \delta(wb) \geqslant \delta(w) + \delta(awb). \tag{8.37}$$

(iv) $\delta(\varepsilon) = 1$.

8.9 The aim of the following exercise is to show that there exist a uniformly recurrent set F, and a bifix code $X \subseteq F$ which is not F-thin.

Let F be the Thue–Morse set, which is the set of factors of a fixed-point of the substitution f defined by $f(a) = ab$, $f(b) = ba$. Set $x_n = f^n(a)$ for $n \geqslant 1$. Show that the set $X = \{x_n x_n \mid n \geqslant 1\}$ is a bifix code included in F which is not F-thin.

8.10 Let $A = \{a, b\}$ and let $X = \{a^{|u|}bu \mid u \in A^*\}$.

1. Show that X is a maximal prefix code which is not suffix.
2. Show that $Y = X \setminus A^*X$ is a maximal bifix code which is not maximal prefix.

8.11 Let $Z \subseteq A^+$ be a code, and let d be a positive integer. Show that the following conditions are equivalent.

1. Z is a group code of degree d.
2. $Z^* = H \cap A^*$ for some subgroup of finite index d of $FG(A)$.
3. Z^* is recognized by a group automaton on the alphabet A with d states.
4. For some morphism $\varphi : A^* \to G$ onto a group and some subgroup K of index d of G, the equality $Z^* = \varphi^{-1}(K)$ holds.

8.9.3 Section 8.5

8.12 Let $\mathcal{A}$ be a finite automaton on the alphabet A. Show that if $F \subseteq A^*$ is a recurrent set, then $\hat{\varphi}_A(J(F)) \subseteq J_A(F)$, where $\hat{\varphi}_A$ is the unique continuous extension to $\widehat{A^*}$ of the transition morphism φ_A of the automaton $\mathcal{A}$.

8.13 Prove the assertions made in Example 8.5.6.

8.14 Prove Lemma 8.5.7.

8.15 Let $Z \subseteq A^+$ be a group code with minimal automaton $\mathcal{A} = (Q, i, i)$, and let $v \in A^*$ be such that $\delta_Z(v) = d_Z$. Denoting by P_v the set of suffixes of v with no prefix in Z, show that the mapping $p \in P_v \mapsto i \cdot p \in Q$ is bijective.

8.16 Let $X \subseteq A^+$ be an F-complete bifix code, where $F \subseteq A^*$. Show that if $u \in \widehat{A^*}$ is an element of $\overline{F}$ such that $\hat{h}_X(u)$ is idempotent, then $u \in \overline{X^*}$, where $\hat{h}_X$ is the unique continuous morphism $\widehat{A^*} \to M(X^*)$ extending the syntactic morphism $h_X : A^* \to M(X^*)$.

8.17 Let $F \subseteq A^*$ be a uniformly recurrent set, and let $Z \subseteq A^+$ be a recognizable complete bifix code. Set $X = Z \cap F$. Show that if $d_X(F) = d_Z$ then $\hat{h}_Z(J(F)) \subseteq J_Z$.

8.9.4 Section 8.6

8.18 Give a new proof of the Finite Index Basis Theorem (Theorem 8.4.2) based on Proposition 8.6.2 and on the Cardinality Theorem (Theorem 8.4.1).

8.9.5   Section 8.7

8.19 Let $F \subseteq A^*$ be a recurrent set, and let X be a finite F-maximal prefix code contained in F. Show that the mapping $w \in \overline{F} \mapsto \delta_X(w) \in \{1, \ldots, d_X(F)\}$ is a continuous mapping, for the topology of $\overline{F}$ induced from that of $\widehat{A^*}$, and for the discrete topology of $\{1, \ldots, d_X(F)\}$.

8.9.6   Section 8.8

8.20 Consider a uniformly recurrent set $F \subseteq A^*$ and a recognizable complete bifix code $Z \subseteq A^+$. Set $X = Z \cap F$.

1. Let K be any maximal subgroup of $J(F)$. Show that there is a surjective morphism $\mu : \hat{h}_Z(K) \rightarrow \hat{h}_X(K)$ such that the diagram

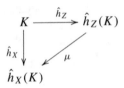

   commutes.
2. Suppose that $h_Z(A^+)$ is contained in J_Z (which happens if Z is a group code). Show that if $\hat{h}_X(K) = G_X(F)$ and the groups $G_X(F)$ and G_Z have the same cardinal, then Z is F-charged. (This improves the implication (iii) $\Rightarrow$ (i) in Theorem 8.8.1.)

8.10   Solutions

8.10.1   Section 8.2

8.1 Let us for example verify the first part of the converse. Let $x, y \in X = RA \setminus R$. If x is a prefix of y, set $y = xv$. If v is nonempty, set $v = wa$ with $a \in A$. Then $xw \notin R$ and thus $y \notin RA$, a contradiction.

8.2 Suppose that M is a nontrivial right unitary submonoid of A^* and set $N = M \setminus \{\varepsilon\}$. The nonempty set $X = N \setminus NN$ generates M. It is a prefix code since $u, uv \in X$ implies $v \in M$ and thus $v = \varepsilon$.

Conversely, suppose that X is a prefix code. Let $u, uv \in X^*$. We show that $v \in X^*$ by induction on $|u|$. The base case $|u| = 0$ holds trivially. Suppose that $|u| > 0$. Set $uv = x_1 \cdots x_n$, with $x_1, \ldots, x_n \in X$. Then $u = x_1 \cdots x_{i-1} p$ and

$v = s x_{i+1} \cdots x_n$ with $x_i = ps$, for some $i \in \{1, \dots, n\}$. In case $p = \varepsilon$, we are done. If $p \neq \varepsilon$ then the length of $u' = x_1 \cdots x_{i-1} \in X^*$ is smaller than that of u, and $u'p \in X^*$, thus $p \in X^+$ by the induction hypothesis. From $ps \in X$ we then get $p \in X$ and $s = \varepsilon$, since X is a prefix code. This concludes the inductive step. Hence, X^* is right unitary.

8.3 Suppose that X is right F-complete. Take any path $i \to q$ in $\mathcal{A}(X^*)$ such that i appears only at the beginning of the path. Let u be the label of the path. Let $a \in A$ be such that $ua \in F$. Because X is right F-complete, there is $v \in A^*$ such that $uav \in X^*$. Since $q \cdot av = i \cdot uav = i$, we must have $q \cdot a \neq \emptyset$.

Conversely, suppose that the second condition in the equivalence holds. We prove by induction on the length of $w \in F$ that w is a prefix of X^*. It is true if w is the empty word. Next, let $w = w'a$ with $a \in A$ and assume that w' is a prefix of X^* as induction hypothesis. Let $q \in Q$ be such that $i \cdot w' = q$. Set $w' = xu$ with $x \in X^*$ and $u \in A^* \setminus XA^*$. Then $ua \in F$ and the word u labels a path $i \to q$ in $\mathcal{A}$ such that i appears only at the beginning of the path. Therefore, by hypothesis, we have $q \cdot a \neq \emptyset$. This implies that ua is a prefix of X and thus the conclusion.

8.10.2 Section 8.3

8.4 The first part is true for the set of functions from any monoid into K. Set $S^+ = S^* - 1$. Then $S^*S = S^+$ and thus $S^*(1 - S) = (1 - S)S^* = 1$.

8.5 For $X \subseteq A^+$ and $w \in A^*$, $((\underline{X})^*, w)$ is the number of factorizations of w in words of X. This proves the desired equivalence.

8.6 The first equation expresses the fact that every word $w \in A^*$ has a unique factorization $w = xu$ with $x \in X^*$ and $u \in U$, which is clearly true for a prefix code X. The second equation results from the first one by taking the inverses of both sides and multiplying them by $\underline{U}$ on the left.

8.7 Set $L = L_X$, $U = A^* \setminus XA^*$ and $V = A^* \setminus A^*X$. By definition of the indicator, we have $L = \underline{V}\ \underline{X^*}\underline{U}$. Since X is prefix, we have by Exercise 8.6, the equality $\underline{A^*} = \underline{X^*}\underline{U}$. Thus we obtain $L = \underline{V}\underline{A^*}$. Multiplying both sides on the right by $(1 - \underline{A})$, we obtain Eq. (8.33). If X is suffix, we have by the symmetric of Eq. (8.32), the equality $\underline{X} - 1 = (\underline{A} - 1)\underline{V}$. Substituting the value of $\underline{V}$ from Eq. (8.33), we obtain Eq. (8.34).

8.8 Assume first that X is a bifix code and let L be its indicator. Then, $L(1 - \underline{A})$ is a characteristic series by Eq. (8.33), which proves Eq. (8.36). The proof of Eq. (8.35) is symmetric. Next, it follows from Eq. (8.34) that

$$-(\underline{X}, awb) = (L, awb) - (L, aw) - (L, wa) + L(w)$$

and thus (8.37) follows.

Conversely, let $L = \delta$ and set $S = (1 - \underline{A})L$. Note that S only takes integer values. Moreover, $(S, \varepsilon) = (L, \varepsilon) = 1$ and for $a \in A$ and $w \in A^*$, we have $(S, aw) = (L, aw) - (L, w)$ and thus $0 \leqslant (S, aw) \leqslant 1$ by Eq. (8.35). Hence, there is a set $U \subseteq A^*$ such that $S = \underline{U}$. Note that $\varepsilon \in U$ and, if $a, b \in A$, then we have

$$(S, awb) = (L, awb) - (L, wb) \leqslant (L, aw) - (L, w) = S(aw),$$

thereby showing that $awb \in U$ implies $aw \in U$. Thus, U is prefix-closed. By Exercise 8.1, the set $X = UA \setminus U$ is a prefix code and $1 - \underline{X} = \underline{U}(1 - \underline{A})$ by (8.32). Symmetrically, the series $T = L(1 - \underline{A})$ is the characteristic series of a suffix code Y and $1 - \underline{Y} = (1 - \underline{A})\underline{V}$. It follows that

$$1 - \underline{X} = \underline{U}(1 - \underline{A}) = (1 - \underline{A})L(1 - \underline{A}) = (1 - \underline{A})\underline{V} = 1 - \underline{Y}.$$

Thus, $X = Y$ is bifix and $L = L_X$ by (8.34).

8.9 Note first that $x_{n+1} = x_n \bar{x}_n$ where $u \to \bar{u}$ is the substitution defined by $\bar{a} = b$ and $\bar{b} = a$. Note also that $u \in F$ if and only if $\bar{u} \in F$.

We have $X \subseteq F$. Indeed, for $n \geqslant 1$, $x_{n+2} = x_{n+1}\bar{x}_{n+1} = x_n\bar{x}_n\bar{x}_nx_n$ implies that $\bar{x}_n\bar{x}_n \in F$ and thus $x_nx_n \in F$. Next X is a bifix code. Indeed, for $n < m$, x_m begins with $x_n\bar{x}_n$, and thus cannot have x_nx_n as a prefix. Similarly, since x_m ends with $\bar{x}_nx_n$ or with $x_n\bar{x}_n$, it cannot have x_nx_n as a suffix. Finally any element of F is a factor of a word in X. Indeed, any element u of F is a factor of some x_n, and so also of $x_nx_n \in X$.

8.10

1. X is clearly prefix. It is not suffix since, for example, $b, abb \in X$.
2. The set $Y = X \setminus A^*X$ is clearly a maximal suffix code and a bifix code. Since Y is strictly included in X, it is not maximal prefix.

8.11 (1) $\Leftrightarrow$ (2) This follows immediately from Proposition 8.3.18.

(2) $\Rightarrow$ (3) The Stallings automaton of H has d states and recognizes $H \cap A^*$ (cf. Proposition 4.3.3).

(3) $\Rightarrow$ (4) Suppose that $\mathcal{A}$ is a group automaton with d states recognizing Z^*. Then, $\mathcal{A}$ is the Stallings automaton of a subgroup H of $FG(A)$ with index d (cf. Proposition 4.3.3). Note that $Z^* = H \cap A^*$. Let $\psi : FG(A) \to M(\mathcal{A})$ be the extension to $FG(A)$ of the transition morphism φ of $\mathcal{A}$. That is, ψ is the representation of H on the right cosets of H. The subgroup $K = \psi(H)$ satisfies $H = \psi^{-1}(K)$, and so K has index d in $M(\mathcal{A})$. As φ is the restriction of ψ to A^*, it follows that $Z^* = H \cap A^* = \varphi^{-1}(K)$.

(4) $\Rightarrow$ (2) Let $\psi : FG(A) \to G$ be the unique group morphism extending φ, and let $H = \psi^{-1}(K)$. Then the index of H in $FG(A)$ is the index of K in G. Moreover, since φ is the restriction of ψ to A^*, one has $H \cap A^* = \varphi^{-1}(K) = Z^*$.

8.10.3 Section 8.5

8.12 Let $u \in F$ be such that $\mathrm{rank}_A(u) = \mathrm{rank}_A(F)$. Let $v \in J(F)$. Then u is a factor of v, and so $\mathrm{rank}_A(F)$ is greater than or equal to the rank of $\hat{\varphi}_A(v)$. But $\hat{\varphi}_A(v) = \varphi_A(w)$ for some $w \in F$. Therefore, the rank of $\hat{\varphi}_A(v) = \varphi_A(w)$ is $\mathrm{rank}_A(F)$. This shows that $\hat{\varphi}_A(v) \in J_A(F)$.

8.13 For each $t \in H$, consider the function $r(t) : q \in \mathrm{Im}(H) \mapsto q \cdot t \in \mathrm{Im}(H)$. One clearly has $r(st) = r(s)r(t)$ for every $s, t \in H$. Let e be the idempotent of H. If $h \in H$ and $q \in Q$, then $q \cdot h = q \cdot he$, whence $r(e)$ is the identity on $\mathrm{Im}(H)$. If s is an inverse of t in H, then $r(s)r(t) = r(st) = r(e) = r(ts) = r(t)t(s)$ is the identity on H, and so $r(t)$ is indeed a permutation. Suppose that s, t are distinct elements of H. Then there is $q \in Q$ such that $q \cdot s \neq q \cdot t$, that is, $(q \cdot e) \cdot s \neq (q \cdot e) \cdot t$. Since $q \cdot e \in \mathrm{Im}(H)$, we conclude that $r(s) \neq r(t)$. Therefore, the morphism r is injective.

Suppose that H' is another maximal subgroup in the same $\mathcal{D}$-class of H. Let e' be the neutral element of H'. We know that there are $x, y \in T$ such that $xy = e$ and $e' = yx$, and that $\varphi : u \in H \mapsto yux \in H'$ is an isomorphism, as seen in the proof of Proposition 3.6.11. Let $\beta : \mathrm{Im}(H) \to \mathrm{Im}(H')$ be given by $\beta(q) = q \cdot x$. Note that the function β is well defined, as $qx = q \cdot ex = q \cdot xyx = q \cdot xe'$. Also, β is a bijection with inverse $\beta' : \mathrm{Im}(H') \to \mathrm{Im}(H)$ given by $\beta'(q) = q \cdot y$. Finally, if $u \in H$ and $q \in \mathrm{Im}(H)$, then $\varphi(q \cdot u) = q \cdot ux = q \cdot eux = q \cdot x(yux) = \beta(q) \cdot \varphi(u)$ holds. Therefore, the pair (β, φ) is an isomorphism of permutation groups.

8.14 Set $\varphi = \varphi_A$. Let $\tilde{e} \in A^*$ be such that $\varphi(\tilde{e}) = e$. Let $v, v' \in \mathrm{Stab}(I)$. Suppose that $\pi_I(v) = \pi_I(v')$. Let $q \in Q$. As $I = Q \cdot \tilde{e}$, we have $q \cdot \tilde{e}v = q \cdot \tilde{e}v'$. This establishes $\varphi(\tilde{e}v) = \varphi(\tilde{e}v')$. In particular, for $k = \mathrm{Card}(I)$ we have $\varphi(\tilde{e}) = \varphi((\tilde{e}v)^{k!})$, since $\pi_I(\tilde{e})$ and $\pi_I((\tilde{e}v)^{k!})$ are equal to id_I. This shows that $\varphi(\tilde{e}v) \in V$. Therefore, ρ_I is a well defined map from $\pi_I(\mathrm{Stab}(I))$ to V. The fact that e is the neutral element of V entails that ρ_I is a morphism of groups.

Let $v \in A^*$ be such that $\varphi(v) \in V$. Then, we have $q \cdot v = q \cdot v\tilde{e}$ for every $q \in Q$, thus $I \cdot v \subseteq I$. And if $v' \in A^*$ is such that $\varphi(v')$ is the inverse of $\varphi(v)$ in H, then $q \cdot \tilde{e} = (q \cdot v'\tilde{e}) \cdot v$ for every $Q \in Q$, thus $I \subseteq Iv$. This shows that $v \in \mathrm{Stab}(I)$, and so ρ_I is surjective.

Finally, if $\pi_I(v) \neq \pi_I(v')$, then $q \cdot v \neq q \cdot v'$ for some $q \in I$. But $q = q \cdot \tilde{e}$, thus $e\varphi(v) \neq e\varphi(v')$, establishing that ρ_I is an isomorphism.

8.15 It is an immediate consequence of Proposition 8.5.10, since the syntactic monoid of Z^* is a finite group, thus $\mathrm{Im}_Z(w) = Q$ for every $w \in A^*$.

8.16 Suppose first that $u \in F$. Since X is left F-complete, there is $v \in A^*$ such that $vu \in X^*$. Because X is prefix, the initial state i of $\mathcal{A}(X^*)$ is its unique terminal state, whence $i = i \cdot vu$. As $h_X(u)$ is idempotent, one has $i \cdot u = i \cdot vu^2 = i \cdot vu = i$, thus $u \in X^*$.

Finally, suppose that $u \in \overline{F}$. Let $(u_n)_n$ be a sequence of elements of F converging to u. By continuity of $\hat{h}_X$, for all sufficiently large n one has $h_X(u_n) = \hat{h}_X(u)$. Hence, for all such n one has $u_n \in X^*$ by the previous paragraph, yielding $u \in \overline{X^*}$.

8.17 Let $u \in J(F)$. Consider a sequence (u_n) of words of A^* converging to u. The set X is a finite F-complete bifix code by Corollary 8.3.21. Because X is finite, for all sufficiently large n the word u_n is not an internal factor of X, and so $\delta_X(u_n) = d_X(F)$ by Theorem 8.3.16. Since $d_X(F) = d_Z$, we have $h_Z(u_n) \in J_Z$ by Proposition 8.5.10. We conclude that $\hat{h}_Z(u) \in J_Z$, by continuity of $\hat{h}_Z$.

8.10.4   Section 8.6

8.18 Consider the setting of Proposition 8.6.2, supposing that F is a tree set. Let $d' = d_X(F)$. Note that $d' \leqslant d$ (cf. Corollary 8.3.22). By the Cardinality Theorem, the equality $\text{Card}(X) - 1 = d'(\text{Card}(A) - 1)$ holds. On the other hand, by Schreier's Index Formula, H has a basis of cardinality $d(\text{Card}(A) - 1) + 1$, and so the cardinality $d'(\text{Card}(A) - 1) + 1$ of its generator set X is greater than or equal to $d(\text{Card}(A) - 1) + 1$. This shows that $d' \geqslant d$, whence $d' = d$ and X is a basis of H. (The equality $d_X(F) = d$ is deduced in Corollary 8.6.6 more generally whenever F is connected.)

8.10.5   Section 8.7

8.19 Let $(w_n)_n$ be a sequence of elements of $\overline{F}$ converging to $w \in \overline{F}$. We want to show that $\delta_X(w_n) = \delta_X(w)$ for all sufficiently large n. As the elements of A^* are isolated in $\widehat{A^*}$, if $w \in \overline{F} \setminus A^*$, then $w_n = w$ for all sufficiently large n, and we are done. Suppose that $w \notin A^*$. Then, since X is finite, for all sufficiently large n the word w_n is not an internal factor of X, thus $\delta_X(w_n) = d_X(F)$ by Theorem 8.3.16. On the other hand, we showed in Proposition 8.7.5 that $\delta_X(w) = d_X(F)$.

8.10.6   Section 8.8

8.20

1. Recall that X is a finite F-maximal bifix code by Corollary 8.3.21. Let $u, v \in K$. Suppose that $\hat{h}_X(u) \neq \hat{h}_X(v)$. We only need to show that $\hat{h}_Z(u) \neq \hat{h}_Z(v)$. Let e be the idempotent of K, and let P_e the set of suffixes of e which have no prefix in X (which is the set of suffixes of e which have no prefix in Z, as $X = Z \cap F$ and e belongs to the factorial set $\overline{F}$). Set $\mathcal{A}(X^*) = (Q_X, i_X, i_X)$ and $\mathcal{A}(Z^*) = (Q_Z, i_Z, i_Z)$. As in the first steps of the proof of Theorem 8.8.1 (namely in (8.22)), the image of $\hat{h}_X(e)$ is $\{i_X \cdot p \mid p \in P_e\}$. Since $\hat{h}_X(u) \neq \hat{h}_X(v)$, there is $q \in Q_X$ such that $q \cdot u \neq q \cdot v$. As $u = eu$ and $v = ev$, we may suppose that $q = q \cdot e$ is in the image of $\hat{h}_X(e)$. Therefore, there is $p \in P_e$ such

that $i_X \cdot pu \neq i_X \cdot pv$. Since $u\mathcal{H}e$, we have $pu\mathcal{L}e\mathcal{L}pv$, and so $pu = xs$ and $pv = yt$ for some $x, y \in \overline{X^*}$ and $s, t \in P_e$, thanks to Proposition 8.7.2. Note that $i_X \cdot pu = i_X \cdot s$ and $i_X \cdot pv = i_X \cdot t$. Therefore, we have $s \neq t$. Since $x, y \in \overline{Z^*}$, we also have $i_Z \cdot pu = i_Z \cdot s$ and $i_Z \cdot pu = i_Z \cdot t$. As s, t are distinct suffix-comparable words of F, we know that $i_Z \cdot s \neq i_Z \cdot t$ (cf. Proposition 8.5.10), that is $i_Z \cdot pu \neq i_Z \cdot pv$, whence $\hat{h}_Z(u) \neq \hat{h}_Z(v)$.

2. Note that $\hat{h}_Z(K) \subseteq G_Z$, because J_Z is the unique $\mathcal{J}$-class of $M(Z^*)$ containing idempotents in $h_Z(A^+) = \hat{h}_Z(\widehat{A^+})$. From $\hat{h}_X(K) = G_X(F)$, we obtain $\mu(\hat{h}_Z(K)) = G_X(F)$, and therefore $\text{Card}(G_Z) \geqslant \text{Card}(\hat{h}_Z(K)) \geqslant \text{Card}(G_X(F))$. Since $\text{Card}(G_Z) = \text{Card}(G_X(F))$, this shows that $G_Z = \hat{h}_Z(K)$.

8.11 Notes

For the general theory of codes, see Berstel et al. (2009). The properties of bifix codes presented in Sect. 8.3 are generalizations of properties proved in Berstel et al. (2009). The generalization, already appearing in Berstel et al. (2012), consists in replacing the hypothesis of maximality (or completeness) with respect to the free monoid A^* to a maximality (or completeness) relative to a recurrent set. In particular, Theorem 8.3.16 is Berstel et al. (2012, Theorem 4.2.8).

The proof that if F recurrent and $X \subseteq F$ is an F-thin and F-maximal code, then it is F-complete is in Berstel et al. (2012, Theorem 4.2.2).

Theorem 8.4.1 is proved in Berstel et al. (2012, Theorem 5.2.1) for episturmian sets. The proof for tree sets appears in Berthé et al. (2015, Theorem 3.6) where the Finite Index Basis Theorem (Theorem 8.4.2) is also proved.

The Charged Code Theorem (Theorem 8.6.3) appeared in the manuscript (Kyriakoglou and Perrin 2017) for group codes. Corollary 8.6.6 is proved in the case of a Sturmian set in Berstel et al. (2012, Theorem 7.2.5). In its entirety, Theorem 8.6.3 is proved in Almeida et al. (2020). In fact, a more general result is proved there, holding when F is a recurrent set, and both Z and X are recognizable. The fact that X may no longer be finite makes the proof more complicated. Also in Almeida et al. (2020), one finds a generalization of the material developed in Sect. 8.7 to arbitrary recognizable prefix and bifix codes.

The computation of the eggbox diagrams in Figs. 8.7 and 8.11 was carried out using the software **GAP** (GAP 2020; Delgado et al. 2006; Delgado and Morais 2006).

Bibliography

J. Almeida, *Finite Semigroups and Universal Algebra*. Series in Algebra, vol. 3 (World Scientific Publishing Co., Inc., River Edge, NJ, 1994). Translated from the 1992 Portuguese original and revised by the author

J. Almeida, Dynamics of implicit operations and tameness of pseudovarieties of groups. Trans. Am. Math. Soc. **354**(1), 387–411 (2002a)

J. Almeida, Finite semigroups: an introduction to a unified theory of pseudovarieties, in *Semigroups, Algorithms, Automata and Languages*, ed. by G.M.S. Gomes, Jean-Éric, P.V. Silva (World Scientific, Singapore, 2002b), pp. 3–64

J. Almeida, Profinite semigroups and applications, in *Structural Theory of Automata, Semigroups, and Universal Algebra*. NATO Science Series II: Mathematics, Physics and Chemistry, vol. 207 (Springer, Dordrecht, 2005a), pp. 1–45. Notes taken by Alfredo Costa.

J. Almeida, Profinite groups associated with weakly primitive substitutions. Fundam. Prikl. Mat. **11**(3), 13–48 (2005b)

J. Almeida, Profinite semigroups and applications, in *Structural Theory of Automata, Semigroups and Universal Algebra*, ed. by V.B. Kudryavtsev, I.G. Rosenberg (Springer, New York, 2005c), pp. 1–45

J. Almeida, A. Azevedo, On regular implicit operations. Portugal. Math. **50**, 35–61 (1993)

J. Almeida, A. Costa, Infinite-vertex free profinite semigroupoids and symbolic dynamics. J. Pure Appl. Algebra **213**(5), 605–631 (2009)

J. Almeida, A. Costa, On the transition semigroups of centrally labeled Rauzy graphs. Int. J. Algebra Comput. **22**(2), 1250018, 25 (2012)

J. Almeida, A. Costa, Presentations of Schützenberger groups of minimal subshifts. Israel J. Math. **196**(1), 1–31 (2013)

J. Almeida, A. Costa, A geometric interpretation of the Schützenberger group of a minimal subshift. Ark. Mat. **54**(2), 243–275 (2016)

J. Almeida, A. Costa, Equidivisible pseudovarieties of semigroups. Publ. Math. Debrecen **90**, 435–453 (2017)

J. Almeida, A. Costa, J.C. Costa, M. Zeitoun, The linear nature of pseudowords. Publ. Mat. **63**, 361–422 (2019)

J. Almeida, A. Costa, R. Kyriakoglou, D. Perrin, On the group of a rational maximal bifix code. Forum Math. **32**, 553–576 (2020)

J. Almeida, B. Steinberg, Rational codes and free profinite monoids. J. Lond. Math. Soc. (2) **79**(2), 465–477 (2009)

J. Almeida, P. Weil, Free profinite $\mathcal{R}$-trivial monoids. Int. J. Algebra Comput. **7**, 625–671 (1997)

© Springer Nature Switzerland AG 2020
J. Almeida et al., *Profinite Semigroups and Symbolic Dynamics*, Lecture Notes in Mathematics 2274, https://doi.org/10.1007/978-3-030-55215-2

P. Arnoux, G. Rauzy, Représentation géométrique de suites de complexité $2n + 1$. Bull. Soc. Math. France **119**(2), 199–215 (1991)

L. Balková, E. Pelantová, W. Steiner, Sequences with constant number of return words. Monatsh. Math. **155**(3–4), 251–263 (2008)

B. Banaschewski, The Birkhoff theorem for varieties of finite algebras. Algebra Univers. **17**, 360–368 (1983)

J. Berstel, C. Reutenauer, *Noncommutative Rational Series with Applications*. Encyclopedia of Mathematics and its Applications, vol. 137 (Cambridge University Press, Cambridge, 2011)

J. Berstel, D. Perrin, C. Reutenauer, *Codes and Automata* (Cambridge University Press, Cambridge, 2009)

J. Berstel, C. De Felice, D. Perrin, C. Reutenauer, G. Rindone, Bifix codes and Sturmian words. J. Algebra **369**, 146–202 (2012)

V. Berthé, M. Rigo (eds.) *Combinatorics, Automata and Number Theory*. Encyclopedia of Mathematics and its Applications, vol. 135 (Cambridge University Press, Cambridge, 2010)

V. Berthé, C. De Felice, F. Dolce, J. Leroy, D. Perrin, C. Reutenauer, G. Rindone, Acyclic, connected and tree sets. Monatsh. Math. **176**, 521–550 (2015)

V. Berthé, C. De Felice, F. Dolce, J. Leroy, D. Perrin, C. Reutenauer, G. Rindone, The finite index basis property. J. Pure Appl. Algebra **219**(7), 2521–2537 (2015)

S. Burris, H.P. Sankappanavar, *A Course in Universal Algebra*. Graduate Texts in Mathematics, vol. 78 (Springer, Berlin, 1981)

A. Costa, Conjugacy invariants of subshifts: an approach from profinite semigroup theory. Int. J. Algebra Comput. **16**, 629–655 (2006)

A. Costa, Semigrupos profinitos e dinâmica simbólica. PhD thesis, Univ. Porto, 2007

A. Costa, B. Steinberg, Profinite groups associated to sofic shifts are free. Proc. Lond. Math. Soc. (3) **102**(2), 341–369 (2011)

J.C. Costa, Free profinite locally idempotent and locally commutative semigroups. J. Pure Appl. Algebra **163**, 19–47 (2001)

T. Coulbois, M. Sapir, P. Weil, A note on the continuous extensions of injective morphisms between free groups to relatively free profinite groups. Publ. Mat. **47**(2), 477–487 (2003)

M. Delgado, J. Morais, SgpViz: A GAP package to visualize finite semigroups. http://www.gap-system.org/Packages/sgpviz.html

M. Delgado, S. Linton, J. Morais, Automata: A GAP package on finite automata. http://www.gap-system.org/Packages/automata.html

F. Dolce, D. Perrin, Neutral and tree sets of arbitrary characteristic. Theor. Comp. Sci. **658**(part A), 159–174 (2017)

F. Dolce, D. Perrin, Eventually dendric shifts. Ergodic Theory Dynam. Syst. (2020). https://doi.org/10.1017/etds.2020.35

X. Droubay, J. Justin, G. Pirillo, Episturmian words and some constructions of de Luca and Rauzy. Theor. Comp. Sci. **255**(1–2), 539–553 (2001)

F. Durand, A characterization of substitutive sequences using return words. Discret. Math. **179**, 89–101 (1998)

F. Durand, D. Perrin, *Dimension Groups and Dynamical Systems* (Cambridge University Press, Cambridge, 2020, to appear)

F. Durand, B. Host, C. Skau, Substitutional dynamical systems, Bratteli diagrams and dimension groups. Ergodic Theory Dynam. Syst. **19**(4), 953–993 (1999)

F. Durand, J. Leroy, G. Richomme, Do the properties of an S-adic representation determine factor complexity? J. Integer Seq. **16**(2), 30 (2013). Article 13.2.6

S. Eilenberg, *Automata, Languages and Machines*, vol. A (Academic, New York, 1974)

S. Eilenberg, *Automata, Languages and Machines*, vol. B (Academic, New York, 1976)

R. Engelking, *General Topology* (Heldermann, Berlin, 1989)

N. Pytheas Fogg, *Substitutions in Dynamics, Arithmetics and Combinatorics*, ed. by V. Berthé, S. Ferenczi, C. Mauduit, A. Siegel. Lecture Notes in Mathematics, vol. 1794 (Springer, Berlin, 2002)

M.D. Fried, M. Jarden, *Field Arithmetic*. Ergebnisse der Mathematik und ihrer Grenzgebiete. 3. Folge. A Series of Modern Surveys in Mathematics [Results in Mathematics and Related Areas. 3rd Series. A Series of Modern Surveys in Mathematics], vol. 11, 3rd edn. (Springer, Berlin, 2008). Revised by Jarden

F.R. Gantmacher, *The Theory of Matrices*, vols. 1, 2 (Chelsea, New York, 1959) Translated from the Russian original

GAP, *GAP* – Groups, Algorithms, and Programming, version 4.11. The GAP Group, 2020. http://www.gap-system.org

M. Gehrke, Stone duality, topological algebra, and recognition. J. Pure Appl. Algebra **220**(7), 2711–2747 (2016)

M. Gehrke, S. Grigorieff, J.-É. Pin, Duality and equational theory of regular languages, in *ICALP 2008, Part II*. Lecture Notes in Computer Science, vol. 5126 (Springer, Heidelberg, 2008), pp. 246–257

A. Glen, J. Justin, Episturmian words: a survey. RAIRO Inf. Théor. et Appl. **43**, 403–442 (2009)

F.Q. Gouvêa, *p-adic Numbers: An Introduction*, 2nd edn. Universitext (Springer, Berlin, 1997)

M. Hall Jr., Coset representations in free groups. Trans. Am. Math. Soc. **67**(2), 421–432 (1949)

M. Hall Jr., A topology for free groups and related groups. Ann. Math. (2) **52**, 127–139 (1950)

G. Hansel, Une démonstration simple du théorème de Skolem-Mahler-Lech. Theoret. Comput. Sci. **43**, 91–98 (1986)

T. Harju, M. Linna, On the periodicity of morphisms on free monoids. RAIRO Inf. Théor. et Appl. **20**, 47–54 (1986)

K. Henckell, J. Rhodes, B. Steinberg, A profinite approach to stable pairs. Int. J. Algebra Comput. **20**(2), 269–285 (2010)

R.P. Hunter, Some remarks on subgroups defined by the Bohr compactification. Semigroup Forum **26**, 125–137 (1983)

R.P. Hunter, Certain finitely generated compact zero dimensional semigroups. J. Aust. Math. Soc. Ser. A. Pure Math. Stat. **44**(2), 265–270 (1988)

S. Ito, Y. Takahashi, Markov subshifts and realization of β-expansions. J. Math. Soc. Jpn. **26**, 33–55 (1974)

J. Justin, L. Vuillon, Return words in Sturmian and episturmian words. Theor. Inf. Appl. **34**(5), 343–356 (2000)

I. Kapovich, A. Myasnikov, Stallings foldings and subgroups of free groups. J. Algebra **248**, 608–668 (2002)

D.E. Knuth, *The Art of Computer Programming*. Volume 2: Seminumerical Algorithms, 3rd edn. (Addison-Wesley, Reading, MA, 1998)

P. Kurka, *Topological and Symbolic Dynamics* (Société Mathématique de France, Paris, 2003)

R. Kyriakoglou, D. Perrin, Profinite semigroups. Technical report, 2017. arXiv:1703.10088 [math.GR]

G. Lallement, *Semigroups and Combinatorial Applications* (Wiley, New York/Chichester/Brisbane, 1979). Pure and Applied Mathematics (A Wiley-Interscience Publication, Chichester)

H. Lenstra, Profinite Fibonacci numbers. Nieuw Archief voor Wiskunde **6**, 297–300 (2005)

D. Lind, B. Marcus, *An Introduction to Symbolic Dynamics and Coding* (Cambridge University Press, Cambridge, 1995)

M. Lothaire, *Algebraic Combinatorics on Words* (Cambridge University Press, Cambridge, 2002)

A. Lubotzky, Pro-finite presentations. J. Algebra **242**(2), 672–690 (2001)

R.C. Lyndon, P.E. Schupp, *Combinatorial Group Theory*. Classics in Mathematics (Springer, Berlin, 2001). Reprint of the 1977 edition

W. Magnus, A. Karrass, D. Solitar, *Combinatorial Group Theory*, 2nd edn. (Dover Publications, Inc., Mineola, NY, 2004)

S. Margolis, M. Sapir, P. Weil, Irreducibility of certain pseudovarieties. Commun. Algebra **26**(3), 779–792 (1998)

S. Margolis, M. Sapir, P. Weil, Closed subgroups in pro-**V** topologies and the extension problem for inverse automata. Int. J. Algebra Comput. **11**(4), 405–445 (2001)

J.C. Martin, Minimal flows arising from substitutions of non-constant length. Math. Syst. Theory **7**, 72–82 (1973)

J.D. McKnight Jr., A.J. Storey, Equidivisible semigroups. J. Algebra **12**, 24–48 (1969)

B. Mossé, Puissances de mots et reconnaissabilité des points fixes d'une substitution. Theor. Comp. Sci. **99**(2), 327–334 (1992)

B. Mossé, Reconnaissabilité des substitutions et complexité des suites automatiques. Bull. de la S. M. F. **124**, 329–346 (1996)

Ana Moura. Representations of the free profinite object over DA. Int. J. Algebra Comput. **21**, 675–701 (2011)

J.R. Munkres, *Topology* (Prentice Hall, Upper Saddle River, 1999)

N. Nikolov, D. Segal, Finite index subgroups in profinite groups. C. R. Acad. Sci. Paris Sér. A. **337**, 303–308 (2003)

K. Numakura, Theorems on compact totally disconnected semigroups and lattices. Proc. Am. Math. Soc. **8**, 623–626 (1957)

J.-J. Pansiot, Decidability of periodicity for infinite words. RAIRO Inf. Théor. et Appl. **20**, 43–46 (1986)

J.-É. Pin, *Varieties of Formal Languages* (Plenum, London, 1986). English translation

J.-É. Pin, Polynomial closure of groups and open sets of the hall topology. Theor. Comp. Sci. **169**, 185–200 (1996)

J.-É. Pin, C. Reutenauer, A conjecture on the Hall topology for the free group. Bull. Lond. Math. Soc. **23**, 356–362 (1991)

N. Pippenger, Regular languages and stone duality. Theory Comput. Syst. **30**, 121–134 (1997)

I.F. Putnam, *Cantor Minimal Systems* (American Mathematical Society, Providence, 2018)

M. Queffélec, *Substitution Dynamical Systems—Spectral Analysis*. Lecture Notes in Mathematics, vol. 1294, 2nd edn. (Springer, Berlin, 2010)

J. Reiterman, The Birkhoff theorem for finite algebras. Algebra Univers. **14**, 1–10 (1982)

C. Reutenauer, Une topologie du monoïde libre. Semigroup Forum **18**(1), 33–49 (1979)

J. Rhodes, B. Steinberg, Closed subgroups of free profinite monoids are projective profinite groups. Bull. Lond. Math. Soc. **40**, 375–383 (2008)

J. Rhodes, B. Steinberg, *The q-Theory of Finite Semigroups*. Springer Monographs in Mathematics (Springer, New York, 2009)

L. Ribes, P. Zalesskii, *Profinite Groups*. Ergebnisse der Mathematik und ihrer Grenzgebiete. 3. Folge. A Series of Modern Surveys in Mathematics, vol. 40, 2nd edn. (Springer, Berlin, 2010)

J.R. Stallings, Topology of finite graphs. Invent. Math. **71**, 551–565 (1983)

B. Steinberg, On the endomorphism monoid of a profinite semigroup. Portugal. Math. **68**, 177–183 (2011)

T. Tao, Open question: effective Skolem-Mahler-Lech theorem (2007). https://terrytao.wordpress.com/2007/05/25/open-question-effective-skolem-mahler-lech-theorem/#more-34

S.J. van Gool, B. Steinberg, Pro-aperiodic monoids via saturated models. Israel J. Math. **234**, 451–498 (2019)

D.D. Wall, Fibonacci series modulo *m*. Am. Math. Mon. **67**, 525–532 (1960)

P. Walters, *An Introduction to Ergodic Theory*. Graduate Texts in Mathematics (Springer, Berlin, 1982)

B. Weiss, Subshifts of finite type and sofic systems. Monatsh. Math. **77**, 462–474 (1973)

S. Willard, *General Topology* (Dover Publications, Inc., Mineola, NY, 2004). Reprint of the 1970 original [Addison-Wesley, Reading, MA]

Subject Index

accumulation point, 28
acyclic set, 179
admissible
    congruence, 111
        generated, 112
    two-sided fixed point, 147
alphabet, 88
aperiodic semigroup, 1, 118
Arnoux-Rauzy
    set, 176
    shift, 176
    words, 194
asymptotic injectivity, 150, 151
automaton, 90
    complete, 219
    finite, 90, 91
    group, 94, 221
    inverse, 94
    minimal, 91
    permutation, 94, 221
    reduced, 91
    Stallings, 94
    states, 90
    trim, 91

balanced word, 190
balance of two words, 190
basis
    of subgroup, 93
    of submonoid, 116
    of a topology, 26
biextendable
    set, 140, 153, 172
    substitution, 145
    word, 172

bifix code, 221
Binet's Formula, 19
bispecial word, 172
block, 142
block map, 143
bounded, 171

Cantor space, 54
Cauchy sequence, 9, 30
Chacon
    set, 173
    word, 173
chain, 27
characteristic series, 256
Clifford-Miller's Lemma, 65
clopen set, 26
closed
    equivalence, 39
    set, 26
closure, 26
cluster point, 28
code, 116
    bifix, 221
    group, 221
    prefix, 116, 217
    profinite, 116
    suffix, 116, 217
cofinal, 28
compact
    semigroup, 35
    space, 30
compatible family of morphisms, 49
complete
    automaton, 219
    metric space, 30

completion
      of monoid, 104
completion of metric space, 30
complexity, 173
congruence
      of semigroups, 39
      syntactic, 122
conjugacy, 144
connected
      set, 179
      space, 53
connection, 146, 201
conservative, 190
continuous map, 28
convergent net, 27

decidable topological group, 209
decoding function, 116
defining pair, 184
degree
      of bifix code, 225
      of permutation group, 234
dendric
      set, 179
      shift, 179
directed set, 27
directive word, 177
discrete topology, 26
distance
      profinite, 57
      ultrametric, 9

element
      periodic, 142
elementary substitution, 161
empty word, 88
entropy of subshift, 213
episturmian word, 175
      strict, 176
equivalence
      closed, 39
      open, 54
equivalent permutation groups, 234
evaluation map, 59
eventually dendric shift, 194
exponent of periodicity, 148
extension graph, 179

F-complete bifix code, 225
F-degree, 225
F-group of X, 238
F-maximal
      bifix code, 225
      code, 217

F-minimum $\mathcal{J}$-class, 234
F-thin set, 224
factor, 36
      internal, 224
factorial
      number system, 11
      set, 140, 153
            minimal, 140
            periodic, 142
            recurrent, 140
Fibonacci
      morphism, 146
      shift, 146
      word, 146
finite intersection property, 31
finitely presented
      profinite group, 113
      profinite semigroup, 112
first countable space, 62
fixed point, 146
formal series, 255
free
      factor, 95
      group, 93
      monoid, 88
      profinite
            group, 106
            monoid, 96
            semigroup, 104
      semigroup, 88
      submonoid, 116
full shift, 140

golden mean shift space, 141
Green
      Lemma, 65
      relations, 41
group
      automaton, 94, 221
      code, 221
      free profinite, 106
      profinite, 56
      Schützenberger, 45, 46
      topological, 38
      transition, 95
group of X, 238

Hall topology, 107
Hausdorff space, 26
Hopfian semigroup, 60
Hunter's Lemma, 84

ideal, 41
idempotent, 33, 35

index, 46
    of congruence, 96
indicator, 256
induced topology, 26
initial state, 90
internal factor, 224
inverse
    automaton, 94
    limit, 10, 50
    system, 49
irreducible
    shift space, 142
iterated palindromic closure, 176

Justin's Formula, 176

k-bounded set, 171
Kleene's theorem, 89

language, 88
    rational, 89
least period, 142
left
    congruence, 41
    ideal, 41
    return word, 170
left-special word, 172
    strict, 172
Lemma
    Clifford-Miller, 65
    Green, 65
length
    of pseudoword, 99
    of word, 88
lifting property, 124
limit return set, 187

matrix
    associated, 147
    primitive, 147
metric
    profinite, 57
    space, 29
        complete, 30
metrizable, 29
minimal
    automaton, 91
    generating set of submonoid, 116
    set, 140
    topological dynamical system, 18
monoid, 32
    cyclic, 46
    free, 88
    free profinite, 96

monogenic, 46
    transition, 90
morphism
    connecting, 49
    Fibonacci, 146
    primitive, 145
    recognizable, 150
    Thue-Morse, 146
    transition, 91

neighborhood, 26
net, 27
neutral
    set, 173
    word, 173
Nielsen-Schreier Theorem, 93
nilpotent semigroup, 118
nonerasing substitution, 145
norm
    p-adic, 9
    profinite, 8
normal space, 62

odometer, 18
1-block code, 143
one-sided shift
    map, 142
    space, 143
        associated, 143
open
    ball, 29
    equivalence, 54
    map, 102
    set, 26
orbit, 141
orbit of a point, 18
order
    prefix, 217
    suffix, 217

p-adic
    integer, 11
    norm, 9
    number, 12
palindrome word, 176
palindromic closure, 176
parse
    enumerator, 223
    of pseudoword, 247
    of word, 222
period, 46, 142
periodic
    element, 142
    point, 146

set, 142
    shift space, 142
periodicity
    exponent, 148
permutation
    automaton, 94, 221
permutation group, 234
    degree, 234
    equivalence, 234
Perron-Frobenius Theorem, 148
Pisano period, 19
prefix
    code, 116, 217
    order, 217
    transducer, 116
prefix-comparable words, 217
presentation
    finite, 112
    of profinite semigroup, 112
    of semigroup, 112
primitive
    matrix, 147
    substitution, 145
    word, 142
pro-
    G topology, 107
    M topology, 107
product topology, 27
profinite
    code, 116
    completion, 112
    distance, 8, 57
    group, 56
    integer, 9
    metric, 57, 103
    natural integers, 14
    norm, 8
    semigroup, 53
    topology, 107
projective
    limit, 50
    profinite group, 115, 125
    profinite monoid, 115
    profinite semigroup, 114
    system, 49
pseudovariety, 118
pseudoword, 98
    length, 99

quasi-component, 84
quotient, 119

rank
    of a $\mathcal{J}$-class, 64

of a free group, 122
of a free profinite group, 159
of a partial transformation, 64
rational
    language, 89
    operations, 89
Rauzy graph, 180, 192
recognizability constant, 150
recognizable
    K-subset, 135
    language, 89
    morphism, 150
    series, 15
    set, 107
recognized
    by automaton, 90
recurrent
    set, 140, 153
reduced word, 93
regular
    class, 45
    element, 45
residually finite
    group, 56
    semigroup, 54
retract, 114
retraction, 114
Return Theorem, 180
return word, 170
    left, 170
    right, 170
reversal of a word, 175
right
    complete set, 217
    congruence, 41
    dense set, 217
    ideal, 41
    return word, 170
    unitary, 170, 246
    unitary submonoid, 217
right-special word, 172
    strict, 172
ring
    of p-adic integers, 11
    of profinite integers, 10

Schreier's Index Formula, 122, 230
Schützenberger group, 45, 46
    of a shift, 196
semigroup, 32
    A-generated, 37
    aperiodic, 1, 118
    compact, 35
    cyclic, 46

equidivisible, 103
free, 88
free profinite, 104
Hopfian, 60
monogenic, 46
nilpoent, 118
presentation, 112
profinite, 53
residually finite, 54
stable, 43
topological, 34
semiring, 13
series
    recognizable, 15
    recognized, 15
set
    periodic, 142
    recognized, 55
shift
    map, 141
    one-sided, 142
    sofic, 160
    space, 141
        irreducible, 142
        one-sided, 143
        periodic, 142
        transitive, 142
sliding block code, 143
sofic shift, 160
space
    compact, 30
    connected, 53
    first countable, 62
    metric, 29
    metrizable, 29
    normal, 62
    totally disconnected, 54
    zero-dimensional, 54
stabilizer, 235
stable
    semigroup, 43
    submonoid, 161, 246
Stallings
    automaton, 94
    folding, 94
standard word, 176
state
    initial, 90
    terminal, 90
Stone dual, 101
strict
    episturmian word, 176
    left-special word, 172
    right-special word, 172

strong
    word, 173
Sturmian
    set, 168, 173, 176
    word, 173
subbasis, 26
submonoid
    right unitary, 246
    stable, 246
subnet, 28
subshift, 141
subshift of finite type, 213
subspace, 26
substitution, 145
    biextendable, 145
    elementary, 161
    expansive, 145
    nonerasing, 145
    order of, 206
    primitive, 145
    proper, 207
suffix
    code, 116, 217
symbolic dynamical system, 141
syntactic
    congruence, 122
syntactic context, 122

terminal state, 90
Theorem
    Hall, 95
    Kleene, 89
    Nielsen-Schreier, 93
    Perron-Frobenius, 148
    Tychonoff, 31
thin set, 224
Thue-Morse
    morphism, 146
    shift, 146
    word, 146
topological
    A-generated semigroup, 37
    closure, 26
    dynamical system, 18, 141
        minimal, 18
    group, 38
        decidable, 209
    semigroup, 34
    space, 26
topology
    discrete, 26
    induced, 26
    pro-$\mathbf{G}$, 107
    pro-$\mathbf{M}$, 107

totally disconnected, 54
transducer
    prefix, 116
transition
    group, 95
    monoid, 90
    morphism, 91
    semigroup, 37
transitive
    shift space, 142
tree
    set, 179
    shift, 179
Tribonacci
    set, 179
    word, 176
trim
    automaton, 91

ultrametric
    distance, 9, 57
    inequality, 8
uniformly continuous, 30
uniformly recurrent
    pseudoword, 153
    set, 140

variety, 119

weak
    word, 173
word, 88
    balanced, 190
    bispecial, 172
    Chacon, 173
    empty, 88
    equivalent, 93
    Fibonacci, 146
    internal factor, 224
    left-special, 172
    neutral, 173
    palindrome, 176
    prefix-comparable, 217
    primitive, 142
    reduced, 93
    reversal, 175
    right-special, 172
    strong, 173
    Thue-Morse, 146
    Tribonacci, 176
    weak, 173

zero-dimensional, 54

Index of symbols

$\mathcal{A}(L)$, 91
A^*, 88
$\widehat{A^*}$, 96
A^+, 88
$\widehat{A^+}$, 104
$\bar{A}$, 93
$\langle A \mid R \rangle$, 112
$\langle A \mid R \rangle_{\mathsf{M}}$, 112
$\langle A \mid R \rangle_{\mathsf{S}}$, 112
$A^{\mathbb{N}}$, 142
$\mathrm{Aut}(S)$, 58
$A^{\mathbb{Z}}$, 140

B_2, 37
$B_d(x, \varepsilon)$, 29

$C_{i,p}$, 47

$\mathcal{D}$, 42
$D(x)$, 42
$d(x, y)$, 9
$d_+(x, y)$, 13
$\delta_X(w)$, 223
$d_S(u, v)$, 57
$d_X(F)$, 225

$E_F(w)$, 172
$e_F(w)$, 172
$\mathrm{End}(S)$, 58
ε, 88
$\varepsilon(f, s)$, 59
η_L, 92
$E_u(\varphi)$, 150

$\widehat{FG(A)}$, 106
$F(\varphi)$, 145

$F(w)$, 153
$FG(A)$, 93
$f^{[-k,\ell]}$, 143
F_n, 12
$F_n(X)$, 143

G, 118
$G(D)$, 45
$G(F)$, 159
$G_{\mathcal{A}}(F)$, 234
$\Gamma(H)$, 46
$\Gamma_F(x)$, 170
$G_n(F)$, 180
G_X, 238
$G_X(F)$, 238

$\mathcal{H}$, 42
$H(A)$, 94
$H(x)$, 42
H_{ba}, 202
Hg, 38

$I(X)$, 224
$\mathrm{Im}_{\mathcal{A}}(u)$, 232
$i \sim_{u,n} j$, 150

$\mathcal{J}$, 41
$J(F)$, 157
$J(x)$, 42
$J_{\mathcal{A}}(F)$, 234
J_X, 238
$J_X(F)$, 238

K, 120
$\mathrm{Ker}\, f$, 39

© Springer Nature Switzerland AG 2020
J. Almeida et al., *Profinite Semigroups and Symbolic Dynamics*, Lecture Notes
in Mathematics 2274, https://doi.org/10.1007/978-3-030-55215-2

$\mathcal{L}$, 41
$L(x)$, 42
$\leqslant_{\mathcal{J}}$, 41
$\leqslant_{\mathcal{L}}$, 42
$\leqslant_{\mathcal{R}}$, 42
$L_F(w)$, 172
$\ell_F(w)$, 172
$\varprojlim_{i \in I} S_i$, 50
$\varprojlim \mathbb{Z}/n\mathbb{Z}$, 10
$\overleftarrow{L_q}$, 91

M, 118
$M(L)$, 92
$M(\varphi)$, 147
$m_F(w)$, 172
$M_{i,p}$, 46

$\mathbb{N}_\infty$, 31
$\widehat{\mathbb{N}}$, 14
$N_{i,p}$, 14

ω, 14
$\mathrm{ord}_p(x)$, 8

$\mathcal{P}(X)$, 102
$\mathrm{Pal}(w)$, 176
$|\varphi|$, 148
φ_H, 205
$\|\varphi\|$, 148
$p_n(x)$, 173
$\psi_a(b)$, 176
$\psi_{i,j}$, 49
PT_n, 33
PT_X, 33
p_{V}, 120

(Q, I, T), 90
$q \cdot w$, 91
$\mathbb{Q}_p$, 12
q_ρ, 39

$\mathcal{R}$, 41
$R(x)$, 42
$r(x)$, 8
$r_+(x, y)$, 14
$\mathrm{rank}_A(u)$, 232

$\mathrm{rank}_A(F)$, 232
$\mathcal{R}_F(x)$, 170
$\mathcal{R}'_F(x)$, 170
$R_F(w)$, 172
$r_F(w)$, 172
ρ_I, 89
ρ_S, 104
ρ_x, 46
$r_S(u, v)$, 57

S, 118
S/ρ, 39
S^1, 34
$(S_i, \psi_{i,j}, I)$, 49
σ_A, 141
S_n, 33
s^ω, 47
$\mathrm{Stab}(I)$, 235
S_X, 33

$T(H)$, 46
T_n, 33
T_X, 33

$\overline{U}$, 26
$u^{-1}L$, 91

$\mathcal{V}A^+$, 119

$|w|$, 88
$\overleftarrow{w}$, 157
$\overrightarrow{w}$, 157
$\widetilde{w}$, 175

$X(a, b)$, 202
$X(F)$, 141
$X(\varphi)$, 145
X^*, 37
X^+, 36
$|x|_\pi$, 8
$[x]_\rho$, 39

$\widehat{\mathbb{Z}}$, 9
$(\mathbb{Z}_+)_\infty$, 31
$\mathbb{Z}/n\mathbb{Z}$, 10
$\mathbb{Z}_p$, 11

LECTURE NOTES IN MATHEMATICS

 Springer

Editors in Chief: J.-M. Morel, B. Teissier;

Editorial Policy

1. Lecture Notes aim to report new developments in all areas of mathematics and their applications – quickly, informally and at a high level. Mathematical texts analysing new developments in modelling and numerical simulation are welcome.

   Manuscripts should be reasonably self-contained and rounded off. Thus they may, and often will, present not only results of the author but also related work by other people. They may be based on specialised lecture courses. Furthermore, the manuscripts should provide sufficient motivation, examples and applications. This clearly distinguishes Lecture Notes from journal articles or technical reports which normally are very concise. Articles intended for a journal but too long to be accepted by most journals, usually do not have this "lecture notes" character. For similar reasons it is unusual for doctoral theses to be accepted for the Lecture Notes series, though habilitation theses may be appropriate.

2. Besides monographs, multi-author manuscripts resulting from SUMMER SCHOOLS or similar INTENSIVE COURSES are welcome, provided their objective was held to present an active mathematical topic to an audience at the beginning or intermediate graduate level (a list of participants should be provided).

   The resulting manuscript should not be just a collection of course notes, but should require advance planning and coordination among the main lecturers. The subject matter should dictate the structure of the book. This structure should be motivated and explained in a scientific introduction, and the notation, references, index and formulation of results should be, if possible, unified by the editors. Each contribution should have an abstract and an introduction referring to the other contributions. In other words, more preparatory work must go into a multi-authored volume than simply assembling a disparate collection of papers, communicated at the event.

3. Manuscripts should be submitted either online at www.editorialmanager.com/lnm to Springer's mathematics editorial in Heidelberg, or electronically to one of the series editors. Authors should be aware that incomplete or insufficiently close-to-final manuscripts almost always result in longer refereeing times and nevertheless unclear referees' recommendations, making further refereeing of a final draft necessary. The strict minimum amount of material that will be considered should include a detailed outline describing the planned contents of each chapter, a bibliography and several sample chapters. Parallel submission of a manuscript to another publisher while under consideration for LNM is not acceptable and can lead to rejection.

4. In general, **monographs** will be sent out to at least 2 external referees for evaluation.

   A final decision to publish can be made only on the basis of the complete manuscript, however a refereeing process leading to a preliminary decision can be based on a pre-final or incomplete manuscript.

   Volume Editors of **multi-author works** are expected to arrange for the refereeing, to the usual scientific standards, of the individual contributions. If the resulting reports can be forwarded to the LNM Editorial Board, this is very helpful. If no reports are forwarded or if other questions remain unclear in respect of homogeneity etc, the series editors may wish to consult external referees for an overall evaluation of the volume.

5. Manuscripts should in general be submitted in English. Final manuscripts should contain at least 100 pages of mathematical text and should always include

   – a table of contents;
   – an informative introduction, with adequate motivation and perhaps some historical remarks: it should be accessible to a reader not intimately familiar with the topic treated;
   – a subject index: as a rule this is genuinely helpful for the reader.
   – For evaluation purposes, manuscripts should be submitted as pdf files.

6. Careful preparation of the manuscripts will help keep production time short besides ensuring satisfactory appearance of the finished book in print and online. After acceptance of the manuscript authors will be asked to prepare the final LaTeX source files (see LaTeX templates online: https://www.springer.com/gb/authors-editors/book-authors-editors/manuscriptpreparation/5636) plus the corresponding pdf- or zipped ps-file. The LaTeX source files are essential for producing the full-text online version of the book, see http://link.springer.com/bookseries/304 for the existing online volumes of LNM). The technical production of a Lecture Notes volume takes approximately 12 weeks. Additional instructions, if necessary, are available on request from lnm@springer.com.

7. Authors receive a total of 30 free copies of their volume and free access to their book on SpringerLink, but no royalties. They are entitled to a discount of 33.3 % on the price of Springer books purchased for their personal use, if ordering directly from Springer.

8. Commitment to publish is made by a *Publishing Agreement*; contributing authors of multiauthor books are requested to sign a *Consent to Publish form*. Springer-Verlag registers the copyright for each volume. Authors are free to reuse material contained in their LNM volumes in later publications: a brief written (or e-mail) request for formal permission is sufficient.

Addresses:
Professor Jean-Michel Morel, CMLA, École Normale Supérieure de Cachan, France
E-mail: moreljeanmichel@gmail.com

Professor Bernard Teissier, Equipe Géométrie et Dynamique,
Institut de Mathématiques de Jussieu – Paris Rive Gauche, Paris, France
E-mail: bernard.teissier@imj-prg.fr

Springer: Ute McCrory, Mathematics, Heidelberg, Germany,
E-mail: lnm@springer.com

Printed in the United States
By Bookmasters